RED BOOK OF MARINE ENGINEERING

QUESTIONS AND ANSWERS

Volume I

RED BOOK
OF
MARINE ENGINEERING

QUESTIONS AND ANSWERS

William B. Paterson

Part One — Third Assistant

Part Two — Second Assistant

Multiple-Choice Questions and Answers
and
A New Section on Automation

Fourth Edition

CORNELL MARITIME PRESS

CENTREVILLE MARYLAND

ISBN 0-87033-088-8

Library of Congress Catalog Card Number: 76-153141

Manufactured in the United States of America

First edition, 1961. Fourth edition, 1971; third printing, 1984

INTRODUCTION

The purpose of this publication is to make available to personnel of the American Merchant Marine specimen questions and answers which will serve as a guide to appropriate study for licenses in the various grades of marine engineer. The samples should be regarded only as typical questions which indicate the scope of each examination. Candidates will not be asked the same number of questions published under the respective grades and titles, nor are the examining officers precluded from using questions not contained in this publication.

The material is grouped in accordance with the grade of license to which the various parts apply. Under each grade there are questions for a license as engineer of steam vessels and as engineer of motor vessels; also questions that are common to both examinations. In studying for the higher grades of license, similar subject matter shown for the lower grades should be carefully reviewed. Candidates for the lower grades should also acquaint themselves with the material given for the higher grades, although they will not be required to show the same degree of knowledge.

Applicants for a license as engineer of steam vessels shall be required to demonstrate their ability to analyze boiler water by actual chemical testing.

Prior to sitting for a license examination, an applicant must meet other requirements. A brief summary of these follows:

Age and Experience

Applicants must be at least 21 years of age, with the exception of Third Assistant Engineers, who must be at least 19 years of age. Minimum qualifying experience is required for each grade of license. These requirements are set forth in detail in CG-191, "Rules and Regulations for Licensing and Certificating of Merchant Marine Personnel."

Citizenship

All applicants for an original, renewal, or raise of grade of license must be citizens of the United States, native born, or fully naturalized. This must be established by acceptable documentary evidence. Persons not able to prove American citizenship will not be examined for an original license.

Applications

Form CG-866 (License Application) may be obtained either by written request or personal application to any Officer in Charge, Marine Inspection, U. S. Coast Guard. It must be completed in all respects. All statements of sea service made therein must be supported by documentary evidence, issued by responsible persons, officers, or organizations. When the application has been completed, it must be presented personally by the applicant at a Marine Inspection Office. Each applicant for an original license is required to have a written endorsement from a master, engineer, and one other licensed officer of a vessel on which he has served.

Physical Examination

Upon acceptance and approval of his application, the candidate will be sent to one of the offices of the U. S. Public Health Service for a physical examination and for an examination in First Aid.

For original license as engineer the applicant must have, either with or without glasses, at least 20/30 vision in one eye and at least 20/50 in the other. The applicant who wears glasses, however, must also be able to pass a test without glasses of at least 20/50 in one eye and at least 20/70 in the other. The color sense will be tested by means of the "Stillings" test, but any applicants who fail this test will be eligible if they can pass the "Williams" lantern test.

Reexamination and Refusal of Licenses

Any applicant for license or endorsement who has been duly examined and refused may come before the same Officer in Charge, Marine Inspection for reexamination at any time thereafter that may be fixed by such Officer in Charge, Marine Inspection, but such time shall not be less that 1 month from the date of his last failure. In the case of another failure, he will not be reexamined until after a lapse of at least 6 months from date of last failure.

A candidate who has been duly examined and refused a license by an Officer in Charge, Marine Inspection shall not be examined by any other Officer in Charge, Marine Inspection until 1 year has elapsed from the date of the last refusal without the sanction of the Officer in Charge, Marine Inspection that refused the applicant.

CONTENTS

SECOND ASSISTANT ENGINEER

PREFACE

This book is a guide for study and not one which, if remembered word for word, will get a man the license he wants. One should recall his own knowledge of situations in answering questions rather than try to "remember what the Red Book said."

The Red Book can serve the student well as a "refresher" to review his written and mental notes on class and text material.

Men preparing to sit for their examinations can use the material presented here in self-examination and as a stimulus to thinking—revealing the weak spots in their knowledge and the gaps they must fill.

Always bear in mind the fact that no guide book can be a substitute for a sound and ready understanding of the subject.

One must pass a First Aid Examination before being eligible to sit for an original license. The questions and answers in Chapter 1 are similar to those given in such examinations.

The information on pages 327 to 452 should be studied by applicants for all grades.

Read the questions carefully. Some of them are written in a negative form and will catch you off guard if you do not read carefully. Answers to the questions will be found at the end of each section.

I wish to extend my appreciation to the following men who gave their time and knowledge in checking much of the material in this book: John M. Wellener—Electricity; Edward Novak—Boilers; Harold G. Kropp—Reciprocating Engines; Walter P. Timlin—Auxiliary Machinery; Harold W. Eastman—Motor; James H. Fitzpatrick—Reciprocating Engines; Thomas L. Della Torre—Mathematics; William H. Paton—General Rules and Regulations; John Paterson—Turbines; William E. Mooney and Frank Krustangel—General Questions; Louis Tesar—Electricity; Gilbert Antonecchia, Engineering Instructor, Seaman's Church Institute(NYC)—General Questions and Diesel; William Van Loo—Electricity and General Questions; Captain Pierre R. Becker, Director of the National Maritime Union Pension and Welfare Plan Upgrading and Retraining School. Mr. Thomas A. McEnery, Head of the Engine Department, and Messrs. Frank Boland, Charles Scott and Harry Smith, instructors in the above school. Also Messrs. A. Mandell and E. Jordan, instructors on the N. Y. C. School Ship John W. Brown; Mr. R. R. Spencer (P. E.), Director School of Engineering M. E. B. A. Dist. 2, Brooklyn, N.Y.; Mr. R. Barone, Director Marine Officers Training Center, State of New Jersey.

Much of the material on Automation was submitted by Mr. Sidney Carlson, Engineering Instructor on the School Ship John W. Brown, New York City Board of Education, and formerly Chief Engineer on the automated ship S. S. Ponce De Leon.

W. B. P.

THIRD ASSISTANT ENGINEER

FIRST AID

Q. What is first aid?
A. The immediate, temporary care given in case of accident or sudden illness, before the services of a physician can be obtained.

Q. What is a "Bill of Health?"
A. A necessary document setting forth the sanitary history and conditions of the ship, cargo, crew, passengers and the last port of call. It is issued in this country by the port health doctor and in a foreign country by the American Consul. It is not necessary in coastwise shipping. Its use helps to control the spread of disease from one country to another.

Q. Describe 3 ways of purifying drinking water aboard ship.
A. 1. Distillation.
2. 1/4 teaspoon of chloride of lime to 50 gallons of water, stir well and let stand 1/2 hour before using.
3. 1 tablespoon of iodine to a barrel of water, stir well and let stand 1/2 hour before using.

Q. How would you treat dysentery?
A. Treat as a highly contagious disease. Take full precautions to prevent its spread by isolation and disinfection.
Use following treatment:
1. Light diet—force fluids.
2. Rest in bed.
3. One sulfaguanidine tablet (1 gram) twice daily.
4. Boiled milk or bismuth.

Q. What diseases may result from drinking polluted water?
A. Cholera; Dysentery; Typhoid.

Q. How do you prevent the spread of smallpox?
A. Vaccination together with the isolation of active cases.

Q. Name the degrees of burns and the treatment of each.
A. First, second and third degree burns. The same treatments may be used for all degrees; the application of
1. Sterile boric acid ointment.
2. Baking soda.
3. Tannic acid sulfadiazine.
4. Sterile vaseline.
All treatments should be applied with a light bandage. In case of severe pain, morphine may be given.
Note: Sulfathiazole by mouth is also used: two tablets every 4 hours.

Q. What is shock? Give symptoms and treatment.

A. Shock is a condition in which the blood becomes pooled and stagnated in the vessels of the abdominal cavity because of dilation of the blood vessels due to a shock to the nervous system.

Symptoms of shock are pallor, lower temperature, cold perspiration (shallow), rapid weak pulse, and dilated pupils.

Treatment consists of lowering of head, maintaining body heat, warm drinks when conscious, such as coffee or tea, and if in pain, morphine tartrate injection (except in cases of head injury).

Q. What is first-aid treatment at sea for syphilis and gonorrhea?

A. Syphilis: Sterile dressings consisting of sterile normal saline solution. It is imperative that patient visit doctor upon entrance to port to determine proper quantitative diagnosis.

Gonorrhea: Isolation and disinfection of all personal articles. Instituting of sulfathiazole therapy, consisting of 2 tablets every 4 hours after initial dosage of 4 tablets. Continue dose for 5 days, and wait a week to determine whether or not there is a recurrence. If there is, repeat original treatment.

Q. How would you give first aid in case of a simple fracture?

A. Immobilize the part, apply a traction splint until medical aid can be obtained and treat for shock.

Q. How would you prepare an enema?

A. Shave a few pieces of soap (castile) into a container, add warm water and mix well until enough is dissolved to make good suds. Have patient lie on back or left side with knees drawn up, apply a light coating of vaseline to enema tube and gently insert about 3" or 4" of tube into the rectum. Elevate can to about hip height to allow portion to flow through tube, ridding tube of air and any foreign particles. Instruct patient to breathe through mouth to avoid cramps. In event of cramps, stop until pain ceases.

Q. Under what conditions is an injection of morphine not given to an injured person?

A. Morphine should not be given to anyone suffering from a head injury, having difficulty in breathing—or in undiagnosed internal bleeding when medical assistance is near, unless badly needed. Morphine is not given if respiration is below 12.

Q. What information should be forwarded when requesting medical advice by radio?

A. Message should be as short and concise as possible. Should contain patient's symptoms, complaints, temperature, pulse, respiration, age, sex, whether there are any similar cases aboard ship or were at last port of call. The treatment already administered and reaction to same should be included.

Q. What is the normal pulse rate, temperature, and respiration?

A. Normal pulse: 72. Normal temperature: 98.6°F.; 37°C. Normal respiration: 16.

Q. In what book would you enter any notes on sickness or injury aboard ship?
A. Master's log.

Q. Name three methods of preventing the spread of typhoid fever aboard ship.
A. Isolation, cleanliness and inoculation.

Q. What precautions would you take before fumigation? After fumigation?
A. Clear everyone off the ship. See that all open foodstuffs, cigarettes, etc., are removed. Post signs and post guards at the gangplank of ship. After fumigation, send man wearing a gas mask to ventilate the ship thoroughly. No one is allowed aboard for 12-24 hours.

Q. Explain how you would reduce a high fever in a patient.
A. Cooling sponge baths, alcohol sponges, fever-reducing drugs (aspirin).

Q. Explain the different types of wounds and the treatment of each.
A. 1. Abrasion: Scraping of the skin, with capillary bleeding. Cleanse area and dust with sulfa powder; apply pressure bandage.
2. Laceration: Torn or jagged type of wound, usually severe bleeding. Cleanse surrounding area, dust sulfa powder after removing all foreign particles and loose tissues; irrigate with boiled salt water; apply sterile gauze bandage.
3. Puncture: A stab or perforation made by a sharp or rounded instrument. Encourage bleeding by light pressure around wound. Remove obvious foreign particles, dust sulfa powder into depth of wound and apply wet dressings. Give tetanus shots if available.
4. Incision: Clean-cut wound, usually severe bleeding. Control bleeding, cleanse surrounding area, dust sulfa powder, apply pressure bandage.

Q. How would you prevent the spread of malaria?
A. In any suspected area, give prophylactic doses of quinine or atabrine to all crew members. Protect both sick and well from mosquitoes. (Because of the scarcity of quinine, atabrine is being used today in many instances.)

Q. What treatment would you give an unconscious person?
A. 1. Red Unconsciousness: Elevate the head slightly, apply cold compresses to the head and keep perfectly quiet. Do not give stimulant.
2. White Unconsciousness: Wrap in blankets, elevate the feet

and lower the head. Inhalation of stimulant while unconscious; stimulants by mouth when there is a return to consciousness.

3. Blue Unconsciousness: Artificial respiration at once, conserve the body heat, shock position and stimulants when patient has returned to consciousness.

Q. Where is a tourniquet applied?

A. It should be placed between the heart and the wound, using any material at hand (tie, strip of shirt, rubber tourniquet, etc.).

In case of severe bleeding, one should be familiar with points of compression on the body so that the flow of blood can be stopped temporarily. Figure 1A shows the arterial system. Figure 2 shows where to apply pressure to stop bleeding.

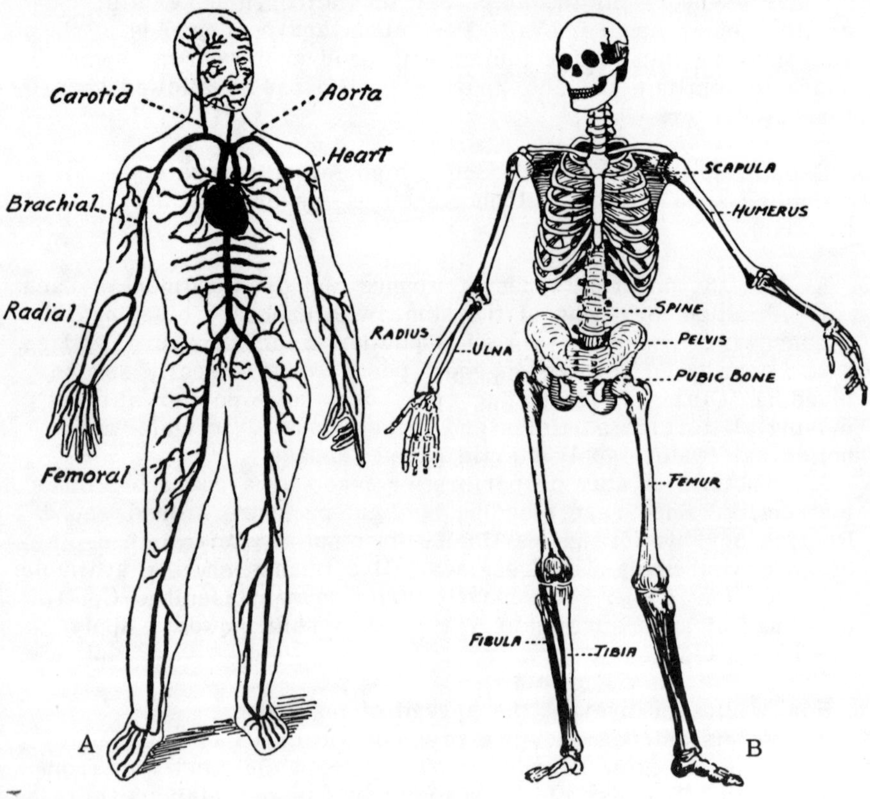

Fig. 1. A. Arterial system; B. Human skeleton.

Q. What is the function of the bones?

A. They make up the framework of the body; act as levers for the action of the muscles; protect delicate structures of the body. Red blood cells are formed in the marrow of the bones. Figure 1B.

Q. How would you treat a man with appendicitis if you were 4 days from port?
A. Complete rest in bed. Fowler's position. Light diet. No cathartics. Sulfa drugs by mouth, giving 2 tablets every four hours.

Q. How would you stop a nosebleed?
A. Apply cold compresses to back of neck and bridge of nose. Lay patient down. In severe cases, pack nose lightly with gauze dipped in adrenalin.

Q. How would you treat an apparently drowned person?
A. Clear mouth of obstructions. Apply artificial respiration. Cover with blankets. After recovery, treat for shock.

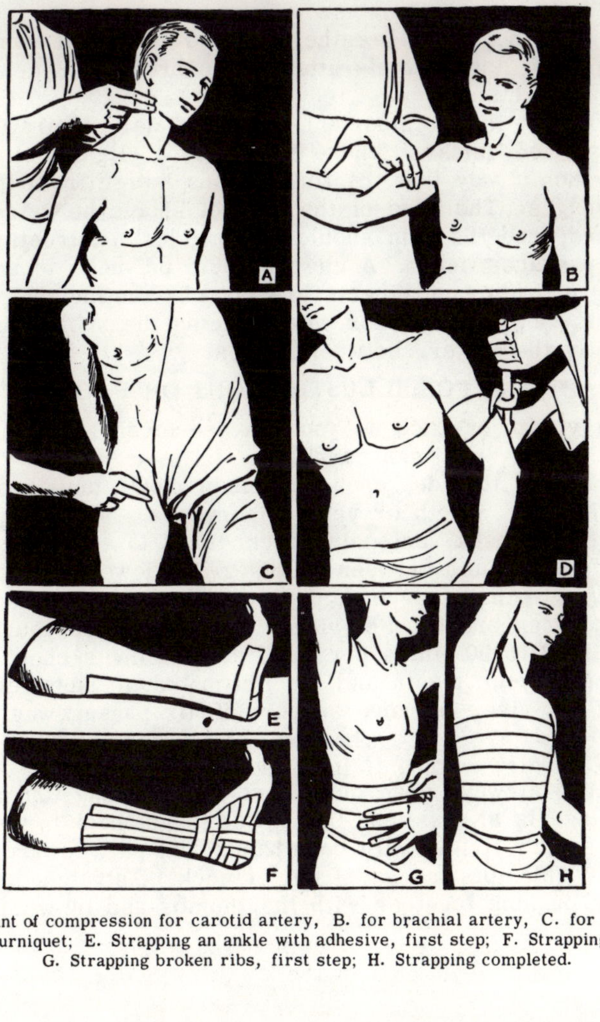

Fig. 2. A. Point of compression for carotid artery, B. for brachial artery, C. for femoral artery; D. Windlass tourniquet; E. Strapping an ankle with adhesive, first step; F. Strapping completed; G. Strapping broken ribs, first step; H. Strapping completed.

GENERAL PRINCIPLES OF MANUAL ARTIFICIAL RESPIRATION

Time is of prime importance. Seconds count. Do not take time to move the victim to a more satisfactory place; begin at once. Do not delay resuscitation to loosen clothes, warm the victim, or apply stimulants. These are secondary to the main purpose of getting air into the victim's lungs.

Begin artificial respiration and continue it rhythmically and without interruption until spontaneous breathing starts or the victim is pronounced dead.

As soon as the victim is breathing by himself, or when additional help is available, see that the clothing is loosened (or removed, if wet) and that the patient is kept warm, but do not interrupt the rhythmic artificial respiration to accomplish these measures.

If the victim begins to breathe on his own, adjust your timing to assist him. Do not fight his attempts to breathe. Synchronize your efforts with his.

Remember, it is all-important that artificial respiration, when needed, be started quickly. There should be a slight inclination of the body in such a way that the fluid drains better from the respiratory passages. The head of the subject should be extended, not flexed forward, and the chin should not sag lest obstruction of the respiratory passages occur. A check should be made to ascertain that the tongue or foreign objects are not obstructing the passages. These aspects can be cared for when placing the subject into position or shortly thereafter, between cycles.

CAPTIONS FOR ILLUSTRATIONS ON PAGE 7

View 1. Lay victim on back so you can see face. Move injured man carefully.

View 2. Turn head to side, open and clean mouth and throat with cloth or your fingers.

View 3. Pull head back, extending neck. Hold lower jaw up.

View 4. With left thumb between teeth, grasp lower jaw at midline, and lift lower teeth higher than uppers. Close nose with right hand.

View 5. Take deep breath. Put victim's mouth inside your lips and blow into it 12 to 20 times per minute. Watch his chest. When it rises, remove your mouth and let victim exhale passively. If his chest does not rise, improve support of air passageway and blow harder.

View 6. Use an airway tube if it is available.

View 7. To use airway, open mouth and insert along tongue until flange is at his lips. Don't push tongue back in throat.

View 8. Grasp jaw firmly with both hands and pull upward, extending neck so chin juts out and front of neck is stretched. Close nostrils by pushing together with the thumbs and blow into airway.

1

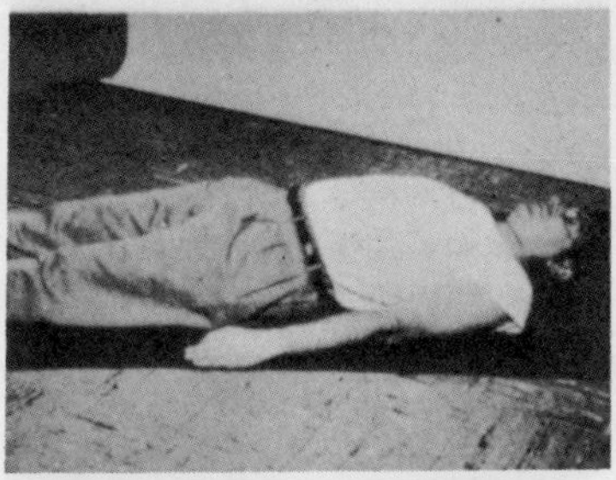

2

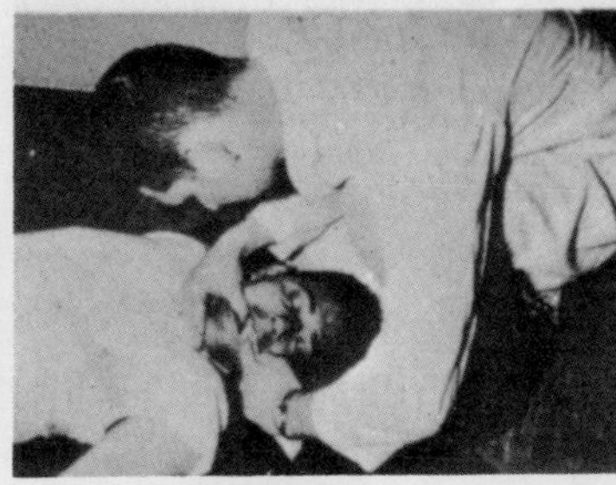

3

4

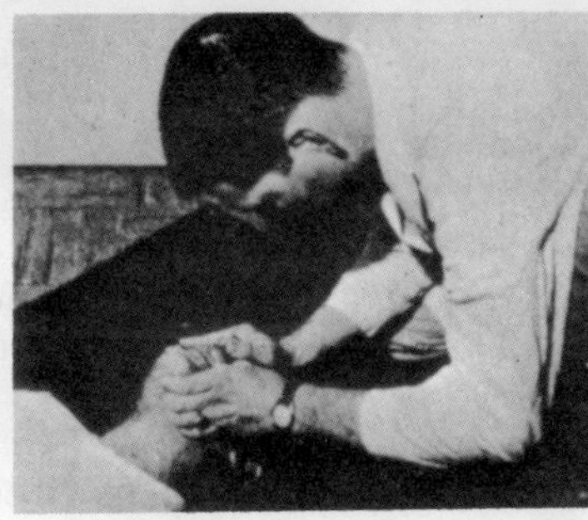

5

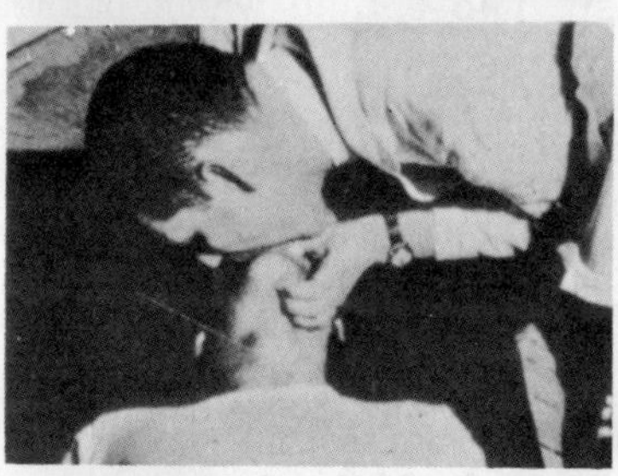

6

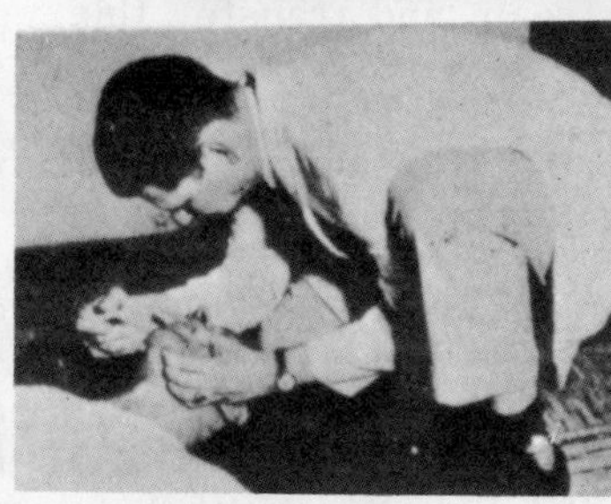

7

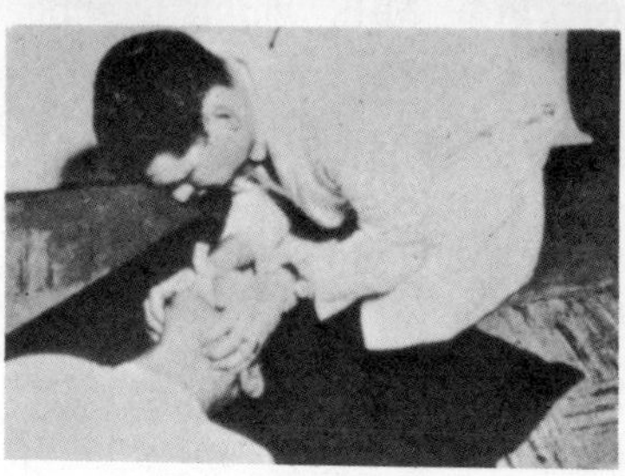

8

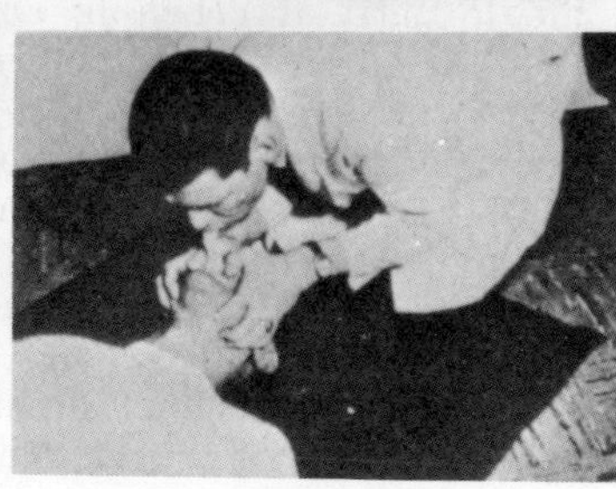

Mouth-to-Mouth Resuscitation.
(See caption on page 6)

Q. What visual signs are present in a conscious person if the back or neck is broken?
A. If there is spinal cord damage, there is usually paralysis below the point of the break.

Q. What stimulants are given in case of severe internal bleeding?
A. None.

Q. What is the function of the nerves? Name the two types.
A. Transmission of impulses to and from the brain. Sensory and motor.

Q. What is ship sanitation?
A. Concerns mechanical cleanliness: water, food, ventilation, and protection against vermin.

Q. How would you prevent the spread of typhus fever? How is it spread?
A. Isolation, inoculation, disinfection and personal cleanliness. Germ carried by body lice and bedbugs.

Q. Describe the function of the kidneys.
A. They separate and remove waste material and poisons from the blood stream and transfer this waste into a fluid which passes from the body as urine.

Q. What is the treatment for diarrhea?
A. 1. Ascertain the cause; isolate if contagious (dysentery, cholera, typhoid).
2. If caused by irritating substance, administer a cathartic to remove it.
3. Administer soft diet and forced fluids.
4. Give three tablets of bismuth subcarbonate three times a day before meals until diarrhea is controlled. Milk which has been boiled 20 minutes may also be used for this purpose.

Q. Describe first-aid treatment for severe head bleeding.
A. Elevate the head, apply pressure over the wound with sterile compress and digital pressure at the proper pressure point. Treat for shock and care for the wound after bleeding has stopped.

Q. What are the symptoms of diphtheria?
A. 1. High fever together with prostration.
2. Greyish membrane adhering to the mucous membrane of the throat; this may interfere with breathing.

Q. If a man failed to pass urine, how would you relieve or treat him?
A. 1. Apply warm water to the genitals.
2. Keep hands of person in warm water.

3. As a last resort, a catheter of soft rubber should be used. The catheter should be boiled and the parts thoroughly washed before insertion. This should be done gently. If not successful, do not use force; instead, try at a later time.

MULTIPLE-CHOICE QUESTIONS & ANSWERS

1. The proper way to stop a slow oozing bleeding after a cut is to apply:
   A. a tourniquet above the cut
   B. a Band-Aid
   C. gentle pressure at the place of injury
   D. a tourniquet below the cut

2. The condition of shock is a direct result of:
   A. any injury however minor
   B. diminished and inadequate blood circulation
   C. excessive hunger
   D. a blow on the head

3. Penicillin should be administered:
   A. whenever a patient requests it
   B. for nausea and vomiting
   C. only on the advice of a physician
   D. for a bad cold

4. Shock is an extremely dangerous situation. A patient in shock should be placed in which of the following positions?
   A. head up, feet up 12-18 inches
   B. head up, feet down 12-18 inches
   C. head down, feet up 12-18 inches
   D. arms above the head

5. The most important factor in treating a severe ankle sprain is:
   A. rest of the affected part
   B. exercise of the joint
   C. hot epsom salt application
   D. immediate immersion in ice water

6. The easiest safe way to sterilize a syringe and needle is to:
   A. boil for one hour
   B. boil for 5 minutes
   C. dip in alcohol
   D. leave in alcohol overnight

7. If you think a patient has acute appendicitis the first thing to do is:
   A. give the patient an enema
   B. place an ice bag on the abdomen
   C. radio for advice
   D. stop all food and liquids for 12 hours

8. In the event of pain and tenderness of the right lower portion of the abdomen accompanied by nausea and vomiting, it would be diagnosed as:
   A. anemia
   B. apoplexy
   C. acute appendicitis
   D. amoebic dysentery

9. The diet of a sick person should be watched closely. When a high fever is present he should be given a:
   A. meat diet
   B. low fat diet
   C. liquid diet
   D. sugar free diet

10. A black and blue spot on the skin indicates:
    A. the patient is anemic
    B. nothing at all
    C. the patient has been poisoned
    D. there has been bleeding under the skin

11. When giving an injection it is important to first clean the skin with alcohol:
    A. because the shot will hurt less
    B. to reduce the bacterial contamination of the skin
    C. so that no oil will be carried into the muscle
    D. to make the injection easier

12. The plunger of a syringe is always pulled back after the needle is inserted:
    A. to determine if a blood vessel has been punctured
    B. to delay the whole operation
    C. to keep air out of the syringe
    D. in order to see if the syringe is working

13. One of the simplest and safest methods of reducing body temperature when a patient has a high fever is to:
    A. give him a tepid sponge bath
    B. give him an ice water shower
    C. put him on a liquid diet
    D. A and C

14. A mild sprain of the ankle is best treated by:
    A. having the patient walk on it so it will not get stiff
    B. applying hot water compresses
    C. keeping the patient off his feet
    D. B and C

15. The most useful drug to reduce a low-grade temperature is:
    A. aspirin
    B. paregoric
    C. bicarbonate of soda
    D. aromatic spirits of ammonia

16. The basic unit of structure of the body is the:
    A. muscle
    B. cell
    C. tissue
    D. bone and joint network

17. The large bone of the thigh is called the:
    A. clavicle C. sternum
    B. femur D. radius

18. The bones at a joint are held together by:
    A. ligaments C. a central nervous system
    B. sutures D. cartilage

19. The brain is covered by the meninges lining and:
    A. lymph C. blood
    B. bacteria D. cerebro-spinal fluid

20. The Eustachian tube connects the:
    A. stomach and liver C. ear and throat
    B. throat and stomach D. kidney and prostate gland

21. The total number of ribs in the body is:
    A. eight C. sixteen
    B. twenty-four D. twenty

22. The blood vessels are called the:
    A. auricles, capillaries and veins
    B. arteries, capillaries and veins
    C. auricles, ventricles and capillaries
    D. arteries, auricles and veins

23. The heart is composed of:
    A. cardiac bones C. cardiac muscle
    B. scar tissue D. blood cells

24. Malaria is most common in which following part of the world:
    A. Alaska C. Argentina
    B. Australia D. Central Africa

25. The early signs of tuberculosis of the lungs are:
    A. fatigue, sweating and a hacking cough
    B. coughing up blood
    C. weight loss of from ten to thirty pounds
    D. loss of appetite

26. When a sore develops on the penis while at sea you should:
    A. give immediate treatment with penicillin
    B. give sulfadiazine daily for a week
    C. radio for advice and try to have the patient removed from the ship
    D. do nothing except follow simple hygienic measures until reaching port

27. The most common venereal diseases are:
   A. syphilis, gonorrhea and buboes
   B. syphilis, gonorrhea and chancroid
   C. syphilis, gonorrhea and crabs
   D. gonorrhea, chancroid and beriberi

28. Asphyxia is a condition that results:
   A. when breathing stops for any reason
   B. after a head injury
   C. when a patient goes into shock
   D. after exposure to overheating

29. In giving artificial respiration the first thing to do is:
   A. call for a mechanical resuscitator
   B. begin at once, do not wait
   C. loosen the clothes, warm the patient and give mild stimulants
   D. remove the patient to a comfortable place

30. The symptoms by which the necessity for artificial respiration may be recognized are:
   A. wheezing type breathing and coughing
   B. coughing up water after immersion
   C. blueness of the skin and suspension of breathing
   D. inability of patient to get his breath in a lying position

31. When a patient has a foreign body in the eye which you cannot remove you should:
   A. apply boric acid ointment
   B. wash with warm water
   C. bandage the eye and keep it closed
   D. attempt to remove with a small pair of tweezers

32. If a shipmate developed chills and fever every other day after having been in a West African port, the most likely disease would be:
   A. yellow fever
   B. pleurisy
   C. malaria
   D. cholera

33. When a shipmate is in an obviously severe depression you should:
   A. put him in a locked room and let no one enter
   B. watch him closely for suicidal tendencies
   C. pay no attention to him—maybe he will snap out of it
   D. tell him jokes to try and take his mind off his troubles

34. The large bone in the leg is called the:
   A. femur
   B. fibula
   C. tibia
   D. humerus

35. The large bone in the lower arm is called the:
   A. femur
   B. ulna
   C. fibula
   D. radius

36. The blood leaving the heart passes to the:
   A. arteries
   B. veins
   C. capillaries
   D. lungs

37. The small center part of the eye is called the:
   A. iris
   B. pupil
   C. retina
   D. cornea

38. The body is protected against germs by the:
   A. red cells
   B. plasma
   C. white cells
   D. none of the above

39. Yellow fever and malaria are transmitted by:
   A. flies
   B. mosquitos
   C. body contact
   D. any of the above

40. Typhus is transmitted by:
   A. body lice
   B. flies
   C. mosquitoes
   D. any of the above

41. Smallpox can be prevented by:
   A. medication
   B. injection
   C. vaccination
   D. A or C

42. Cholera and typhoid are transmitted by:
   A. mosquitoes
   B. swallowing cholera germs in food or water
   C. flies
   D. body contact

43. Blood cells are formed in the:
   A. kidneys
   B. lungs
   C. bone marrow
   D. heart

44. Oxygen is distributed to the blood cells by way of the:
   A. red corpuscles
   B. white corpuscles
   C. spleen
   D. liver

45. Salt water should not be used for laundering (because of contamination) when the ship is:
   A. riding out a storm
   B. under way
   C. in the South Pacific
   D. in port

46. It is important to the health of the crew to keep the rat population of the ship as low as possible because rats:
    A. eat the crew's food
    B. are fire hazards
    C. carry insects that spread disease
    D. carry infectious diseases

47. The most effective way to carry out rat control is to:
    A. fumigate the ship monthly
    B. spread rat poison around the ship
    C. set rat traps
    D. prevent rats from coming aboard as much as possible

48. The best chemical for disinfecting drinking water is:
    A. sodium chloride C. lime
    B. chlorine D. sodium bicarbonate

49. Aromatic spirits of ammonia is used:
    A. for a severe cough
    B. as a stimulant for faintness
    C. as a mild sedative
    D. to relieve pain

50. The normal oxygen content of air is approximately:
    A. 20% B. 15% C. 25% D. 30%

51. The normal oxygen content of the human body is approximately:
    A. 20% B. 65% C. 50% D. 10%

52. Codeine should be locked in a special safe because it:
    A. is a powerful stimulant C. is a narcotic
    B. deteriorates easily D. is a dangerous poison

53. A yellowish discoloration of the skin frequently follows the use of:
    A. aspirin C. codeine
    B. atabrine (quinacrine) D. aromatic spirits of ammonia

54. Pentobarbital sodium is useful for:
    A. a stimulant C. a quick acting sedative
    B. relief of pain D. cure of gonorrhea

55. The average rectal temperature is higher than the oral temperature by:
    A. 1° B. 2° C. 3° D. 4°

56. The normal body temperature taken orally is:
    A. 90° B. 94.6° C. 98.6° D. 99.5°

57. The best treatment for a first-degree burn is the application of:
A. an ice pack
B. a cold compress
C. vaseline
D. A or B

58. The normal pulse beat is:
A. 50 - 70 beats per minute
B. 70 - 90 beats per minute
C. 90 - 110 beats per minute
D. 110 - 120 beats per minute

59. The usual dose of morphine sulphate is:
A. 1/5 grain B. 1/2 grain C. 1 grain D. 1/4 grain

60. Ships pass through quarantine to check on:
A. passports
B. dangerous contagious diseases
C. possible rat conditions
D. all of the above

61. A temporary relief for a toothache is:
A. peanut oil
B. aspirin
C. oil of cloves
D. aromatic spirits of ammonia

62. A good treatment for diarrhea is:
A. a cleaning out with castor oil
B. small doses of paregoric
C. light or fluid diet
D. all of the above

63. The symptoms of heat exhaustion are:
A. fainting
B. rapid and feeble pulse
C. skin cold and clammy
D. all of the above

64. The treatment for heat exhaustion is:
A. wrap patient in blankets
B. give stimulants such as tea or coffee
C. put to bed for rest
D. all of the above

65. The symptoms of sunstroke are:
A. dizziness
B. high temperature
C. skin hot and dry
D. all of the above

66. The treatment for sunstroke is:
A. give patient a cold bath
B. apply ice pack to head
C. give cold drinks
D. all of the above

67. The worst classification of a burn is a:
    A. first-degree burn
    B. second-degree burn
    C. third-degree burn
    D. none of the above

68. The best cure for frostbite is to:
    A. rub the affected parts with ice water
    B. rub the affected parts with warm water
    C. let it cure by itself
    D. none of the above

69. The signs of a bone fracture are:
    A. deformity
    B. extreme pain at the break
    C. swelling and discoloration
    D. all of the above

70. In a second-degree burn the affected area:
    A. turns blue
    B. blisters
    C. turns red
    D. bleeds profusely

ANSWERS TO FIRST AID QUESTIONS

1. C	15. A	29. B	43. C	57. D
2. B	16. B	30. C	44. A	58. B
3. C	17. B	31. C	45. D	59. D
4. C	18. A	32. C	46. D	60. B
5. A	19. D	33. B	47. D	61. C
6. B	20. C	34. C	48. B	62. D
7. C	21. B	35. D	49. B	63. D
8. C	22. B	36. A	50. A	64. D
9. C	23. C	37. B	51. B	65. D
10. D	24. D	38. C	52. C	66. D
11. B	25. A	39. B	53. B	67. C
12. A	26. C	40. A	54. C	68. A
13. D	27. B	41. C	55. A	69. D
14. D	28. A	42. B	56. C	70. B

STEAM ENGINES

RECIPROCATING—CONSTRUCTION

Q. What is a single-acting engine?
A. It is one in which work is done on only one side of the piston, having only one working stroke to a revolution.

Q. What is a double-acting engine?
A. It is an engine in which work is done on both sides of the piston, having two working strokes to a revolution.

Q. What is a condensing engine?
A. It is one which exhausts into a condenser.

Q. What is a non-condensing engine?
A. It is one which exhausts into the atmosphere.

Q. What is a triple-expansion engine?
A. It is a three-cylinder engine in which there are three stages of expansion.

Q. What is a quadruple-expansion engine?
A. It is a four-cylinder engine in which there are four stages of expansion.

Q. What is a four-cylinder triple-expansion engine?
A. It is a four-cylinder engine in which there are three stages of expansion. The steam leaving the I. P. divides up, and one half goes to the forward L. P. while the other half goes to the after L. P.

Q. How are the by-pass valves connected?
A. They are run directly from the main steam line to the I. P. and L. P. receivers.

Q. What is the purpose of a balance piston?
A. Relieves the eccentric of some of the weight of the valve and valve gear.

Q. What is the purpose of an assistant cylinder?
A. Relieves the eccentric of some of the weight of the valve and valve gear and balances the inertia force of the valve gear on the downstroke.

Q. What is meant by lap on a valve? What kinds of lap are there?
A. Steam lap is that part of the valve which overlaps the admission

edge of the port when the valve is in mid-position.

Exhaust lap is that part of the valve which overlaps the exhaust edge of the port when the valve is in mid-position.

Q. State which sections of the crankshaft of a triple expansion engine are subject to the greatest torsional stress and explain why.

A. The sections of shaft that fit into the webs. All the twisting (torsional) motion imparted to the cranks from the crossheads is located at these points. The section with the greatest torsional stress would be the part secured into the bull gear of the jacking engine.

RECIPROCATING — OPERATION

Q. How would you stand by with engines required for immediate use?

A. 1. Boiler pressure about 25 lbs. below safety valve setting.

2. Steam maintained at throttle.

3. Auxiliaries running slowly.

4. Links moved through full throw every few minutes.

5. Engine drains open.

6. Vacuum at about 10".

7. With direct-connected air or circulating pumps, use auxiliary pumps.

Q. Before turning engine over in warming up, what precautions are taken?

A. 1. See that jacking gear is disengaged.

2. Get permission from bridge.

3. Pass warning to stand clear.

4. Rock engine a few times without going over dead centers to ease out water caused by condensation.

5. Roll engine carefully over dead centers to get rid of any excess condensation.

Q. How is total lead changed with link motion, and how is valve given more lead on one end than on the other?

A. For total lead, eccentric is changed on shaft. To change lead on one end of valve, remove or add shims under foot of valve stem.

Q. What is the effect of adjusting the links?

A. Power in the several stages of expansion is equalized, thus giving a smoother-running engine.

Q. What is the effect obtained by linking-in an engine?

A. Shortens the valve travel, thereby giving earlier cut-off and less power as more work must be done by expansion.

Q. What is the effect obtained by linking-out an engine?

A. Increases the valve travel, thereby giving later cut-off and more power as steam is let into the cylinder for a longer period of the stroke.

Q. What is the purpose of the by-pass valves on a reciprocating engine?

A. 1. To turn engine off dead center if caught in that position while maneuvering.

2. They are used for warming up the engine.

3. They may be used in emergency to obtain extra power.

Q. How is the horsepower of an engine increased?

A. Increase steam pressure or link out on the H. P. (Later cut-off for less expansion and more power.)

Q. How is the eccentric set in relation to the crank for an inside valve?

A. 90° minus lap and lead behind the crank in the direction of rotation.

Q. How is the eccentric set in relation to the crank for an outside admission valve?

A. 90° plus lap and lead ahead of the crank in the direction of rotation.

Q. Explain in detail how you would set a slide valve (for small engines such as generators, circulating pumps, etc.).

A. 1. Square the valve.

a. Rotate engine with eccentric fixed in any position and note the amount the valve uncovers the steam ports at each end. Alter the length of the valve stem until these port openings are equal. (Small engines usually have threaded valve stems.) Valve is now squared. Note total lead.

2. Place engine on one dead center.

3. Determine direction in which engine is to run and, if an outside valve, turn full part of eccentric ahead of the crank in that direction until the desired port opening is obtained (approximately 1/2 of total lead).

4. Secure eccentric.

5. Place engine on opposite dead center and see if lead is the same. If not, make it so by altering length of valve stem.

6. Lead is usually made slightly more on bottom end to allow for cushioning and the angularity of the connecting rods.

In most large engines, the eccentrics are keyed to the shaft and in these cases it is only necessary to have a little more lead on the bottom by altering the length of the valve stem. This is done by inserting or removing shims from beneath the foot of the valve stem.

Q. What would you look for if the vacuum began to drop while under way? Give remedy for each trouble.

A. There are many things which may cause this to happen. The following are listed according to the ease with which each can be accomplished in a short time.

a. See if circulating pump is running. If stopped, and trouble appears to be mechanical, immediately line up the auxiliary circulating pump or the fire pump to the main condenser. [Stop main engine first (notify bridge) to prevent overheating of condenser.]

b. If circulating pump is running, check to see if it is pumping. If not, the suction may be plugged. Operate the steam line to the sea-suction stool.

c. Check overboard discharge. If it is a 2-pass condenser and the discharge is cool, the division plate probably has carried away.

d. If both the main circulating pump and cooling water are in perfect order, the trouble must lie in the air pump. It may be air-bound, or possibly some of the valves may have carried away. To check the inside of the pump, it would be necessary to secure the engine and remove the pump inspection plates.

Q. When is water most likely to get into cylinder, and what is done about it?

A. When preparing to start engine; keep all cylinder and valve-chest drains open until all water is blown out.

Q. How would you take leads on a main engine crankpin?

A. 1. Mark nuts and studs for position.

2. Secure pulley end of chain fall to eyebolts in bottom of cylinder head. Screw eyebolts into tapped holes in top of crankpin bearing bolts. Attach hook end of chain fall to these eyebolts.

3. Remove stud nuts.

4. Lower bottom half of bearing clear of crankpin.

5. Lay strips of lead wire about 2" apart in bottom half of bearing. A spot of Albany grease on the ends will hold them in place.

6. Pull up bottom half and replace stud nuts.

7. Detach chain fall.

8. Tighten up stud nuts to original marks.

9. Attach chain fall again, remove stud nuts, and lower bottom half of bearing again.

10. Remove the leads in order and place on large board or piece of cardboard in correct positions.

11. Measure thickness of the leads with a micrometer.

12. High spots can be readily detected by excessive flattening of leads.

13. Scrape high spots; go over oil grooves.

14. Wipe out well and douse with oil.

15. Secure as in steps 6, 7, and 8.

Q. How is a broken eccentric sheave repaired?
A. Repair temporarily with bolts. If necessary, bore and tap holes for them.

Q. What checks would you make on an engine after completing an overhaul job?
A. 1. Check job performed to see that everything is in its proper place.
2. See that all tools, chain falls, etc., are removed.
3. Jack engine through 2 or 3 revolutions to see that everything is clear.
4. Disengage jacking gear.

RECIPROCATING—GENERAL

Q. What is back pressure?
A. It is the pressure of the steam on the exhaust side of the piston, which tends to hold the piston back.

Q. Why is the I. H. P. of an engine always more than the S. H. P.?
A. Some of the work developed in the cylinders is used to overcome the friction of running the engine.

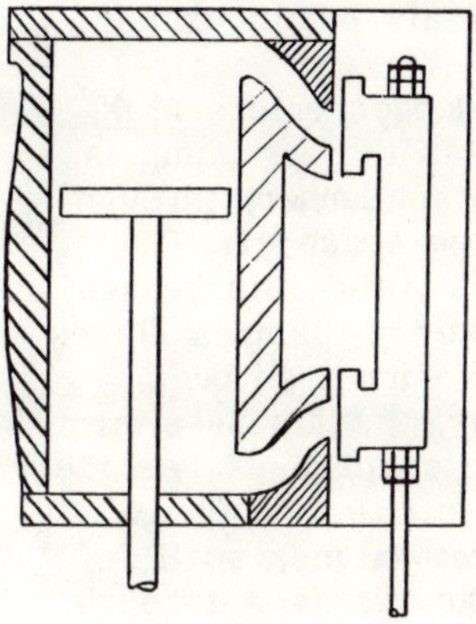

Fig. 3. D-slide valve.

Q. Make a free-hand sketch of a D-slide valve in position in an engine.
A. Figure 3.

Q. What is "mean effective pressure"? How is its amount ascertained?
A. It is the average effective pressure throughout the stroke. It is found by taking an indicator card and measuring the results.

Q. How is mean effective pressure (M. E.P.) computed from an indicator card?
A. Divide the working area of the card into 10 equal spaces. Measure the height at the center of each space (as shown by dotted lines on the sketch in Fig. 4). Separate scales (rulers) are furnished with indicator sets with the inches broken down into divisions corresponding to the scales of the various springs. That is, a scale with 80 divisions to the inch would be used to measure the card which was taken with an 80-lb. spring and 1" on the scale (or 80 markings) would represent 80 lbs. of pressure. Add the 10 pressures obtained in this manner and divide the total by 10. This will give you the M. E. P. of the card.

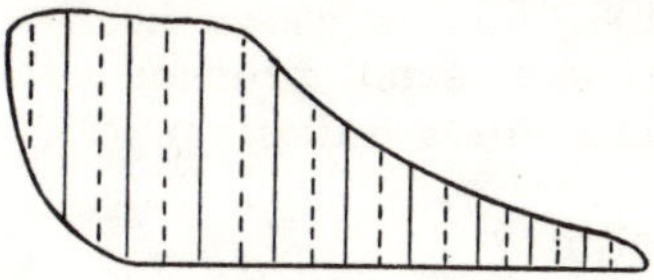

Fig. 4. Indicator card divisions.

Q. What is meant by the dead center of an engine?

A. The points in the revolution of the crankshaft, at which the piston rod, connecting rod, and the axes of the crank webs are all in a straight line.

Q. What is a butterfly valve? What is its purpose?

A. A butterfly valve is similar to a damper employed in furnace pipe lines and stacks. It is obvious that this type of valve will not be steamtight. It is used essentially for throttling the engine in a seaway. Some butterfly valves are operated from the engine governor.

Q. Describe the construction and operation of an air-pump "Kinghorn" valve.

A. It consists of 3 flat circular-shaped discs of different diameters. The three discs, A, B and C, fit over a stud which is screwed into the valve deck V and are held in place by coil spring S and lock nut L. The two bottom discs, B and C, have small holes (approximately 1/4") drilled in them so that some water will flow between the discs and form a seal which prevents air leakage, Figs. 5 and 6.

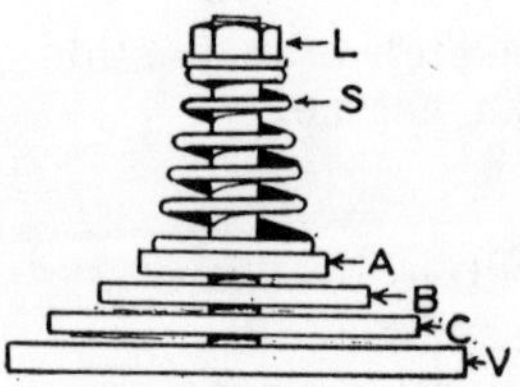

Fig. 5. Kinghorn valve.

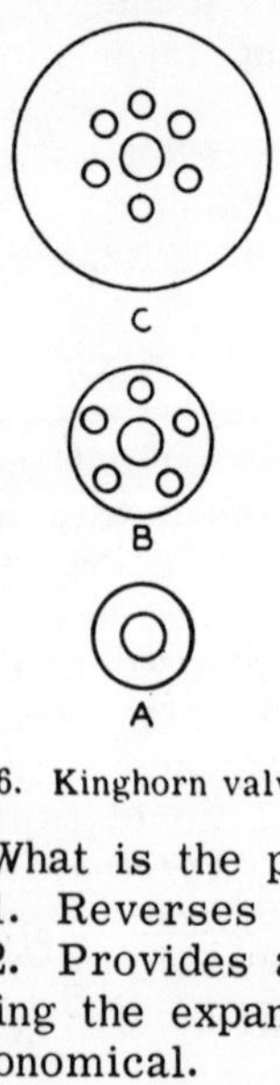

Fig. 6. Kinghorn valve discs.

Q. Sketch a bucket-type air pump; explain operation.

A. Water from the condenser enters at X and flows by gravity (as the air pump is below the level of the condenser) through the foot valves C to the space #1. The plunger or bucket Z is fitted with bucket valves B. As the bucket or plunger comes down, the water in #1 cannot go back through the foot valves, so it passes through the bucket valves B into space #2. As the plunger comes up, the water cannot go back through the bucket valves, so it is forced through the head valves A through passage Y to the hot well or feed tank, Fig. 7.

Q. Why are piston valves used on high-pressure steam instead of slide valves?

A. They are better balanced valves. There is very little friction, for the valve is practically floating, and is not pressed against a seat as is a slide valve.

Q. What is the purpose of the "Stephenson Link?

A. 1. Reverses the engine.

2. Provides a means of adjusting the stroke of the valve by using the expansion of steam to its fullest extent, which is more economical.

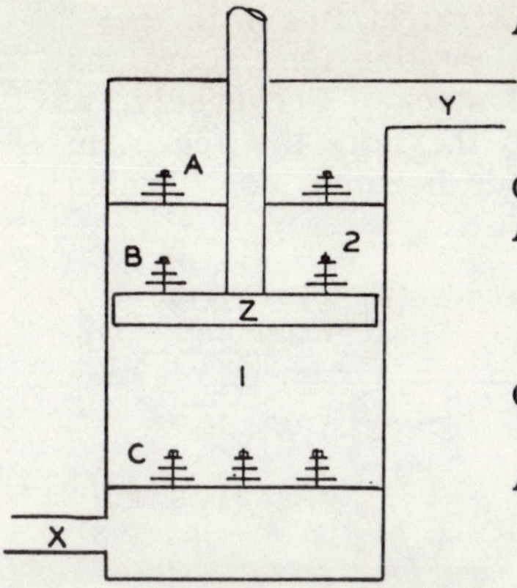

Fig. 7. Bucket-type air pump.

Q. What is the purpose of an eccentric?
A. To convert the rotary motion of the shaft to the reciprocating motion of the valve.

Q. What is throw of the eccentric?
A. It is equal to the distance from the center of the eccentric disc to the center of the crankshaft.

Q. What is the purpose of a thrust bearing?
A. To transmit the thrust of the propeller to the ship and prevent the main engine from being forced out of line.

Q. Why do the cylinders in a multiple-expansion engine vary in size?
A. They must increase in size from the H. P. to the L. P. as the pressure drops from cylinder to cylinder and there is, therefore, an increase in the volume of steam. It is also necessary that each cylinder operate at approximately the same horsepower. With reduced pressures, the size of the piston must be increased to obtain this equality.

Q. Has a four-cylinder triple-expansion engine any advantage over a three-cylinder one? If so, what?
A. Yes, in theory. Less weight and space for same power; better steam balance and distribution.

Q. What damages might occur if water should get into a cylinder?
A. 1. Bent rod.
2. Broken cylinder head.
3. Broken piston.
4. Frictional damage due to overheating, as water would destroy lubrication.

Q. Sketch an Edwards type air pump. Explain operation.
A. Air and condensate enter at A and flow either directly into top of plunger B (depending on its position) or into lower chamber E. The piston going up forces the water through the head valves D. On coming down, the water in E is forced up around curved channel, as shown by arrows, to the top of plunger. This water is removed on the upstroke. Inlet water enters through openings C, Fig. 8.

Q. What is the purpose of the expansion slot?
A. It permits percent of cut-off to be altered while engine is running. In Fig. 9, it will be noted that when the adjusting screw is turned so that the block in the slot moves to the right, the drag

links pull the quadrant to the right. In doing so, the eccentric rod (left) becomes closer to being in a straight line with the valve stem. When the two are in a straight line (one above the other), the valve will be making its full stroke. Conversely, as the block in the expansion slot moves to the left, the eccentric rod is pushed to the left and the two rods become less "in line", thus shortening the travel of the valve.

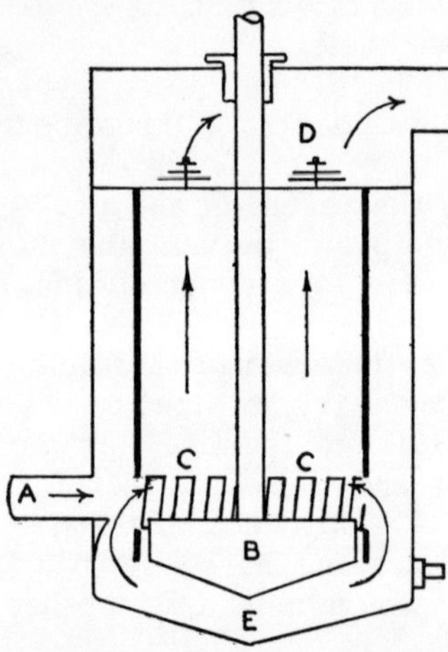

Fig. 8. Edwards air pump.

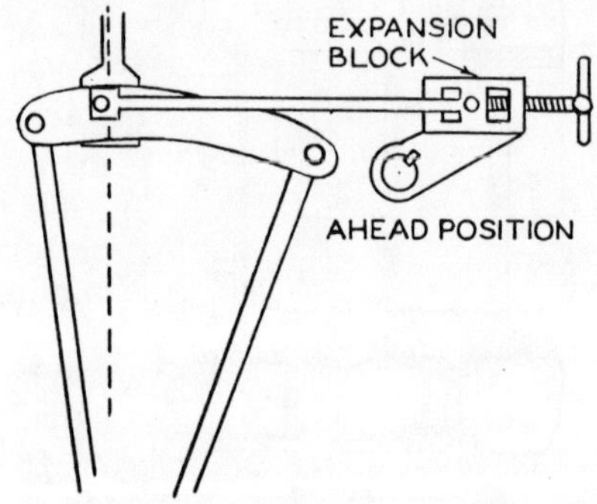

Fig. 9. Expansion block in ahead position.

Q. What are the troubles that may be indicated by the overflowing of the hot well?

A. 1. Chronometer valve is stuck or levers which operate the valve are stuck.

2. Overflow bucket in hot well which controls the chronometer valve may be stuck.

3. Drains from heat exchangers, superheaters, heating systems, etc., may be open or leaking.

4. Traps discharging to hot well may be broken.

Q. Why is it not desirable to use highly superheated steam in a reciprocating engine?

A. This type of steam is too dry. Saturated steam has sufficient moisture in it to help act as a lubricant between the rings and cylinder walls.

Q. Sketch a piston valve.

A.

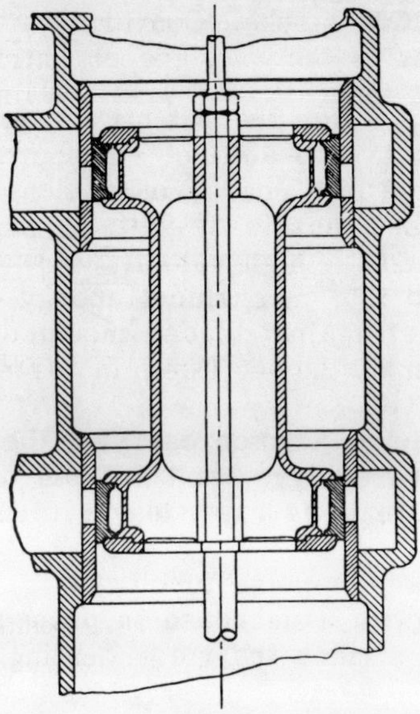

Q. List some causes of knocks in the L. P. engine.

A. 1. Piston rod and connecting rod out of line.
2. Too much clearance in slipper guide.
3. Too much clearance in cross head or crank pin bearings.
4. L. P. crank out of line.
5. Too much clearance on air pump rod guide.

TURBINES

CONSTRUCTION

Q. What is a steam turbine?

A. A steam turbine is a heat engine in which the potential energy of steam is changed into useful work in two distinct steps:

1. The available energy is converted into energy of motion, kinetic energy, by steam expansion in a nozzle or suitable passage, from which the steam emerges at a high velocity.

2. This kinetic energy is converted into mechanical energy or useful work by directing the steam jet against blades mounted on a revolving rotor, or by the reaction of the jet itself in the expanding passage if the passage revolves.

A steam turbine consists essentially of a rotor carrying the blades, the casing in which the rotor revolves, and nozzles or stationary blades through which the steam is expanded or directed.

Q. What is the operating principle of the impulse turbine?

A. The steam expands in the nozzles with a decrease in pressure, but an increase in velocity before it strikes the rotor blading.

Q. What is the operating principle of the reaction turbine?

A. The steam expands in the rotor blading, thus producing work. The principle of the reaction turbine is seen in a rotary lawn sprinkler where the force of the water kicks back, turning the sprinkler. A set of stationary curved blades direct steam into a set of blades on the rotor, where it expands, giving a similar kickback effect and thus producing work.

Q. What is the difference between an impulse and a reaction turbine?

A. In the impulse type, the steam expands in the nozzles, losing its pressure but gaining velocity; in the reaction type, the steam expands in the blading.

Q. What is a stage in a turbine?

A. A stage refers to that part of the unit in which a single pressure or velocity drop takes place.

Q. What is "tip clearance?"

A. The distance between the tip of the blade and the casing.

Q. What is "axial clearance?"

A. The distance between adjacent rows of blades or a row of blades and a diaphragm, depending on the type of turbine.

Q. What is the purpose of binding wires on blading?

A. Binding wires are used to stiffen blading and cut down on blade vibration, **Fig. 10.**

Q. What is the purpose of "shrouding?" How is it installed?

A. Shrouding is a strip of metal secured to the ends of a section of blading. Shrouding adds strength to the blades, lessens vibration, prevents steam leakage over blade tips, prevents wet steam being being thrown against casing by centrifugal force of rotor, which causes pitting in the casing, **Figs. 11 and 12.**

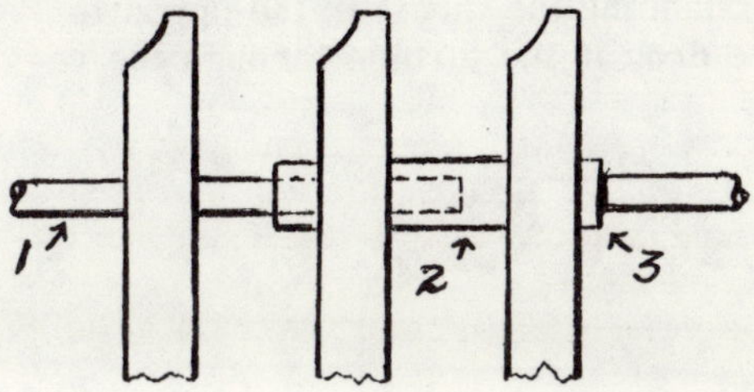

Fig. 10.

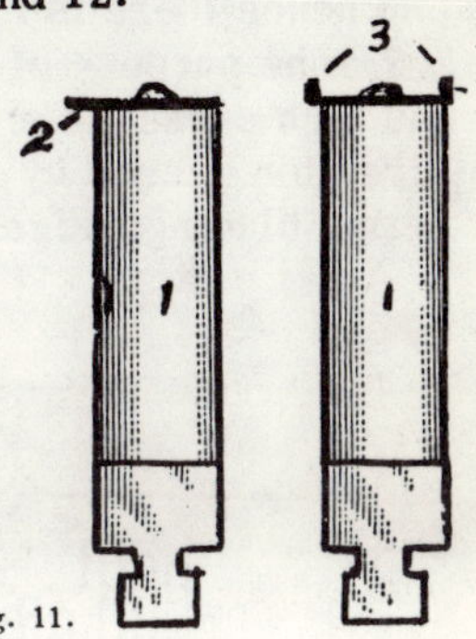

Fig. 11.

Fig. 10. Junction tube between segments, (1) binding wire; (2) junction tube; (3) wire welded to tube.

Fig. 11. Forms of shrouds, (1) blade; (2) flat shroud; (3) channel shroud.

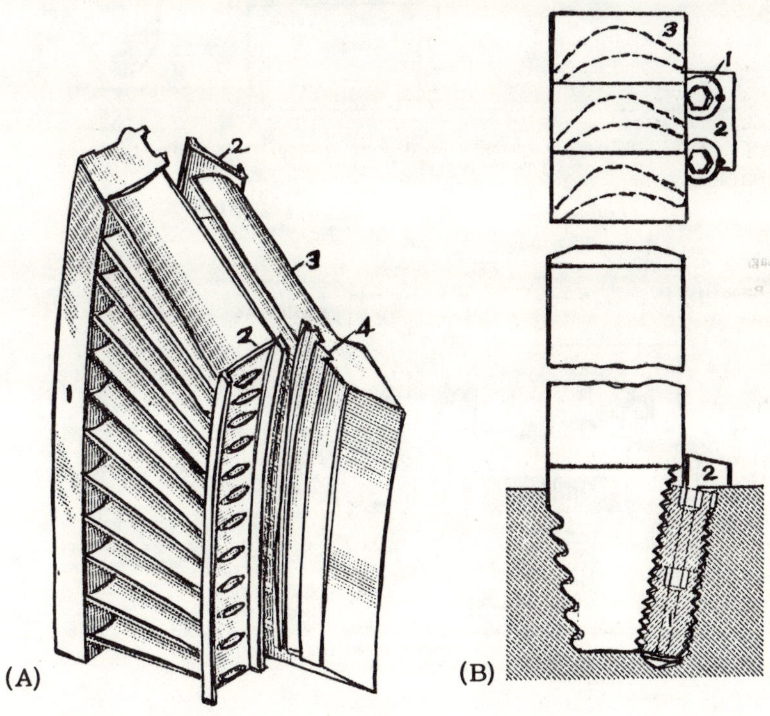

Fig. 12. A. Segments of fixed and moving turbine blades, (1) casing section; (2) shroud; (3) rotor section; (4) serrations. B. Securing last blade in position on rotor by setscrews, (1) setscrew; (2) wedge piece; (3) blade.

Q. (a) Why isn't it necessary to have dummy pistons on impulse turbines?

(b) What purpose does the dummy piston serve on reaction turbines?

A. 1. Dummy pistons are not necessary on impulse turbines due to the fact that the pressure on both sides of the moving wheels is the same and hence there is no axial thrust transmitted to the shaft.

2. The purpose of the dummy piston is to produce a thrust toward the high pressure end counter-balancing the thrust in the opposite direction caused by the pressure drop in the turbine through the reaction blading, Figs. 13 and 14.

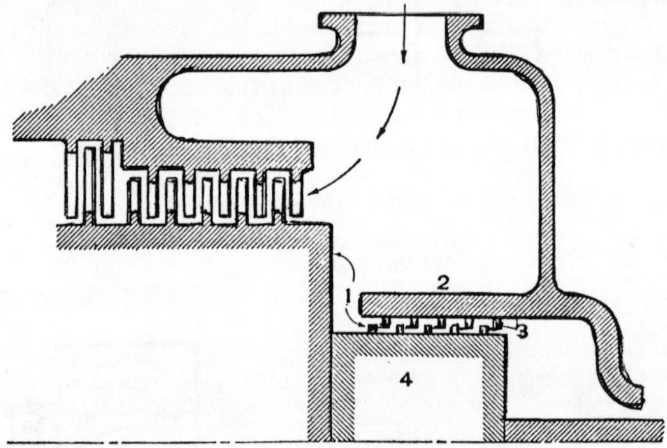

Fig. 13. Balance, or dummy piston, (1) steam thrust; (2) dummy cylinder; (3) dummy packing; (4) dummy piston.

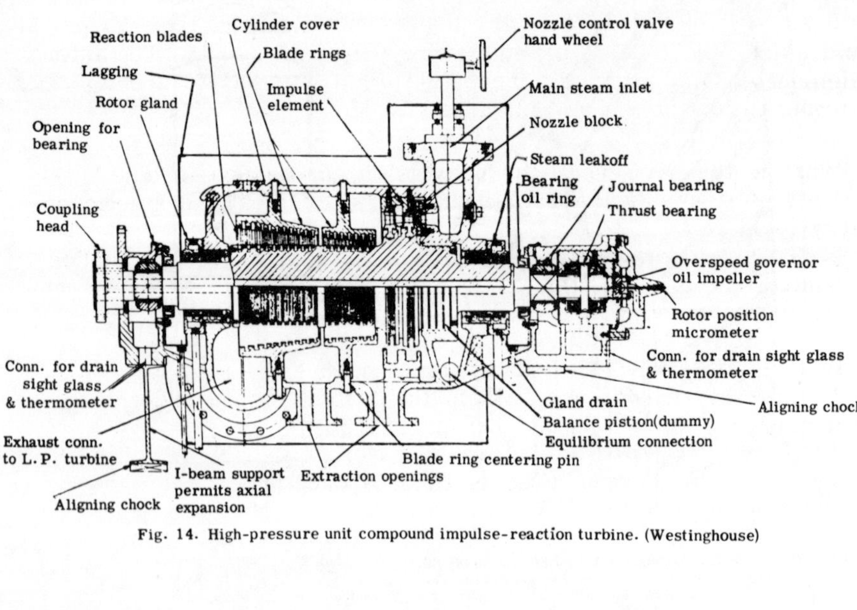

Fig. 14. High-pressure unit compound impulse-reaction turbine. (Westinghouse)

Q. What is "axial clearance" and how does its importance differ in the impulse and reaction type turbines?

A. The clearance between blades and nozzle or between different stages of blades. So as not to allow the high velocity steam jets to spread in the impulse, the blade wheel should be set comparatively close to the nozzle. In the reaction type, the velocity is much lower and the jet fills the whole periphery. The axial clearances in this type may therefore be more liberal.

Q. What is the distinguishing difference between impulse blading and reaction blading?

A. In the impulse blading the cross-sectional area of the steam passages between the blades remains approximately the same. In the reaction blading the passages between adjacent blades must be in the form of correctly shaped nozzles in order that the steam may expand the correct amount without turbulence.

Q. What is the purpose of reduction gears?

A. To reduce the high, efficient speed of a turbine to the low, efficient speed of a propeller.

Q. What is the purpose of double reduction gears?

A. Where the necessary speed ratios between the turbine and propeller are high and an unnecessary large main or bull gear would be required, the double reduction unit would handle it.

Q. In a cross-compound turbine, where is the astern turbine usually installed? How does horsepower of the astern turbine generally compare with the ahead turbine?

A. It is built on the same rotor as the L. P. turbine, the L. P. and astern units being assembled in the same casing. The astern element usually produces about 75-80% of the normal ahead torque at 50% of the normal ahead RPM.

Q. What is the purpose of the nozzles in an impulse turbine?

A. 1. To change the thermal (heat) energy of steam to kinetic energy (velocity).

2. To direct the jet of steam from the nozzles against blades mounted on a wheel which is free to rotate. Thus the kinetic energy is converted into work.

Q. What is the purpose of gland-steam seals on a turbine?

A. To prevent steam leakage from the high-pressure end and to prevent air leaking into the low-pressure end.

Q. What are the 3 main types of flexible couplings?

A. Claw, gear and pin.

Q. (a) What is meant by "pressure stage"?
(b) What is a "velocity stage"?

A. (a) A pressure stage may be defined as any combination of a nozzle (or set of nozzles) with one or more rows of blades, in which a single pressure drop occurs between the inlet to the nozzle and the exit from the last row of blades.

(b) A velocity stage is one in which the speed at which steam flows while going through one set of vanes is reduced without reduction of pressure, Figs. 15 and 16.

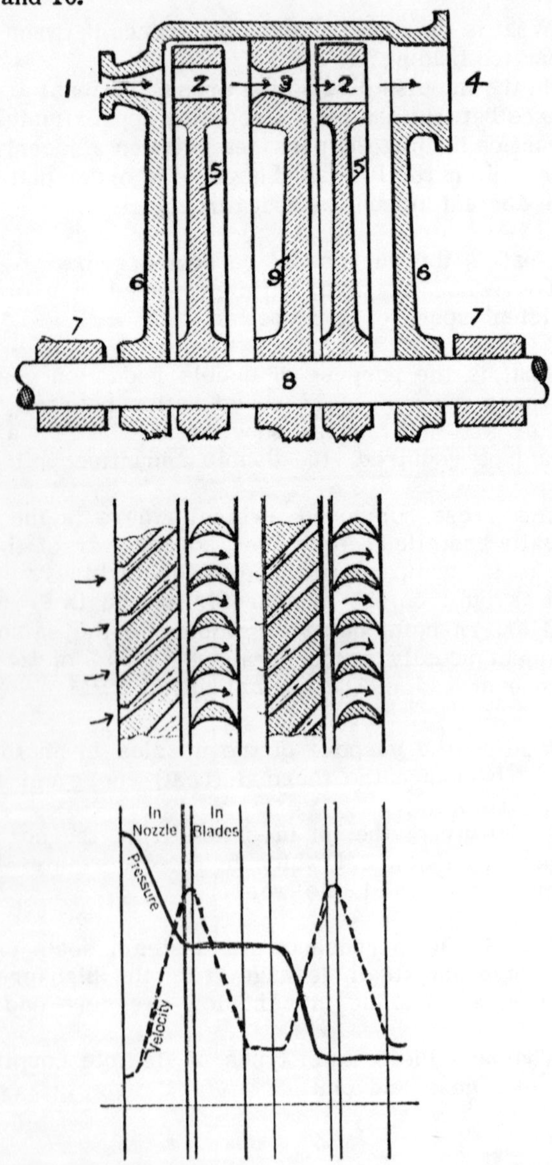

1. Steam-inlet nozzle
2. Blade
3. Diaphragm nozzle
4. Exhaust
5. Wheel
6. Casing
7. Bearings
8. Shaft
9. Diaphragm

Fig. 15. Pressure-velocity diagram of Rateau turbine.

Q. Where is the thrust bearing usually located in a turbine installation?

A. Usually on the hot end (steam inlet end). On most installations this would be on the free or forward end, the other (after) end would be connected to reduction gear or motor shaft.

Q. What is the purpose of stationary blades in a reaction turbine?

A. The reaction turbine has a series of stationary blades for the purpose of directing steam flow to the next set of moving blades.

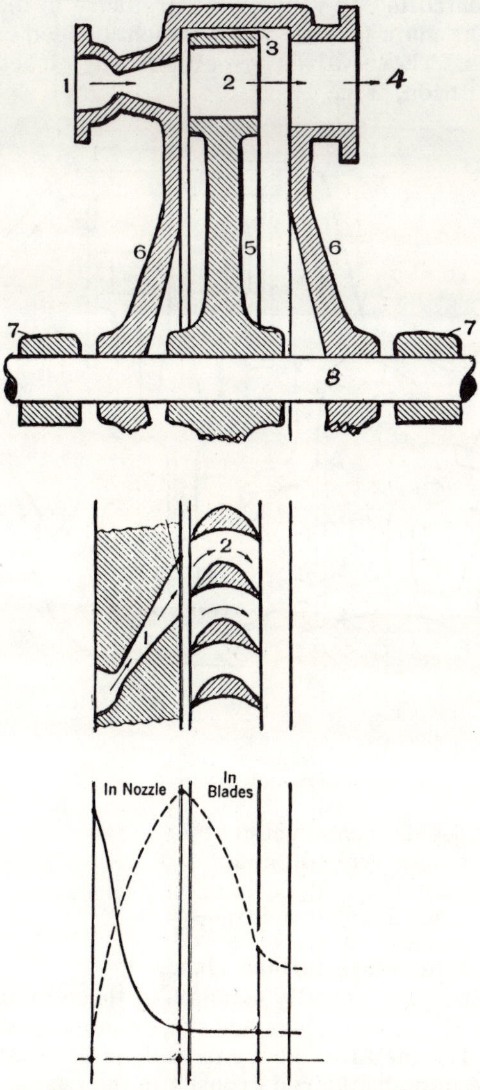

1. Steam-inlet nozzle
2. Blade
3. Radial clearance
4. Exhaust
5. Turbine wheel
6. Casing
7. Bearing
8. Shaft

Fig. 16. Action of steam flowing through impulse turbine.

Q. Describe a steam chest and valve arrangement for admitting steam to the nozzle block chamber of a steam turbine.

A. Steam is admitted to the steam chest through a throttle valve direct from the main steam line. The nozzles for the first stage are arranged in groups, each being controlled by an individual hand nozzle valve. By opening or closing the hand nozzle valve, steam is either shut off or admitted to the group of nozzles which it controls. With a given pressure in the steam chest, the amount of steam flowing through the turbine depends upon the number of hand nozzles opened which are usually arranged radially on the upper half of the steam chest. Speed may be controlled by varying the pressure in the main steam chest by use of the main throttle and additional speed control is given by hand nozzles. These valves are either fully closed or fully opened during operation, Fig. 17.

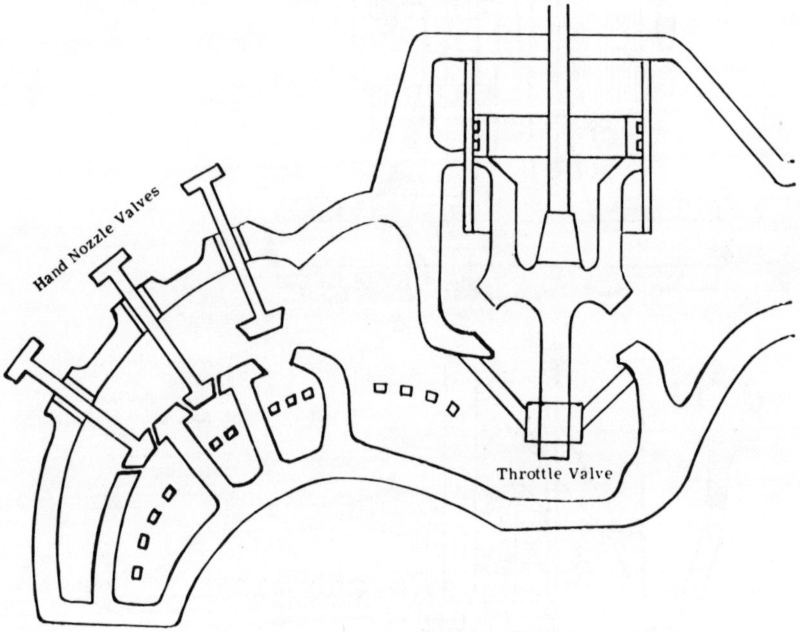

Fig. 17. Throttle valve and hand nozzles.

Q. How are gland-steam seals constructed?

A. 1. Usually of the labyrinth type packing.
2. Auxiliary units of some machines have carbon rings.
3. Combinations of 1 and 2 are sometimes fitted.

Q. Describe a typical high pressure turbine gland.

A. A high pressure turbine gland usually consists of three or more rows of metallic labyrinth packing rings and two or more rows of carbon packing rings. The metallic labyrinth packing is usually in four segments which fit into shouldered grooves in the casing and are held out, close to the rotor, by springs and kept from turning by dowels or stop pins. There is usually a leak-off line leading from between the last two rows of labyrinth packing back to one of the later stages of the turbine. Between the last row of labyrinth packing and the carbon packing a gland sealing line is led. The carbon rings are in segments and are held in place much the same as the labyrinth rings. A leak-off line usually leads from before the last carbon ring to the gland exhaust condenser, Fig. 18.

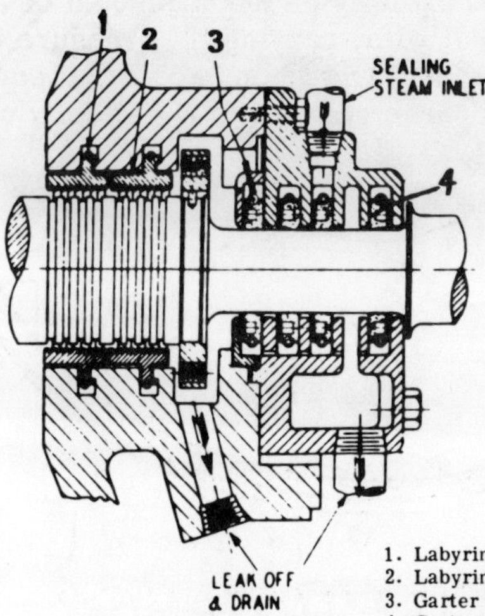

1. Labyrinth seal retainer spring
2. Labyrinth seal retainer spring
3. Garter spring on carbon ring
4. Carbon ring

Fig. 18. Combination of labyrinth seal and carbon ring packing.

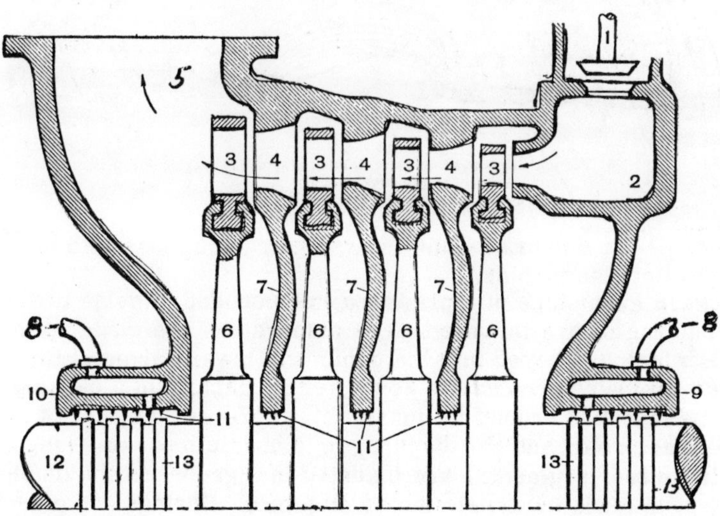

1. Throttle valve
2. Steam chest
3. Blades
4. Nozzles
5. Exhaust
6. Rotor wheels or disks
7. Diaphragms
8. Gland-steam inlet
9. High-pressure gland
10. Low-pressure gland
11. Labyrinth-seal strips
12. Rotor shaft
13. Collars

Fig. 19. Steam seal labyrinth.

Q. What is the purpose of diaphragms and in what type of turbine are they found?

A. Diaphragms are found in impulse turbines. They are used to hold the nozzle blocks and prevent steam leakage between the stages, **Fig. 20.**

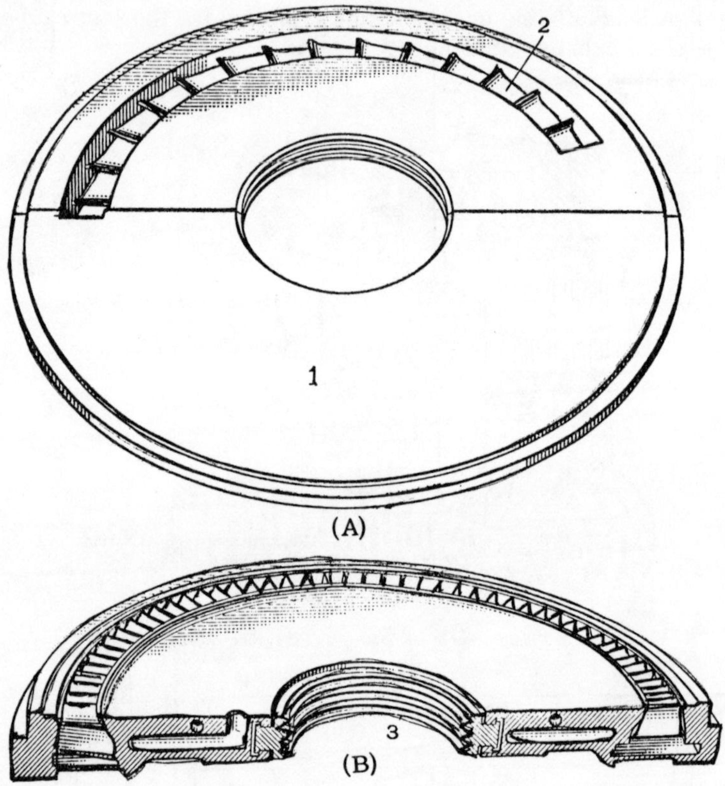

Fig. 20. Turbine diaphragm. View A, (1) diaphragm; (2) nozzles. View B, (3) labyrinth packing.

Q. Describe the diaphragm and interstage packing installed in a multistaged impulse turbine.

A. Between each stage of a pressure-compounded impulse turbine nozzle diaphragms are installed. The diaphragms are made up in two halves which are keyed into the upper and lower halves of the casing. Since a pressure drop exists across each diaphragm a packing seal must be installed around its inner web to prevent the flow of steam between the rotor and the diaphragm. Labyrinth packing rings, usually in four segments, are inserted in a groove in the diaphragm and kept in place by dowel pins or stop pins. Springs are placed behind each of the packing rings to maintain their close clearance with the shaft, Fig. 20.

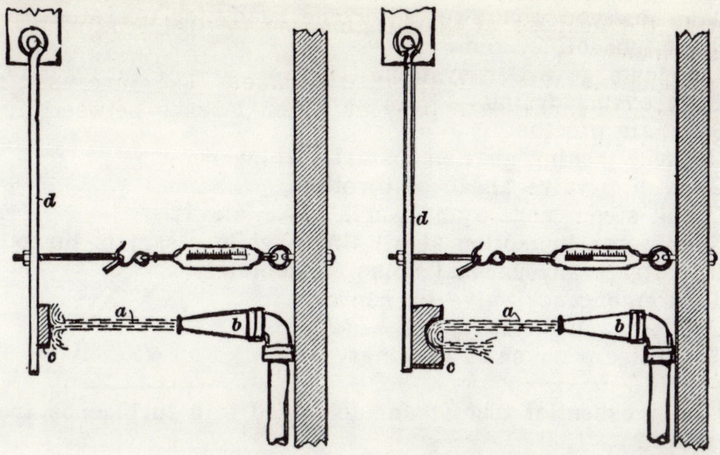

a. Water jet b. Nozzle c. Block d. Movable arm

Fig. 21. Diagram to explain impulse and reaction forces.

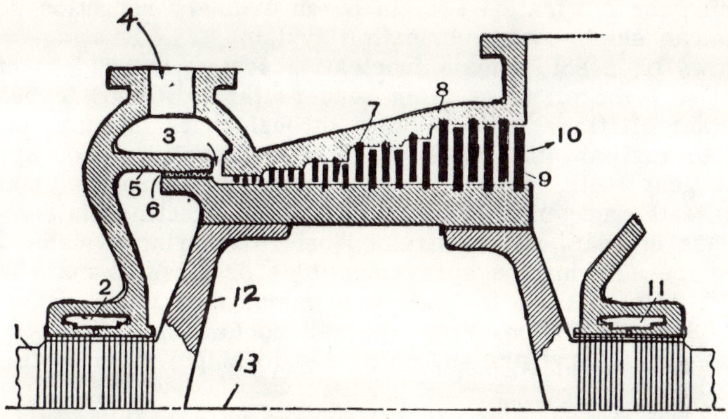

1. Shaft
2. Gland-steam seal
3. Steam chamber
4. Steam inlet
5. Dummy cylinder
6. Dummy packing
7. Stationary blades
8. Casing
9. Moving blades
10. Exhaust
11. Gland-steam seal
12. Hollow rotor
13. Rotor axis

Fig. 22. Major parts of axial flow turbine.

TURBINES—OPERATION

Q. Name three methods of determining the quantity of steam consumed by a turbine.

A. The quantity of steam consumed by the turbine is generally determined by one of the following methods: (1) By measuring the condensate. (2) By measuring the feed water. (3) By a steamflow meter.

The first method—that of weighing the condensate— will, generally, result in greater accuracy than will any of the other methods. Consequently, where practicable, it should be used.

Q. Explain how you would get a turbine ready for operation.

A. 1. Start lube oil system.
2. Lubricate governor system.
3. Open casing drains.
4. Start air ejector.
5. Connect jacking gear and start turning rotor.
6. Open stop valve ahead of throttle.
7. Crack steam nozzle and warm rotor slowly.
8. Keep rotor in motion at all times while steam is turned on.
9. Investigate any unusual noise immediately.
10. Trip emergency valve by hand.
11. Check bearing clearances.
12. Turn steam on shaft packings.

Q. Why is it essential that steam delivered to a turbine be free from water?

A. Because wet steam would soon pit, warp and break the blading in the turbine and loosen the blading.

Q. Describe the lubrication system on an ordinary reduction gear connected to ship's main lubrication system.

A. Oil flows from ship's main lubrication system to headers on gear units. Some oil flows from headers through tubes to bearings. Other oil flows from headers through tubes to spray nozzles or to auxiliary manifolds which have spray pipes for spraying over gear teeth, the oil being directed to the zone of mesh or gear teeth engagement, and spread by a directing plate across entire face of gear. Oil is strained before entering system, and again before entering the spray manifolds or pipes to main bearings. Oil falls to sump for return to main cooling and cleaning system. Some reduction gears are self-contained, have their own lubricant pumps, and are not connected to ship's main system.

Q. Explain the operation of the controls which permit the steam-driven lube oil pump to automatically start in the event of failure of the electrically-driven lube oil pump, and continue the lubrication of the main propulsion turbine.

A. Low pump discharge pressure acts on a pressure regulating valve in the steam line to the stand-by pump causing it to start if the lube oil pump discharge pressure drops below normal.

Q. The lube oil pump which supplies oil to the main bearings and governors of a turbine-generator set is operated through gearing connected to the reduction gear shaft. How is oil pressure provided for the speed controls and bearing lubrication before the unit is started up?

A. On main unit: by means of lube oil pump or gravity tank.
On generators: by means of a hand operated pump.

Q. What pressure is usually carried on a gland-sealing system?

A. 2 to 3 lbs. gage pressure.

Q. Sketch a gravity oil system as used in connection with a turbine.
A. Figure **23**.

Q. Sketch a force-feed oil system as used in connection with a turbine. Explain operation.
A. When the pump is running, part of the oil goes to the bearing and part goes through a small hole X in check valve C to the gravity tank. The oil overflows into a pipe from the gravity tank and runs through sight glass D.

The high-pressure alarm A is set for a specified pressure, at which point an alarm rings and a light flashes to warn the engineer. If the pumps should stop, the oil in the gravity tank will flow back down through valve C and set off a similar alarm, Fig. **24**.

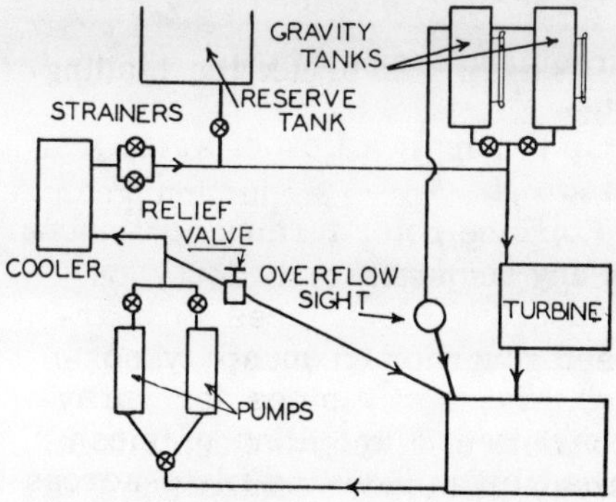

Fig. 23. Turbine gravity oil system.

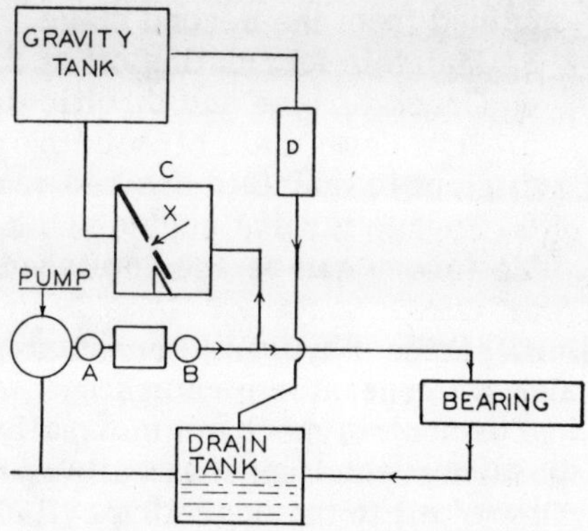

Fig. 24. Turbine force-feed oil system.

Q. What are the possible effects of excessive wear in the main bearings of a steam turbine?
A. The rotor will drop. In the case of the reaction type turbine, where the tip clearance is very small, this could be disastrous.

Q. Describe briefly the operation of an overspeed control governor on a main propulsion turbine.
A. The overspeed control valve is usually held open against spring pressure by oil at constant pressure supplied by the lubricating oil service pumps. This oil is supplied to the operating cylinder through a pilot valve which, when actuated by the governor, will shut off the oil flow and drain the operating cylinder allowing the spring pressure to close the overspeed control valve. The pilot valve may be actuated hydraulically by a small centrifugal oil pump directly connected to the turbine shaft or by mechanically connected speed governor weights.

Q. How should a gland seal operate?
A. There should be a slight wisp of steam issuing from both ends of the shaft at all times.

Q. Explain how you would shut down an auxiliary turbine if you found a badly overheated bearing.

A. If a turbine has to be shut down due to an overheated bearing it should be slowed down but kept turning over at a low speed until the bearing and journal have cooled sufficiently. Otherwise the bearing metal will freeze to the shaft and make repairs much more difficult.

Q. What is the procedure, with respect to a geared turbine vessel, when coming to a stop and the orders are to stand by for an indefinite period, being ready to get under way within 15 minutes.

A. 1. Open the recirculating valve from the deaerating feed tank to the condenser. It may also be necessary to open the recirculating valve from the main air ejector to the condenser.

2. Secure the first stage of the air ejector and maintain vacuum as obtained from the second stage.

3. Maintain lubricating oil at the required temperature.

4. Crack turbine and throttle drains.

5. Slow down main circulating pump to supply a flow of water just sufficient to maintain desired vacuum.

6. Engage turning engine and keep turbine rotor turning continuously.

The above can be accomplished in any sequence.

Q. Describe a hydraulic constant-speed governor frequently used to control generator turbines.

A. The control mechanism usually consists of a weighted centrifugal governor which mechanically operates a pilot valve controlling the flow of oil to the operating cylinder. The spring-loaded operating cylinder in turn controls the amount of opening or closing of the turbine nozzle valves. The oil pump and governor are mounted on the low-speed gear shaft of the turbine reduction gear. With an increase in turbine speed the governor weights move outward and draw the pilot valve downward; with a reduction in turbine speed the weights move in and push the pilot valve upward. The position of the pilot valve with respect to the pilot-valve bushing determines the flow of oil to the operating cylinder and therefore, the position of the operating valves that admit steam to the turbine.

Q. Outline the correct procedure to secure a turbine installation.

A. 1. Start auxiliary condensate system and transfer auxiliary exhaust and makeup feed to the auxiliary condenser.

2. Use hand tripping device to shut steam off main turbine, then close throttle and bulkhead stops.

3. Secure the main air ejector and the main condensate pump.

4. Shut off gland seal steam and open turbine drains.

5. Rotate main turbine with turning engine until cool, then secure turning engine and the main lubricating oil pumps.

6. When main condenser has cooled off secure main circulating pump.

Q. Outline the correct procedure to secure a turbine installation.
A. 1. Start auxiliary condensate system and transfer auxiliary exhaust and makeup feed to the auxiliary condenser.
2. Use hand tripping device to shut steam off main turbine, then close throttle and bulkhead stops.
3. Secure the main air ejector and the main condensate pump.
4. Shut off gland seal steam and open turbine drains.
5. Rotate main turbine with turning engine until cool, then secure turning engine and the main lubricating oil pumps.
6. When main condenser has cooled off secure main circulating pump.

Q. Name at least two means of reducing the speed of a turbine to the proper speed of the propeller.
A. Reduction gears or turboelectric drive through a generator and motor.

Q. Outline the correct procedure to secure a turbine installation.
A. Secure air ejectors. Open vacuum breakers. Secure all steam stops. Open all drains. Keep condensate pump running until all lines are dry. Engage jacking engine and run for 2 or 3 hours. Secure lube system. Shut down circulator when condenser is cool.

Q. Trace the flow of steam from the throttle through a cross-compound steam turbine unit when admitted for ahead power and when admitted for astern power. What provision is made for operating either the H. P. or L. P. turbine independently in order to meet an emergency?

A. For ahead power, high pressure steam from the ahead maneuvering valve enters the high pressure turbine at the forward end of the cylinder and flows toward the aft end. The exhaust from the high pressure turbine enters the low pressure turbine in the aft end of the cylinder and flows toward the forward end. The exhaust from the low pressure turbine passes downward directly into the condenser. Part of the ahead steam is under control of three hand-operated nozzle control valves, which are located in the high pressure turbine cylinder. In operation, the nozzle control valves should be opened only as required to develop the desired power. Operating with more than the required number of hand valves open will necessitate throttling, will result in greater steam consumption, and will tend to cause wire drawing and valve seat erosions at the maneuvering valve.

In addition to normal operation, either turbine may be operated, for ahead power only, independently of the other turbine of that unit, by installing the emergency piping.

For astern power, high pressure steam from the astern maneuvering valve enters the low pressure turbine at the forward end of the cylinder and flows through the astern blading which is located in the exhaust end of the cylinder. It then passes downward through the main exhaust opening into the condenser.

Q. If there is no steam issuing from the L. P. gland seal, what is likely to occur?

A. Air will leak into the unit and destroy the vacuum.

Q. Explain how you would continue to operate a modern steam turbine at sea, maintaining a vacuum, if an air leak developed that couldn't be corrected.

A. If an excessive amount of air is leaking into the condenser it will be necessary to use two nozzles for each stage of the air ejector in order to maintain vacuum. However, when this condition occurs you cannot expect to have as much vacuum as normally, since the increased temperature in the air ejector will cause a reduction in the possible vacuum. In order to keep the air ejector cool it may be necessary to take on a great deal of extra-feed, striking the excess down from the deaerating feed tank back to the reserve feed tank, and to recirculate water from the outlet of the air ejector back to the condenser.

Q. A ship has to be towed because the main turbine plant is damaged beyond operation. How would you protect the turbine from further damage while under tow?

A. The shaft must be prevented from turning by:

1. Engaging the line shaft brake if the unit is equipped with one.

2. Lift the full bearing (last one on the line shaft) remove shims, replace top of bearing and secure. This should prevent the shaft from turning. If this doesn't hold, remove the top half of one or two of the spring bearings, insert pads and tighten down bearing covers.

Never use jacking gear to hold turbine; it is too light for this purpose.

Secure rudder in midship position.

TURBINES—MAINTENANCE

Q. What are the two major adjustments of the main propelling turbine?

A. In all main propelling turbines installed in vessels, the two major adjustments are the fixing of the rotor in its proper radial and axial position. The radial position of the rotor is maintained by the main bearings and the axial position by the thrust bearing.

Q. What is pitting and what is its cause and effect on turbine reduction gears?

A. Pitting is the flaking of metal from the surface of the teeth or the loss of metal due to corrosion. Pitting, particularly along the pitch line, may occur in the first few months of service. This pitting, usually slight, ceases after this time and does not seriously affect the operation of the gears. Pitting in older gears is usually caused by corrosion due to water or an acid condition in the lubricating oil and must be stopped immediately before the tooth bearing surface is destroyed.

Q. What is the principle of labyrinth packing?

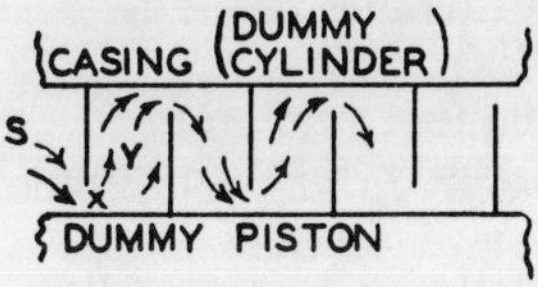

Fig. 25. Simple form of labyrinth packing.

A. There are many types of labyrinth packing, but all work on the principle of pressure drop. The diagram in Fig. 25 shows a simple form of labyrinth packing attached to a dummy cylinder and piston. The steam entering at S must pass through the small passage X. In doing so, it enters a large space Y, and thus the pressure is reduced with a corresponding increase in volume. By the time the steam passes through a designed number of these pressure drops, its pressure is negligible.

Q. If the L. P. turbine compound gage indicates 18" of vacuum, what is the absolute pressure?

A. 18" vacuum = 9 pounds pressure (absolute)
Atmospheric pressure = 15 pounds
15 — 9 = 6 pounds absolute

Q. What causes failure of turbine oil supply?

A. 1. Oil-sump tank level too low for the pump to pick up suction. (Watch level of sump with particular care when the vessel is rolling due to bad weather.)

2. Pump failure. Start another pump.

3. Strainers clogged with dirt. Change and clean suction and discharge strainer.

4. Broken oil line, or valve closed.

Q. Describe the usual lubrication arrangement for reduction gears.

A. The oil is sprayed on through a nozzle to the area where the teeth mesh. It drops to the sump tank in the lower half of the casing and is returned to the lube oil system (the gears do not run in the oil).

Q. What is a bridge gage? How is it used?

A. An instrument used to find the radial position of a crankshaft, or rotor shaft, Fig. 26. The bearing cap is removed and the gage placed on the lower bearing housing. Feelers are inserted between the bridge and the shaft and checked with original clearance stamped on gage.

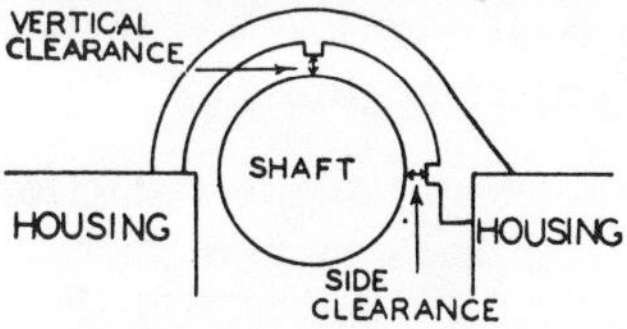

Fig. 26. Bridge gage, showing clearances.

Q. When operating at less than maximum capacity, is it better to use a large number of nozzles with reduced steam pressure or velocity, or to use a smaller number of nozzles?

A. It is more economical to use fewer nozzles and high chest pressure than a large number of nozzles with reduced pressure.

Q. Why is it important, when "warming up" a turbine, to see that sufficient steam is admitted to start the rotor turning immediately?

A. To prevent uneven heating of rotor and consequent warping which would result if steam were not on sufficiently to turn rotor.

Q. Explain how to refit a carbon packing ring if there is excessive steam leakage.

A. Manufacturers usually specify the required radial clearance. Therefore, it will be necessary for the packing to be fitted to a mandrel in order to obtain this proper clearance. For example, if a 4" shaft is to have a carbon gland clearance of .012" (radial), make a mandrel with a 4.024" diameter. Fit the carbons to the mandrel with no clearance but each segment butting. (A rough estimate of .002" to .003" radial clearance per inch diameter of shaft is sometimes used.)

Q. Explain how a single labyrinth gland may be refitted after it has become worn.

A. The single labyrinth packing gland consists of one or more metallic rings which are loosely supported by a shoulder in the packing chamber. Each ring is composed of three or more equal segments which are held together by a garter spring. One of the segments is provided with a stop to prevent the ring from rotating. When first assembled each ring is so machined that the tips of the saw tooth projections hug the shaft. When worn the glands may be expanded in height and drawn out to a feather edge by the use of appropriate hand chisels. The drawing out of the glands should be continued until they come within .005 in. of touching the shaft.

TURBINES—GENERAL

Q. What are some of the principal reasons for the greater efficiency of the steam turbine when compared with the reciprocating engine?

A. 1. Uses higher steam pressures and temperatures and exhausts to lower pressures than any other type of steam engine.
   2. Less space per horsepower.
   3. Less vibration.
   4. Uniform torque.
   5. Greater overload capacity.
   6. More reliable.
   7. Lighter in weight.
   8. Lower lube oil consumption.
   9. No lube oil in exhaust to get into boilers.

Q. Where is the thrust bearing usually installed in a turbine installation?
A. Usually on the free end of the turbine. That is, the end opposite to that connected to the reduction gear or generator.

Q. What are the advantages of the turbo-electric drive?
A. 1. Better maneuverability.
2. Elimination of the astern turbine.
3. Increased economy at reduced powers, especially when two or more generators are provided.
4. If the motor is located in the after end of the ship, there is a considerable reduction of shafting.
5. Full astern power available.

Q. What are the advantages of turbines as compared with reciprocating engines?
A. 1. Require less engine-room space per horsepower.
2. Lighter in weight.
3. Require less attention.
4. Use higher vacuum with better efficiencies.
5. Permit use of steam at higher pressures and temperatures.
6. Fewer reciprocating parts.
7. No rubbing parts such as rings, etc.
8. Exhaust steam free from oil.
9. High overload capacity.
10. Reliability.

BOILERS

FIRE-TUBE —CONSTRUCTION

Q. Describe a stay tube in a Scotch boiler and tell how it is installed.

A. Stay tubes are thicker walled than common tubes in order that they may bear the stress between the two tube sheets, due to the steam pressure acting on the tube sheets. The ends of the tube are upset, that is, the walls are thickened to increase the outside diameter of the tube without decreasing the inside diameter, and the upset ends are threaded. The holes for stay tubes in the tube sheet are also threaded. To install a stay tube, first screw it through the front tube sheet beyond the thread on the tube, and slide the tube through to the back tube sheet. The tube will then screw into both tube sheets. After the tube is screwed in place, it is expanded and the ends are flared, and also may be beaded. Sometimes nuts are fitted to the protruding ends of stay tubes.

Q. Why should rivet holes be drilled rather than punched?

A. Punching weakens the material around the hole. In some instances the holes are punched smaller and then reamed to correct size. The size of the hole that may be punched prior to reaming is regulated by U. S. C. G. regulations.

Q. What difference, if any, is there between a fusible plug marked "inside" on the casing and one marked "outside" on the casing?

A. An inside plug is one that is screwed in from the inside of the combustion chamber (fire side). An outside plug is one that is screwed in from the top of the combustion chamber on the water side, Fig. 27.

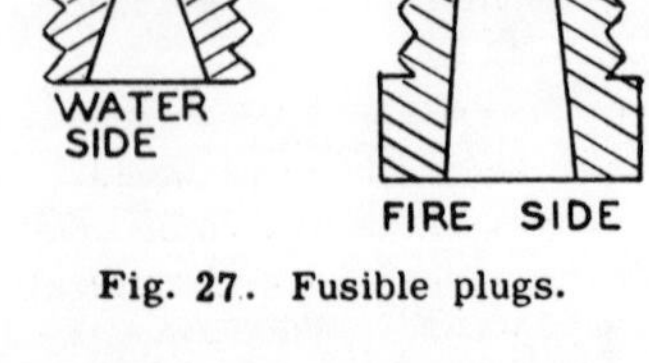

Fig. 27. Fusible plugs.

Q. What is meant by water hammer? What causes it, and how can it be prevented?

A. If there is condensed moisture in a steam line, this may form a slug of water ahead of the steam flowing through the line. This slug of water bangs against elbows, fittings, valves, etc., with a noise that sounds like a hammer hitting metal. Usually, it can be prevented by draining the line thoroughly or opening the valve very slowly and letting the line warm up gradually.

Q. Name the parts on a Scotch boiler.
A. Figures 28 and 29.

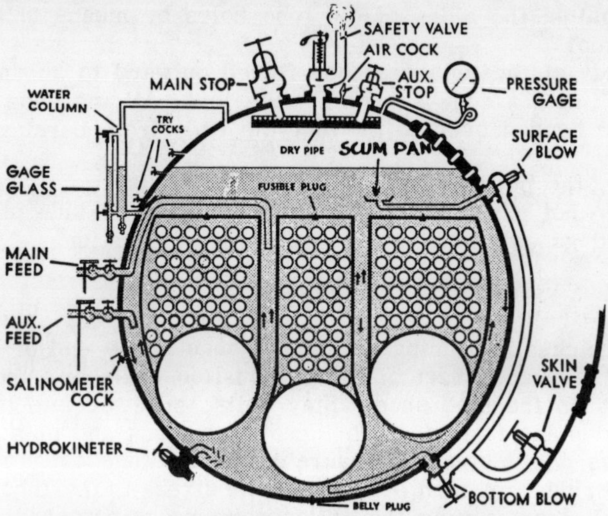

Fig. 28. Front view of Scotch boiler with fittings attached.

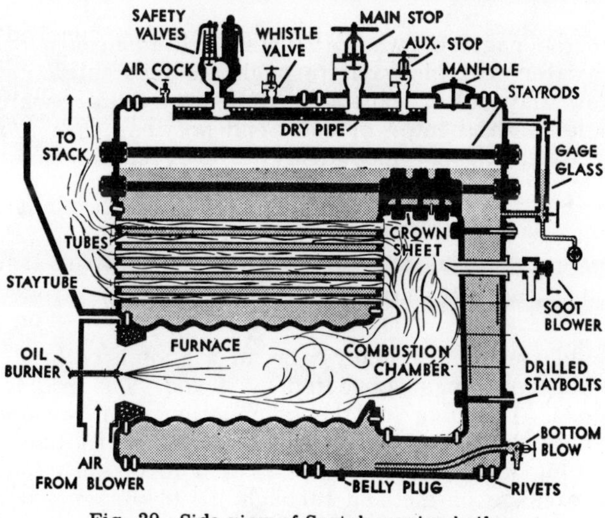

Fig. 29. Side view of Scotch marine boiler.

Q. Describe a plain tube in a Scotch boiler and tell how it is installed.

A. A plain tube is a cylindrical shell of comparatively small diameter and long enough to extend from tube sheet to tube sheet. It may be made of plate formed to a cylindrical shape with a welded seam, or it may be of seamless drawn steel. Its diameter is equal throughout its length and is from 1/32" to 1/16" smaller in diameter than the tube holes in the tube sheets. The tube is of sufficient length so that it extends 3/16" to 1/4" beyond the tube sheets. After the tube is placed in position, it is expanded tightly against the sides of the tube holes by means of an expanding tool.

The ends of the tube are then flared outward to an angle of about 45° with a cone-shaped flaring tool. All the ends of the tubes are beaded over to prevent the ends from burning off.

Q. What is the purpose of furnace bridge walls?
A. To keep hot gases in close contact with under side of furnace. To retard gases and give more perfect combustion.

Q. What is a girder stay? Describe.
A. A bridge built up of plates or structural shapes, separated by distance pieces, spanning the crown sheet. One end of the bridge rests on the back sheet of the combustion chamber, and the other end rests on the tube sheet. Stay bolts, screwed into the crown sheet and made fast to the bridge by washers and nuts, transmit the stress due to steam pressure on the crown sheet to the bridge, and the bridge, in turn, transmits the stress to the tube and back sheets, thus preventing the collapse of the crown sheet.

Q. Explain fully the purpose and requirements of telltale holes in stay bolts.
A. To warn the engineer when the stay bolt is broken or cracked. Steam or water will blow out through the test hole.

All screw stay bolts, except flexible stay bolts, shall have a telltale hole in the center of each end not less than 3/16" in diameter, extending not less than 1/2" beyond the inside of the plate.

FIRE-TUBE—OPERATION

Q. How would you plug a leaky tube in a Scotch boiler?
A. By two tapered plugs connected by a rod. The rod is threaded on both ends. The plug for the combustion chamber end has a threaded hole through it, the threads of which fit those on the rod. The plug for the front end has a hole of sufficient size to allow the passage of the rod through it. In plugging a tube, the rod is inserted in the tube, and a man in the combustion chamber screws the combustion chamber end plug on to the rod and hammers the plug firmly into the end of the tube. The other plug is slid over the front end of the rod and is followed by a washer and a nut. The front end of the rod, which has a square end, is held with a wrench, and the nut is tightened up with another wrench, making a tight joint.

Q. What is a hydrokineter?
A. A fitting located in the bottom of a Scotch boiler. Its principle is that a jet of steam will pick up some of the surrounding water and push it along. The purpose of the hydrokineter is to circulate the water in the boiler while lighting off and raising steam.

Q. When finding water level by test cocks, which one is opened first and what should come out?
A. Bottom cock; water.
In a high-pressure boiler, steam or steam and water will come out of all cocks. Learn to recognize the difference between wet steam from the bottom cock and dry steam from the top one.

Q. Describe the double shut-off method of testing the water-column and gage-glass connections.
A. 1. Close the water connections on column and glass.
2. Open steam connections on column and glass.
3. Open the drain and, if steam blows through strongly, the steam connections are clear.
4. Close steam connections on column and glass.
5. Open water connections on column and glass.
6. Open drain and, if water blows through strongly, the water connections are clear. (Actually, this will appear as very wet steam, due to the fact that the water being under pressure in the boiler will change to vapor upon its sudden change to an area of lower pressure.)

Q. What is the purpose of furnace bridge walls?
A. To keep hot gases in close contact with under side of furnace. To retard gases and give more perfect combustion.

Q. What effect does the height of the bridge wall have on combustion?
A. If too low, the gases escape too rapidly with a consequent loss of heat. If too high, the gases are held in the furnace too long and not allowed to reach the tubes which have the greatest heating surfaces.

Q. How would you get up steam in a Scotch boiler and cut it into a bank of running boilers?
A. 1. Drop the water level in the boiler until it shows about 1" above the lower nut on the gage glass connection.
2. See that the surface and bottom blow valves are closed.
3. Ease up on the main and auxiliary stop valves without raising the discs off their seats.
4. Open the air cock.
5. Circulate the oil in the boiler-front fuel manifold until the proper burning temperature is reached.
6. Light a fire, using the smallest available tip on the burner.

Allow the fire to burn for 15 or 20 minutes and secure it. Repeat this in the other furnaces in rotation. It should take from 8 to 12 hours to get the boiler up to steaming pressure, depending on the size of the boiler. The necessary interval between lighting off and securing fires is best learned by experience.

7. Close air cock when strong jet of steam issues from it.

8. Before opening the steam stop, the line must be drained. All large steam lines are fitted with ample drains. Be sure that they are all open to insure that there is no water in the line, and leave the drains cracked while the stops are being opened. When dry steam issues from all the drains, they may be closed.

9. Open by-pass line to equalize pressure between boiler and line.

10. Open the stop valve slowly.

11. Notify the water tender and fireman that boiler is cut in and have water tender open feed stop and check.

Q. What is the purpose of furnace bridge walls?

A. To keep hot gases in close contact with under side of furnace. To retard gases and give more perfect combustion.

Q. What effect does the height of the bridge wall have on combustion?

A. If too low, the gases escape too rapidly with a consequent loss of heat. If too high, the gases are held in the furnace too long and not allowed to reach the tubes which have the greatest heating surfaces.

Q. How would you shut down and secure a Scotch boiler?

A. 1. Secure the fires and shut the air checks on the furnace fronts.

2. When the boiler has cooled down and is no longer making steam, close the main and auxiliary steam stops.

3. Fill the boiler with water until the gage glass shows 3/4 full and secure the feed stop and check valves.

Q. How would you cut a boiler out of service?

A. 1. Close oil valves and air registers in each burner. (Oil pressure must therefore be reduced).

2. Secure main oil valve to that boiler.

3. Slow down blower.

4. Secure steam stops.

5. Secure feed lines.

6. Remove all burners from boiler, remove tips, and place in kerosene.

7. Open air vent when pressure drops to about 5 lbs. (Prevents formation of vacuum.)

BOILERS

WATERTUBE — CONSTRUCTION

Q. What are waterwalls in a watertube boiler and what is their purpose?

A. They are the generating tubes placed in the furnace, generally arranged so as to form a lining for the furnace. In modern boilers which have waterwalls on all sides and the floor of the furnace, approximately 45-60% of the total heat transferred is through these waterwalls. They also act as protection for the refractory against intense furnace heat.

Q. How is a tube plugged in a watertube boiler?

A. By means of soft iron plugs driven into the tube ends. Boiler pressure keeps the plugs in place.

Q. Describe in detail how to cut out and secure a modern watertube boiler.

A. According to the Marine Service Manual, published by Babcock & Wilcox, the procedure is as follows:

In cutting out a boiler, the fuel supply should be stopped, following which, the air supply can be shut off. On oil-burning installations, as soon as the burners are shut off, they should be removed. As a safety precaution, burners should not be allowed to remain in place any time except when actually in use.

When the boiler no longer requires any feed, the non-return valve (if fitted) has closed, the main steam stop valve should be closed. The pressure should be allowed to drop naturally without the aid of any open vents. The superheater drains, however, should be opened sufficiently to keep all condensate out of the superheater headers. Hastening the cooling of the furnace by allowing large quantities of air to pass through the setting tends toward brickwork difficulties.

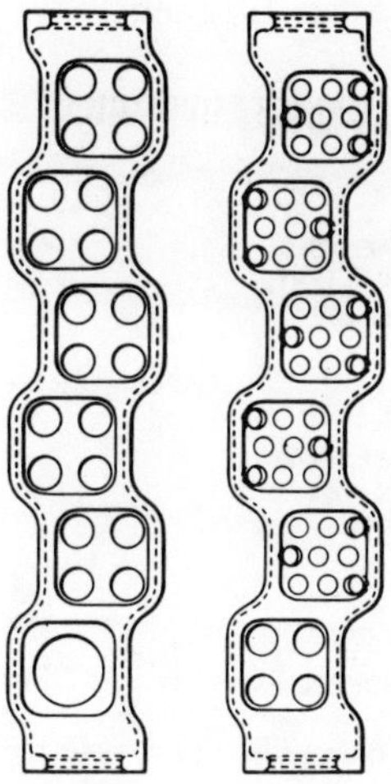

Fig. 30. Tube headers.

When the steam pressure has dropped to 5 or 10 lbs., the drum steam-vent valve (air cock) should be opened to prevent the formation of a vacuum within the boiler by the total condensation of the steam. Any tendency to set up a vacuum within the unit will serve to cause future leakage of any gasket joints which may not have been drawn to a perfect seat or fully followed up.

The boiler should not be emptied until the furnace has cooled to a temperature at which one can enter and remain in the furnace.

Q. What causes panting?
A. Panting is usually caused by:
1. Deficiency of air.
2. Excessive oil temperature.
3. Poor oil-air mixture.

Q. Describe the tube headers used on a sectional header tube boiler. How are they connected to the other pressure parts of the boiler?

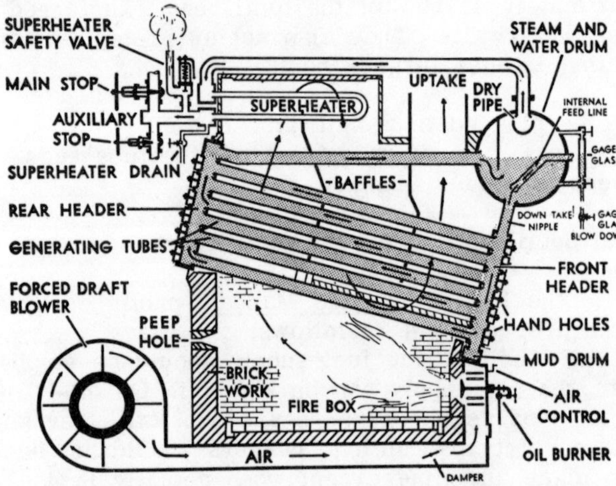

Fig. 31. Side view of B and W straight tube, cross drum watertube boiler—sectional header.

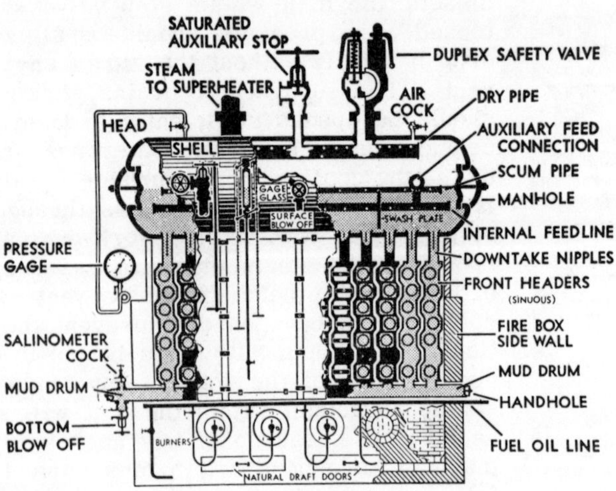

Fig. 32. Front view of B and W straight tube, cross drum watertube boiler.

Note. **Check boiler diagrams in 2nd Assistant Engineer Section.**

A. Tube headers are forged steel boxes (usually sinuous or serpentine in shape), Fig. 30. The rear surface is drilled for tube insertion and the front surface is fitted with a square handhole and fitted handhole plate. The tubes are rolled (expanded) into the tube holes. The hole in the top of the header is connected to the steam and water drum by a long nipple, sometimes referred to as a downtake nipple. It is expanded into both pieces. The hole in the bottom of the header is connected to the mud drum by a short nipple which is expanded into both pieces. Thus, each header is connected individually to the steam and water drum and the mud drum. The small space between each header is calked with asbestos to prevent the gases of combustion from leaking out.

Q. What is the purpose of the gas baffles in a watertube boiler?
A. They are used to divert the path of combustion gases, especially in a three-pass boiler, Fig. 31.

Q. What is the purpose of the steam baffles?
A. To prevent the steam coming back through the return tubes from blowing directly into the water in the steam and water drum setting up a turbulence effect, Fig. 31.

Q. What is the purpose of a furnace in a boiler?
A. It is to provide a space in which the fuel can mix with air and begin burning.

Q. What is the bottom blow valve?
A. A valve connected to the bottom of a boiler, for the purpose of removing sediment, mud, bits of scale, etc., from the bottom of the boiler and also to reduce the concentration of impurities in the boiler water by removing some of the impure water through the bottom blow, and replacing it with pure water, thus diluting the impurities. Figure 32.

Q. What is the feed-stop valve and where is it located?
A. A valve in the feed line next to the boiler. It is used to secure the feed from the boiler when it is necessary.

Q. What is the feed-check valve and where is it located?
A. It is a valve next to the feed-stop valve and is between the feed stop and the feed pump. The valve is made so that it will allow water to flow through it in one direction only. That is, the feedwater can flow into the boiler, but in case the feed pump should stop, the pressure in the boiler cannot force the water back out through the feed check. This valve is also adjustable and is used to regulate the amount of water entering the boiler.

Q. What is a boiler safety valve?
A. It is an automatic valve to prevent the steam pressure within the boiler from exceeding the allowable working pressure.

Q. Describe a dry pipe.

A. A large pipe located at the top of the steam space of a boiler and running parallel to the shell or steam drum. The upper portion of the pipe is perforated with small holes, or slots, along its entire length. The purpose of the dry pipe is to collect the steam evenly from the length of the steam space and to cause the steam to drop most of the moisture that it carries with it, and thereby supply drier steam.

Dry pipes should have a wall thickness at least equal to standard commercial pipe thickness for their diameter. Areas of openings must be at least twice the nominal area of the dry pipe. Openings in pipe should be as near as practicable to the valve; slots in pipe to be not less than 1/4", hole diameters to be not less than 3/8". Drains shall be provided at each end to prevent accumulation of water.

Q. What is a salinometer cock?

A. A small valve placed on the boiler below the water level for the purpose of drawing off samples of the boiler water for testing.

Q. What is an air heater?

A. An arrangement of tubes located in the boiler uptakes. The uptake gases flow over the outer surface of the tubes while the air, which is supplied to the furnace to support combustion of the fuel, flows through the tubes before entering the furnace.

Q. What is an economizer?

A. An arrangement of tubes located in the uptakes to heat feedwater before it enters the boiler. Gases of combustion, which would otherwise be wasted, flow over the tubes and heat the water.

Q. What is a surface blow valve and scum pan ?

A. The surface blow valve is a valve connected to the boiler near the normal steaming water level. The valve is connected to a scum pan or perforated pipe inside the boiler, the purpose of which is to collect the floating impurities on the water, and to a line outside of the boiler, leading overboard. The surface blow is used to remove oil and other floating impurities from the surface of the water.

Q. Name the valves on a boiler.

A. Although the valves on a boiler differ with the type of boiler and with the individual installation, a typical installation will have: safety valve, main stop, auxiliary stop, air-vent cock, steam-gage valve, gage-glass valves, try cocks, main feed-stop, main feed-check, auxiliary feed-stop, auxiliary feed-check, surface-blow valve, bottom-blow valve, whistle valve, drain valve, soot-blower valve, salinometer cock.

Q. What is the purpose of a superheater?
A. To add heat to the steam without raising the pressure of the steam.

Q. What is a desuperheater? Describe a typical desuperheater.
A. A desuperheater is a device for removing the superheat from steam. It usually consists of a coil of piping located in the steam drum of a boiler below the normal steaming water level. A connection is made from the outlet of the superheater to one end of the piping, and the other end of the piping is connected to a line leading to those auxiliaries which cannot use superheated steam. In passing through the desuperheater, the steam gives up its superheat to the water in the drum and emerges as saturated or mildly superheated steam.

BOILERS—OPERATION

Q. What is the principal precaution in the operation of blow-down valves?
A. Do not leave blow-down valve until shut off.

Q. Why are by-pass lines installed around large steam-stop valves?
A. Steam-stop valves over 6'' in diameter are required to have by-pass lines installed around them in order to warm up the line before opening the stop valve.

Q. Where is the air cock located and what is its purpose?
A. The air cock is a small valve located at the very top of the steam space of a boiler. Its purpose is to allow air to escape from the boiler while raising steam or filling the boiler with water, and to allow air to enter the boiler while it is being drained.

Q. When and how would you use the bottom blow?
A. The best time to use the bottom blow to remove solid matter from the boiler is when the boiler is cut out, as most of the heavy matter in the water will have had a chance to settle to the bottom of the boiler and a larger amount of it can be removed by blowing. It is better to give several short blows than one long blow, as more of the solid matter will be removed in that way. If reducing the salinity of the boiler water is the object, the boiler should be filled until the gage glass shows about 3/4 full; the bottom blow is then opened and kept open until the gage glass shows the water level to be just above the bottom nut on the gage-glass connection. This is repeated until the salinity, as indicated by samples taken from the boiler through the salinometer cock, is at the desired point.

Q. How would you cut a boiler out of service?

A. 1. Close oil valves and air registers in each burner unit. (Reduce oil pressure)

2. Secure master oil shut-off valve to boiler.

3. Slow down blower.

4. Secure steam stops.

5. Secure feed lines.

6. Remove all burners from boiler, remove tips, and place in kerosene.

7. Open air vent when pressure drops to about 5 lbs. (Prevents formation of vacuum.)

Q. What would you check if your fires went out in the boiler?

A. 1. See that the fuel-oil pump is operating.

2. Fuel may not be hot enough for combustion.

3. Fuel may be too cold to pump.

4. Check pump and burner strainers.

5. Bunker tank may be empty.

6. Test for water in the fuel.

7. Valve in the line may be closed by mistake.

8. Vent screen in bunker tank may be clogged with dried fuel from a previous overflow, causing a vacuum in the tank.

Q. Should the stress on a stop valve be relieved before steaming the boiler? Explain.

A. Yes. Break valve from seat to prevent valve from freezing when steam expands it.

Q. What is the position of the feed-stop and check valves when the boiler is steaming? When the boiler is cut out?

A. When the boiler is steaming, the feed-stop valve is wide open and the feed-check valve is open only sufficiently to maintain the level of the water in the boiler. When the boiler is cut out, both stop and check valves are closed to prevent any water from entering the boiler.

Q. If the water becomes dangerously low in a Scotch boiler, what would you do?

A. If the water in a Scotch boiler became dangerously low, the fusible plug would melt out. When this happens, secure the fires and feedwater at once; then secure the main and auxiliary steam stops and open the safety valve with the hand-lifting gear; secure the air to the furnaces and allow the boiler to cool gradually. (Call chief engineer.) When the boiler has cooled, carefully examine the crown sheets and upper rows of tubes for sag or any other sign of overheating. If no damage is found, renew the fusible plugs and get up steam again. A report must be submitted to the inspectors in the next port where there is an inspector. If damage was done by overheating, the inspector must first approve any necessary repairs.

Q. What is the most important requirement of any steam boiler? What is the essential difference between a water-tube and a Scotch boiler?

A. A high factor of safety. The Scotch boiler is constructed with large tubes (4-inch) and the gases of combustion pass through the tubes and heat the water which surrounds the tubes. The water-tube boiler is constructed with smaller tubes (usually 2-inch & 1-inch) and the water passes through the tubes and is heated by the gases of combustion which pass around the tubes.

Q. What causes foaming in a boiler?

A. Generally caused by high concentrations of dissolves or suspended solids or both in the boiler water. A fluctuating water level and wet steam are the results.

Q. Name ten instruments or devices used in the modern boiler plant for control of operating conditions.

A. Pressure gages, thermometers, oil meter, manometer, pressure reducing valves, pressure regulating valves, traps, pyrometer, CO_2 recorder, thermostats, check valves, feedwater testing outfit, relief valves, safety valves.

Q. What would you do if a tube ruptured in a Scotch boiler?

A. 1. Secure the fires.

2. Secure the main and auxiliary steam stops.

3. Speed up the blowers to force the escaping steam up the stack if there is any danger to the fireroom personnel.

4. Continue feeding the boiler until the furnace temperature drops so that exposed surfaces will not become overheated.

Q. Why is it desirable to exclude air from boiler feedwater? Why is this more important with high-pressure boilers than with low-pressure boilers?

A. Dissolved oxygen attacks boiler steel at saturation temperatures and pressures. The rapidity of the attack increases sharply with increase in temperature.

Q. How is a feedwater regulator used while maneuvering?

A. If the regulator is in good condition, it will keep a steady water level during maneuvering. However, either the B & W or Bailey regulator may be locked in the wide-open position by a simple operation and the water level controlled by the manually-operated check valve.

Q. How would you use the surface blow?

A. Raise the water level in the boiler to above the scum pan, open the surface blow valve quickly and close it when the water level in the gage glass ceases to drop. Blow out the gage glass and note the condition of the water which appears in the glass after blowing it; if there are any signs of floating matter, repeat the surface blow until the water shows clean.

Q. What are the most common sources of salt-water leakage into the feed system?

A. 1. Leaky condenser tubes.
2. Evaporator carry-over.
3. Leaky evaporator coils.
4. Leaky double-bottom reserve tanks.

Q. What color flame indicates efficient or poor combustion?

A. A yellowish orange, rolling flame indicates good combustion. (At high rates of combustion the flame should be a soft white.)
An incandescent white flame with jagged edges shows excess air.
A reddish, smoky flame indicates insufficient air.

Q. What color smoke indicates (1) efficient combustion; (2) insufficient air; (3) excess air?

A. 1. A light brown haze at the stack usually indicates efficient combustion.
2. Black smoke at the stack usually indicates insufficient air.
3. White smoke at the stack indicates a large amount of excess air.

Q. Why should atomizers be removed from the burner when not being used?

A. Secured atomizers left in a hot furnace will cause overheating of the sprayer plate and baking of carbon on the atomizer tips.

Q. What causes "flarebacks?"

A. A flareback is due to the explosion of a mixture of oil vapor or gas and air in the furnace. A flareback is most apt to occur when lighting off, or when attempting to relight a burner from a hot brick wall.

Q. How can "flarebacks" be prevented?

A. Open air register and let air blow through furnace before attempting to light burner. This will force any explosive gas or vapor mixtures up the stack before lighting off. Always use a torch when lighting off a fire.

Q. What does a sputtering burner indicate?

A. Water in the fuel.

Q. Describe in detail the path of air and gases through a Scotch boiler.

A. The air enters through the air registers at the front of the furnace and mixes with the oil which is sprayed into the furnace. The oil begins to burn in the furnace and passes into the combustion chamber, where burning is completed. The resulting gases pass through the tubes and give up their heat to the water. From the tubes, the gases pass into the uptakes, breechings and stack to the atmosphere.

Q. (a) Explain why boilers require extra feed. (b) What is indicated by a constantly rising water level in a manually fed boiler?

A. (a) Extra feed is required to make up the losses of steam or condensate throughout the vessel. These losses may come from leaking glands, joints, steam relief or safety valves, and boiler tubes; also by overflowing of open heaters and the use of the steam whistle.

(b) A constantly rising water level would indicate a leaking condenser tube or plate, a leaking or stuck open extra feed valve, a jammed or excessively open check valve, or excessive pressure or speed of the feed pumps.

Q. What is the purpose of the surface blow valve?

A. To remove oil and other floating impurities from the surface of the water in the steam drum.

GENERAL

Q. Explain why it is so important to keep air leakage into the furnace from the boiler casings to a minimum.

A. Waste in fuel due to air leakage into the furnace will be in proportion to the air pressure generally used and may amount to 10 per cent or more. The air leaking through boiler casings does not become intimately mixed with the fuel and, therefore, does not aid in the combustion of unburned gases. On the contrary, it has a decided chilling effect on both the gases of combustion and the heating surfaces.

Q. Explain the effect of a cracked regulator diaphragm or bellows upon the boiler water level when operating under the control of the thermo-hydraulic automatic feed system. How can the feed be controlled under these conditions?

A. A cracked diaphragm would cause the loss of the pressure within the generator and the spring of the regulator would close the valve, thereby shutting off the boiler feed. To operate the boiler under these conditions it will be necessary to bypass the regulator or jack it open and control the water level with the hand check valve.

Q. List the precautions to be taken when it's necessary to break a joint on a steam line?

A. The primary concern here is one of safety considerations. The service of which the piping is a part should be secured by physically lashing valves shut and putting warning signs on the valves to prevent inadvertent opening of the valves when the joint or flange is opened. Before the flange is broken careful tests must be made to make certain that pressure has been removed from the line. In slacking off on the flange nuts it is advisable to keep two diametrically opposite nuts and bolts in place until the actual breaking apart of the joint.

Q. Explain why boilers need extra feed. What is indicated by a constantly rising water level in the boiler?

A. In reciprocating drives, more extra feed is needed than in turbine drives due to larger losses through piston rods, valve stems, drains, air pump relief valves when starting, etc. A constantly rising level may be caused by a leak in the condenser (although this should show itself immediately on the salinity indicator), drains open and leading to the hot well, traps not working properly and discharging straight through to hot well.

If there are two or more boilers in the battery and one seems to be getting more water than the others, check the fires.

Q. How does oxygen corrosion usually affect steaming boilers? Where is this type of corrosion most frequently noted?

A. Metal, when taken from the earth as an ore, is some form of iron oxide. In the manufacture of iron, the oxygen is removed and pure iron is obtained. In the presence of oxygen and moisture, this metal has a tendency to return to its original state and corrosion results.

This type of corrosion is known as pitting. It is most frequently found on the furnaces of Scotch boilers and in the lower tubes of water-tube boilers near the front end of the tubes. The theory behind this is the fact that the oxygen bubbles are liberated at this point following a change of direction of the water flowing down the front headers.

Q. What is the operating principle of the electrical salinity indicators placed in various points of the feedwater system?

A. This indicator operates on the principle that salt water is a better conductor of electricity than fresh water. Two metal rods are placed in the hot well, or any other place where a reading is desired, at a prescribed distance apart; the rods being separated form an open circuit. As long as the water remains fresh, the circuit remains open. The presence of salt in the slightest degree immediately makes the water a conductor and passes the electricity through the water from one rod to the other, thus completing the circuit. At this point, a warning bell rings and a red light appears on the salinity indicator control box.

Q. Why should all superheaters have safety valves attached? How are they set in relation to boiler pressure?

A. To protect the superheater from overheating in case excess pressure in the boiler causes the drum safety valves to blow and the surge of steam from the drum out the escape pipe causes a momentary stop in the flow of steam from the drum through the superheater. The superheater safety valve is set about 5 pounds lower than the drum safety valve, thus it will blow first, insuring that there will be a flow of steam through the superheater.

Q. What precaution should be taken when gagging a safety valve?

A. Do not tighten too much, as it may bend the stem. Gags are usually fitted with a thumb screw. Do not use a wrench.

Q. Describe and give the purpose of the drain inspection tank connected into the fuel oil heating system.

A. The trap on the exhaust line from the fuel oil heater leads to the inspection tank and then drains into the hot well. This tank is fitted with two port glasses on opposite sides of the tank. A light is placed in one port and shows the inside of the tank when looked at through the opposite port. The water level is carried about halfway up the port. If there is any appearance of fuel oil on the surface of the water (due to a leak in the heater), the inspection tank is immediately drained to the bilge until the trouble is remedied.

Q. What is the minimum flash point of fuel oil allowed by the regulations to be burned under boilers on vessels subject to inspection? Why is it so important to know the flash point?

A. Below 150°F. This is to prevent inflammable vapors from forming in the storage tanks under ordinary atmospheric conditions.

Q. Define: Flash point; fire point; viscosity.

A. Flash point is the temperature at which an oil will give off an inflammable vapor.

Fire point is the temperature at which an oil will flash and continue to burn.

Viscosity is "internal friction" or "resistance to flow."

Q. What is "priming?" What effect does priming have on machinery?

A. Carryover of entrained water with the steam into the engines. Priming abrades blades in turbines, breaks cylinder heads, piston rods, valves, etc.

Q. What are pitting and grooving?

A. Pitting and grooving are forms of corrosion found in a boiler. Pitting is local corrosion usually found along the water line. Grooving is usually found around seams or where the metal has been bent or strained. Also found around rivet heads.

Q. What are the requirements for feed lines and feed valves on marine boilers?

A. Feed stop valves shall be attached directly to the boiler or its economizer. Where conditions of installation will not permit of direct attachment, forged or cast steel fittings made as short as practicable may be used.

All feed lines shall be provided with a feed check valve adjacent to the feed stop valve. An approved form of feed water regulator may be interposed between the check and the stop valve.

All boilers except small donkey boilers shall be equipped with two connections for supplying feed water, and where practicable, these connections shall be through separate openings.

Q. What is superheated steam?

A. Saturated steam heated to a temperature above that due to its pressure. This is accomplished by passing saturated steam, as it is generated in a boiler, through a series of so-called superheater tubes or coils, which are in most instances heated by the boiler furnace. The reason for the use of superheater tubes or coils is that saturated steam cannot be superheated so long as it remains in contact with the water from which it was generated. To add more heat to the water would merely increase its temperature and pressure and would in no way superheat the steam generated by its evaporation.

Q. Define the three methods of heat transfer.

A. Conduction is the flow of heat within a substance without any actual motion of the substance.

Convection is the transfer of heat through the motion of a heated fluid (liquid or gas).

Radiation is the transfer of heat by waves which carry the heat from a hotter to a colder substance when they are not actually in contact with each other.

Q. What is "boiler horsepower?"

A. The evaporation of 34.5 lbs. of water per hour from and at 212° F.

Q. Why doesn't the cylindrical shell of a boiler need bracing?

A. Due to the cylindrical construction, there is uniform pressure all over.

Q. What is a furnace explosion?

A. The rapid combustion of a mixture of oil vapor and air in a furnace.

Q. Does the water level in the gage glass in a water column 2 feet from the boiler show the true level of the water in the boiler? Why?

A. No. The water in the column is cooler and, therefore, denser (or heavier) than that in the boiler. Hence, water in the glass is lower than that in the boiler.

Q. What is saturated steam?

A. Steam produced by boiling water in a closed vessel has the same temperature as the water, and it is said to be saturated. Saturated steam, then, is steam at the temperature of the boiling point of the water from which it was generated, and may be either wet or dry saturated steam, depending upon whether or not it contains moisture.

Therefore, the term saturated steam refers merely to the fact that the steam is under the same conditions, temperature and pressure, as the water, and does not indicate whether it is wet or dry.

Q. What is the effect of oil or grease in the water in a boiler?
A. Grease in the boiler water will cause foaming and priming when floating on the surface of the water. It will also adhere to the heating surfaces and cause poor heat transfer; and, if the situation becomes bad enough, overheating and damage to the parts affected. Some forms of oil or grease cause corrosion due to chemicals contained in them.

Q. What effect has scale on the heating surfaces of a boiler?
A. Scale in a boiler forms an insulating layer on the water side of the heating surfaces. This insulating layer of scale resists the transfer of heat from the fire and hot combustion gases to the water, and the heat which would normally pass into the water remains in the metal and heats it. If the layer of scale is sufficiently thick and resists the transfer of heat sufficiently, the metal will overheat, become weak and bag, blister or rupture.

Q. How is scale formed?
A. When a steam bubble is formed on a heating surface, all the salts in the bubble will flow toward the hottest part, which is the heating-surface metal. Since slightly soluble salts become insoluble at high temperatures, scale will be formed at the base of the bubble. When the bubble leaves the heating surface, a ring of scales is left. These rings interlace and scale is formed.

Q. How is scale formation prevented in a boiler?
A. By adding a soluble chemical to the boiler water, which will react with the calcium sulphate (most common scale-forming salt) to form a highly soluble salt and a harmless sludge.

For low-pressure boilers (below 200 lbs.), sodium carbonate (in boiler compound) plus calcium sulphate (scale-forming salt) is equal to sodium sulphate (soluble salt) plus calcium carbonate (harmless sludge).

For high-pressure boilers (over 200 lbs.), sodium phosphate (in boiler compound) plus calcium sulphate (scale-forming salt) is equal to sodium sulphate (soluble salt) plus calcium phosphate (harmless sludge).

ELECTRICITY

DIRECT CURRENT

Q. What is a DC generator?
A. One which generates DC current for the external circuit.

Q. Explain how current and voltage are generated in a DC generator.
A. Conductors passing through a magnetic field have a voltage induced in them. Current flows if the circuit is closed.

Q. How are the fields of a DC generator excited?
A. They are excited by the current generated by the machine itself.

Q. What is meant by bad commutation on a DC generator?
A. It is making the connection at the brushes at the wrong time, so that the current taken off is not the maximum voltage.

Q. What is bad commutation? Give ten reasons for it.
A. This means that the brushes are not set at the proper point. It is noted by sparking at the brushes. Reasons: brushes set too far ahead; brushes set too far behind; rough commutator; hard brushes; high mica; short in a coil; open in a coil; bearing low, causing commutator to be off-center; overload; loose brush holders.

Q. Describe a series-wound generator. Give its applications. Make an electrical sketch.
A. A generator which has its field wound in series with the armature, Fig. 33. They are not in general use but may be used for boosting voltage in transmission of DC. The field has a few turns of large wire. The voltage will rise with an increase in load.

Q. Describe and make an electrical sketch of a shunt-wound generator. Give its applications.
A. A generator which has its field wound in parallel with the armature, Fig. 34. It is used for battery charging and sometimes for AC generator excitation. The field consists of a large number of turns of small wire. The voltage drops off with an increase in load.

Q. Describe and make an electrical sketch of a compound-wound generator. Give its applications.
A. One which has two field coils on each pole piece, Fig. 35. One of the coils is made up of a few turns of heavy wire and is wound in series with the armature. The other coil is composed of a great number of turns of small wire and is wound in parallel with the armature. It is used for most all general purposes where

DC is required. The voltage is fairly constant from no load to full load.

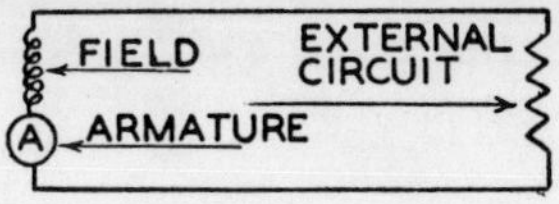

Fig. 33. Series-wound generator.

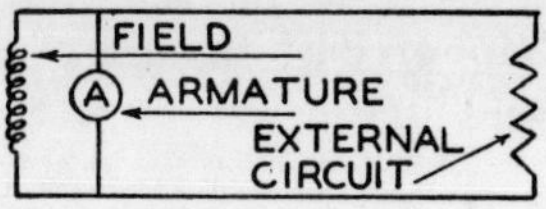

Fig. 34. Shunt-wound generator.

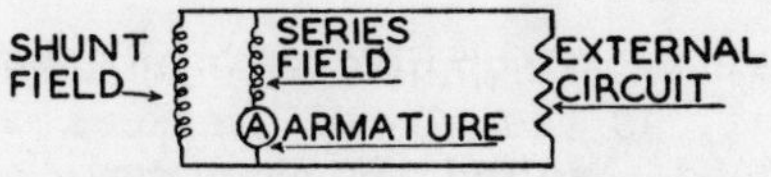

Fig. 35. Compound-wound generator.

Q. Why is it poor practice to use a belt drive when operating a series-wound motor as a prime mover?
A. If belt breaks, the motor will overspeed and run out of control.

Q. What are some causes of hot commutators?
A. Sparking under brushes; poor brush contact; near some hotter part of machine.

Q. How would continuous overload affect a DC generator? How much overload will most generators usually stand? What does "service factor" mean?
A. Continuous overload would heat up the machine. Most generators will stand a 25% overload. Service factor is a number on the name plate of the machine that tells you how much of the rated load the machine will stand. For instance, a service factor of 1.25 means that the machine will operate on a 25% overload.

Q. Why are armature cores laminated? Why are cores used at all?
A. The cores are laminated to cut down on losses by eddy currents. The reason we have cores is to concentrate the magnetic lines of force and to provide a path of low resistance for the flow of lines of force. Of course, the core is used as a means of carrying the armature coils.

Q. What is a rheostat?
A. An instrument for regulating the strength of an electric current by varying the resistance in the circuit.

Q. Explain in detail how to parallel two DC generators, one of which is already on the line.
A. Warm up the machine and bring up to speed with full open throttle. Close circuit breaker and adjust voltage to same as line voltage. Close main switch and adjust the load equally between the two machines.

Q. If two generators were running in parallel, explain how you would cut out one machine and take it off the line.

A. Reduce the load on the machine to be cut out as much as possible. Trip circuit breaker and open the main switch. Secure the cut-out machine.

Q. If several generators were running and connected to a switchboard and the engine of one generator failed, would any injury result?

A. If circuit breakers were supplied with an overload release, no harm would be done, as the circuit breakers would automatically cut out when overload resulted. In case the system was not equipped with overload release, the dead generator would take current from the live ones and try to run as a motor. This would result in excessive sparking and overload, and the live generators might kick out their circuit breakers, stopping all electrically-driven auxiliaries.

Q. Name some generator checkups which should be made at least once a day.

A. Check load condition, commutator condition, lubrication system operation, governor action of prime mover, bearing temperature, and vibration.

Q. A generator is vibrating. What checkup should be made before changing any balance weights.

A. Misalignment, sprung shafting, something chafing the rotating element, foreign matter on shaft, overloading, loose bolts, faulty speed governor.

Q. Why is it necessary to keep armature and field coils clean and dry?

A. If the armature coils and field coils become dirty and damp, they will overheat.

Q. What care should be given a commutator while in operation?

A. It should be kept clean, smooth and free from oil and grease.

Q. What are some causes of failure of a generator to build up?

A. Field connections reversed; brushes not in proper position; wrong direction of rotation; speed too low; field circuit open; not enough residual magnetism; machine short-circuited.

Q. Upon what does voltage of a D.C. generator depend?

A. Speed of armature; strength of field; number of armature conductors.

Q. How could you determine spacing intervals of brushes?

A. Lift all brushes. Place piece of heavy paper around commutator. Mark the edge of each brush on the paper. Lift brushes and remove paper. Measure distances between markings.

ALTERNATING CURRENT

Q. What is the meaning of "alternating current?"
A. It is current that between any two consecutive instants of time is either increasing in strength, decreasing in strength, or reversing in direction.

Q. Describe one cycle of A. C.
A. Starts at zero strength and increases in strength to maximum in one direction, decreases in strength in the same direction to zero, reverses direction, increases in strength to maximum in the new direction, decreases in strength to zero, where it reverses again to the original direction.

Q. What is meant by frequency of A. C. ? What is 60-cycle A. C. ?
A. Frequency means the number of times per unit of time the cycle is repeated. 60-cycle A. C. means that the cycle is repeated 60 times per second.

Q. What is a step-up transformer? What is a step-down transformer?
A. A step-up transformer is one that raises the voltage and decreases the amperes. A step-down transformer is one that lowers the voltage and raises the amperes. The change in volts is directly proportional, as turns in the primary are to the turns in the secondary. The change in amperes is inversely proportional, as the turns in the primary are to the turns in the secondary.

Q. What is an exciter? Why is it used with an A. C. generator?
A. It is a small D. C. generator. It supplies the exciting current to the rotor of the A. C. generator.

Q. What is meant by exciting current?
A. It is current required to create any magnetic field.

Q. What is a power-factor meter and why is it used?
A. It is a meter that states the power factor of the circuit, reading 1.0, .9, .8, etc. It is used to tell us the conditions of our circuit. As power factor may be thought of as % Eff., it is important that the power factor be kept as near to 1.0 as possible.

Q. Explain the theory of a simple A. C. generator.
A. A field of constant polarity (direction) is mechanically moved so as to cut across a varying number of stationary conductors per unit of time. This induces in the conductors a varying and alternating voltage and current.

Q. What is a synchronous motor?
A. It is a motor that turns in synchronization (cycle-for-cycle) with the generator.

Q. With A.C. generators, why are the armature coils made stationary while the field coils revolve?
A. Due to the excessive speed of the rotor, it is difficult to keep the windings secured to the commutator. It is also difficult to keep insulation intact at such high speeds.

Q. What is the purpose of the slip rings and brushes on the revolving field type A.C. generator?
A. To supply D.C. current to field windings on rotor.

GENERAL

Q. Define volt, ampere, ohm, watt, kilowatt.
A. 1. Volt: The unit of electrical pressure. The amount of pressure required to force a current of 1 amp. through a resistance of 1 ohm.

2. Ampere: The unit of current. The amount of current that will flow through a line when acted on by a pressure of 1 volt through a resistance of 1 ohm.

3. Ohm: The unit of resistance. Limits the flow of current to 1 amp. when acted upon by a pressure of 1 volt.

4. Watt: The unit of electrical power.
Watts = volts X amps. W = VA.

5. 1000 watts = 1 kilowatt (Kw.)

6. 746 watts (3/4 Kw.) = 1 horsepower.

Q. What is meant by alternating current and by direct current?
A. Alternating current is a current that changes its direction of flow. It flows in both directions, rising from zero to a maximum intensity and going back to zero again, repeating over and over again according to its frequency. Direct current is a current that flows only in one direction.

Q. What is an electrical circuit?
A. A closed path through which current flows.

Q. What is meant by the "external circuit?"
A. It is that part of the electrical circuit leading from the source of supply back to the source of supply.

Q. What is a series circuit? Sketch.
A. A circuit in which all parts of the circuit are so connected together that there is only one path offered to the flow of current. Figure 36 shows a diagram of a battery, lamp, and an electromagnet all hooked up in series.

Q. What is a parallel circuit? Sketch.
A. A circuit in which the various parts are so connected together that there is more than one path for the flow of current, Fig. 37.

Q. Sketch a combination series-parallel circuit.
A. Figure 38.

Q. Should a series-wound motor, as a prime mover, have a belt drive?

A. No. If the belt should break, the load would be entirely removed and cause it to race at a dangerous speed.

Q. What is an open circuit? What is the effect of an open in a series circuit? In a parallel circuit?

A. An open circuit is one that is broken. It breaks the path for the flow of current. It can be created accidentally or intentionally. A switch is one method of creating an "open" intentionally. It would stop the flow of current in all parts of a series circuit. However, it would stop the flow in a parallel circuit only in that part or branch that had the open in it. The current would continue to flow in the parts of the parallel circuit that were still closed.

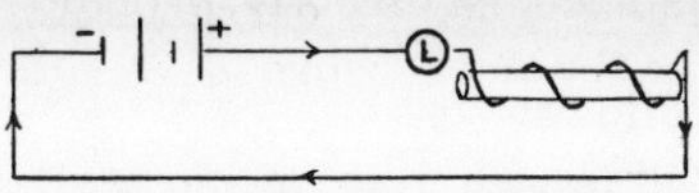

Fig. 36. Series circuit.

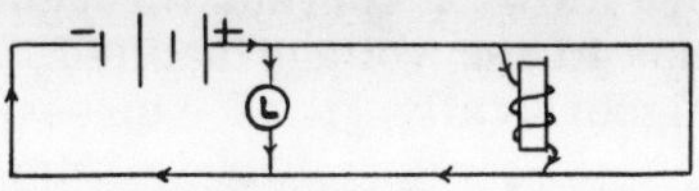

Fig. 37. Parallel circuit.

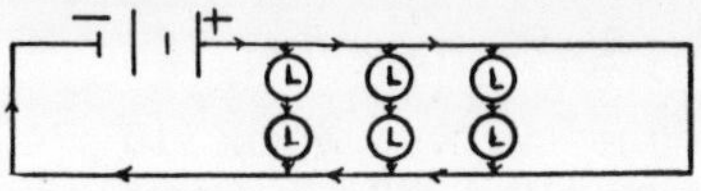

Fig. 38. Series-parallel circuit.

Q. What is a short circuit? What is a dead short? What is the effect of a short in a series circuit? In a parallel circuit, how would you detect a short?

A. A short circuit is a low resistance path that actually shortens the intended path for the flow of current. A dead short is a short circuit that has such low resistance that the circuit is made inoperative. In a series circuit, the short would probably not cause much damage because of the resistance in the rest of the circuit limiting the flow of current. In a parallel circuit, however, the drain of current through the low resistance path would rob the rest of the circuit of current and make it inoperative.

Q. What is a grounded circuit? What is the effect of a ground? What is the effect of two grounds? How would you test for a ground aboard ship?

A. A grounded circuit is one that has come in contact with the earth either by coming in contact somewhere in itself or by coming in contact with a conductor which is in contact with the earth. There is no serious effect when one ground occurs because it is not a completed path for the flow of current. However, if two grounds occur, there is the possible chance of a short circuit. Two grounds give an added unwanted path for the flow of current. Some circuits do use the two grounds to act as the return to the source of supply.

The simplest test for a ground on board ship is to use a test lamp. By attaching one lead of the lamp to one side of the circuit and touching the other lead from the lamp to a metal part of the ship, the light will light if there is a ground on the side of the circuit which the lamp is not attached to. This will not tell where the ground is located but will tell you that there is a ground somewhere in the circuit. To locate the ground, open the circuit at junction boxes and test each section separately.

Q. What is Ohm's Law?

A. A pressure of 1 volt will force a current of 1 ampere through a resistance of 1 ohm.

Q. How is Ohm's Law expressed by formula?

A. The flow of current through a resistance is directly proportional to the voltage applied.

Algebraically, it is expressed as follows:

$I = \frac{E}{R}$ I = current (amps.)

$R = \frac{E}{I}$ R = resistance

$E = IR$ E = voltage

For example, find the current when 110 volts are applied to a resistance of 22 ohms: $110 \div 22 = 5$ amps.

Find resistance when current is 10 amps. and applied voltage is 150 volts: $150 \div 10 = 15$ ohms.

Find voltage when resistance is 15 ohms and a current of 60 amps. must flow through the line: 60 X 15 = 900 v.

Q. Define insulators and conductors.

A. An insulator is a substance that offers a high resistance to the flow of current.

A conductor is a substance that offers a low resistance to the flow of current.

Q. List 3 good conductors and 3 good insulators.

A. Conductors: Aluminum; Copper; Silver. Insulators: Porcelain; Rubber; Mica.

Q. What is the effect of bearing currents in electrical equipment bearings, and how are they prevented?

A. Pit the bearings (especially brass bearings) and sometimes the shaft. Ground frame of machine to ship.

Q. What precautions should be taken when using compressed air to clean out electrical apparatus?

A. It is not good practice to use compressed air, due to moisture. If it must be used, drain thoroughly and let air dry out line before applying to electrical apparatus.

Q. How would you determine the circular mil area of a 1/4" wire?
A. Change fraction of inch to thousandths; Drop decimal point; Square the answer.
1/4" = .250"
250 X 250 = 62,500

Q. What is the major factor affecting insulation on a wire?
A. The voltage of the circuit.

Q. What is the capacity of #14 rubber-insulated wire?
A. 15 amps.

Q. Explain how you would make a splice in a wire.
A. Bare the two ends of the wire by penciling down the insulation. Scrape clean, twist solidly together by wrapping the end of one piece on the body of the other, solder the joint, wrap the joint with rubber insulation to the original thickness, then wrap with friction tape for mechanical protection.

Q. What are the advantages of having several branch circuits from a main switchboard?
A. Better control from the switchboard; better operating advantages, such as testing for grounds; one blown fuse will not cut out all the power.

Q. What is a fuse? Explain its function.
A. A link of low-melting metal placed in series with the line. As it will melt at a predetermined temperature, an overload (amps.) will increase its temperature and cause it to melt, thus breaking the circuit and stopping the flow of current.

Q. What is a permanent magnet?
A. It usually is a piece of steel that has been hardened and placed under the influence of a magnetic field. When removed, it retains some of the magnetism of the field throughout its life.

Q. Define: Lines of force (magnetic); magnetic field.
A. Lines of force is the name given to indicate the strength and direction of a magnetic field. A magnetic field is made up of all the lines of force between two unlike poles of a magnet or series of magnets.

Q. How do like poles of a magnet react to each other? Unlike poles?
A. Like poles repel each other while unlike poles attract each other.

Q. Make a sketch showing lines of force and their direction.
A. Figure 39.

Q. What is residual magnetism?

A. It is the magnetism remaining in a substance after it has been removed from the influence of a magnetic field.

Q. What is permeability?
A. It is the ease with which metals may become magnetized or demagnetized.

Q. What is retentivity?
A. It is the power a metal has for retaining magnetic lines of force.

Q. What is the effect of heating on magnets?
A. It will destroy their magnetic properties.

Q. Show by sketch a field and current-carrying wire.
A. Figure 40.

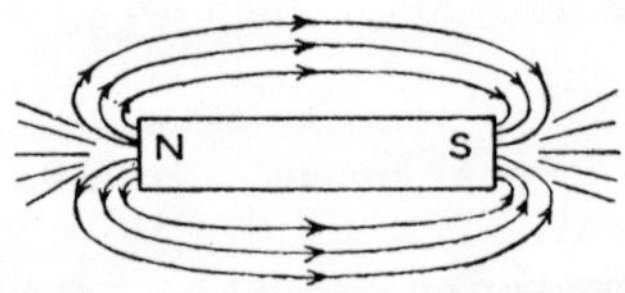

Fig. 39. Direction of lines of force.

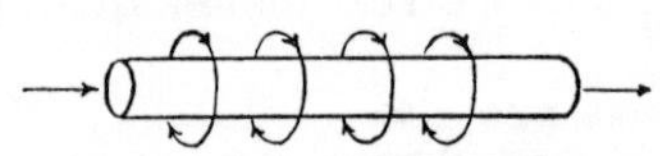

Fig. 40. Field and current-carrying wire.

Q. State the rule for determining the direction of this field.
A. Grasp the wire with the right hand, thumb extended in the direction of the flow of current. The fingers will point in the direction the magnetism is flowing.

Q. What is an electromagnet?
A. It is a piece of soft steel or iron that is magnetized by having a coil of current-carrying wire wrapped around it. When the current is shut off, the iron or steel becomes demagnetized.

Q. State the rule for determining the polarity of electromagnets.
A. Grasp the magnet in the right hand with the fingers extended in the direction of the flow of current. The thumb will then point in the direction of the north pole.

Q. What is the position of the poles in an electromagnet with relation to direction of the flow of current in the coil?
A. The field setup is always at right-angles to the direction of the flow of current.

Q. Why are soft iron cores used in electromagnets?
A. They can be quickly magnetized and demagnetized. In other words, we can control the direction and strength of the field.

Q. What are some causes of sparking at the brushes?
A. Overload; brushes set wrong; poor brush contact; rough commutator; weak field; armature winding broken or short-circuited.

Q. Describe the use of the following instruments and meters.
A. Ammeter: To measure amperes or rate of flow.
Voltmeter: To measure volts or electrical pressure.
Wattmeter: To measure electrical power produced at any instant of time.
Watt-hour meter: To measure electric work done in a given unit of time.
Megger: A small hand-cranked D. C. generator used for testing the effectiveness of insulation.
Magneto: A small hand-cranked A. C. generator used to test for grounds and short circuits.

Q. What are some causes of too low voltage?
A. Overload; speed too low; some reversed poles; some poles short-circuited.

Q. What are some causes of too high voltage?
A. Field too strong; speed too high.

Q. What are some causes of too low speed?
A. Overload; brushes set wrong; excessive friction; too little field resistance; short or ground in armature.

Q. What are some causes of too high speed?
A. Brushes set too far forward; open field circuit; wrong connections; too much field rheostat-resistance.

Q. What are some causes of hot armature coils?
A. Overload; damp windings; short-circuited coils.

Q. What are some causes of hot field coils?
A. Too large field current; damp windings; overload.

Q. What care should the bearings of a D. C. generator receive? Why is this very important aside from mechanical reasons?
A. The bearings should be kept well oiled and clean. Aside from mechanical reasons, if the bearing should run hot and run the metal out of the bearing, it will change the air gap between the armature and the field, thus changing its induction characteristics.

Q. What general care should a generator receive while in operation?
A. It should be kept clean and dry. The bearings should be kept well oiled. The governor should be kept in good condition.

Q. How are generators protected from overload and why is this necessary? What care should these devices receive?
A. Generators are protected by circuit breakers. The contacts should be kept clean and the connections kept tight. The carbon contacts should be renewed if burned very badly.

Q. What is an equalizer? How does it work?

A. An equalizer is a connection between two generators of different capacities running in parallel so that the running load is divided proportionately between the two, Fig. **41**.

Q. Sketch a series motor. Explain its characteristics and its applications.

A. Applications and characteristics: Speed varies with the load; at no-load it overspeeds, at full-load it decreases speed. This motor has a very high starting torque. Used for streetcars, cranes, elevators, locomotives, etc. See Fig. **42**.

Q. Sketch a shunt-wound motor. Explain its characteristics and uses.

A. Characteristics and applications: Constant speed over load range, constant load over speed range, has good control and used mainly on machine tools where variable speed is desired. See Fig. **43**.

Q. Sketch a compound-wound motor. Explain its characteristics and uses.

A. Characteristics and applications: Has desirable features of both above types, i.e., good starting torque, a flexibility of speed control, and constant load-speed application. Used for centrifugal pumps, cargo winches, boat hoists, air-compressor drive. See Fig. 44.

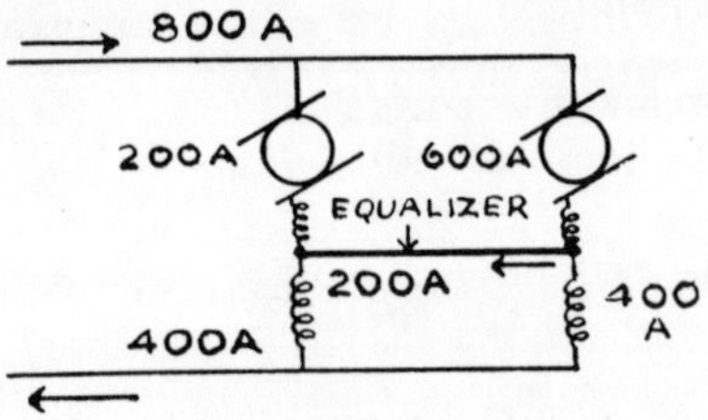

Fig. 41. Equalizer.

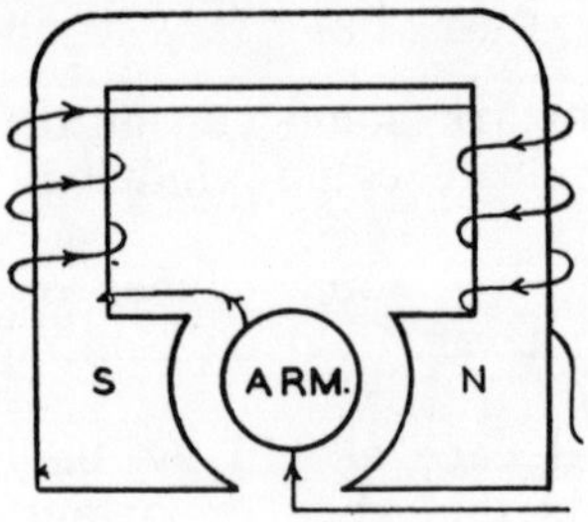

Fig. 42. Series motor.

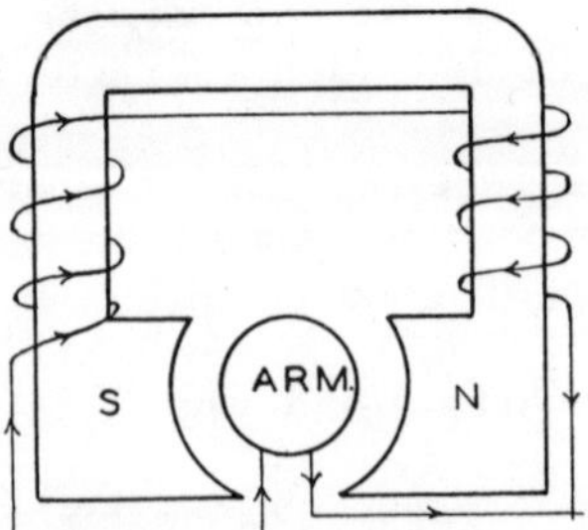

Fig. 43. Shunt-wound motor.

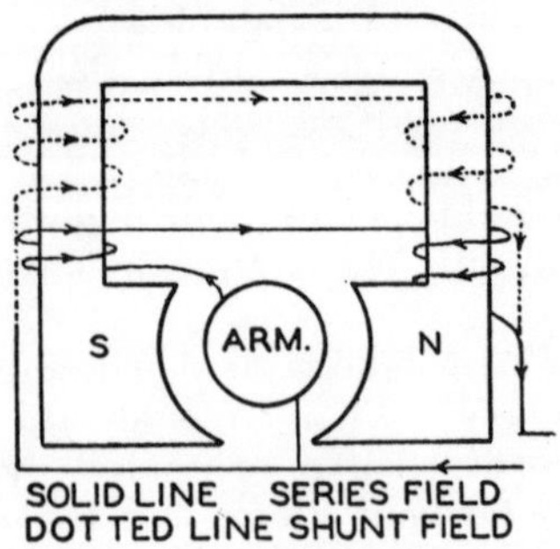

Fig. 44. Compound-wound motor.

Q. What are some causes of hot bearings?
A. Too little or improper oil: dirt in oil; bearings too tight; poor alignment; hot commutators; rough shaft; overload.

Q. What is a motor?
A. A device for converting electrical energy into mechanical energy.

Q. What is a D. C. motor?
A. It is an electrical machine for converting electrical (D. C.) energy to mechanical energy.

Q. Explain the general theory of a D. C. motor.
A. Poles of constant polarity are set up by D. C. in the stator, called the field. Current from the brushes passes into the armature coils at the proper time and in the proper direction to create poles on the armature core. These poles are attracted by the field poles to give rotation.

Q. What is a universal motor? Explain how it is possible for a motor to run on either A. C. or D. C.
A. When the motor is attached to a D. C. line, the current is in the same direction all the time. In an A. C. line, the current is reversing itself every cycle. However, as the current is reversing in the fields as well as the armature, there is no change of relationship between the poles of the armature and the field. Therefore, the motor continues to run in a constant direction. The universal motor must be a series motor.

Q. Give some general rules to follow in the operation of a D. C. motor.
A. Keep the motor clean and the commutator smooth. Keep brushes in good condition, properly spaced and on the neutral point. Don't overload the motor. Use the proper voltage. Use the motor only in the temperature for which it is designed.

Q. Suppose your motor failed to start. Explain the sequence in which you would look for and correct the trouble.
A. First, look for broken leads to the motor. The connections may be hooked up wrong. Field may be too weak. Motor may be overloaded. There may be excessive friction in the bearings.

Q. What would cause a motor speed to be too low?
A. Overload, too little field resistance, brushes set wrong, excessive friction, shorted or open coil in armature circuit.

Q. Why are bars of the commutator insulated from each other? What is generally used for this insulation?
A. The commutator bars are insulated from each other because the difference of potential between them sometimes becomes as high as line voltage. If the bars were not insulated from each

other, short circuits would occur and the machine would not operate. Mica is used for this insulation.

Q. What is undercutting of mica? What is the average depth of undercutting?
A. Undercutting of mica means cutting the mica down below the surface of the commutator. The average depth of this cutting is about 1/32" to 1/16".

Q. What is the effect of high mica between commutator segments?
A. If the mica is higher than the commutator, the brushes will not bear properly on it. The high mica will knock the brushes off the commutator and possibly cause breakage of the brushes.

Q. Explain how you would undercut the mica on a commutator.
A. Use a hacksaw blade and, by drawing in toward the open end of the commutator, the mica can be scraped from between the bars. The blade shaves a thin slice off the top of the mica at each draw. Care must be taken that the saw blade does not jump out of the slot and scar the surface of the commutator.

Q. What would cause a commutator to become badly worn or grooved?
A. If the brushes are too hard, the commutator will become worn excessively. If the brushes are not staggered properly, the commutator will become grooved.

Q. What would cause a motor speed to be too high?
A. Too much resistance in the field, brushes too far forward, wrong connections, open field circuit.

Q. When a motor is operated in a hot place or with too little ventilation, what is the effect upon its insulation?
A. The insulation resistance is increased until the insulation is thoroughly dry, and beyond that point the insulation will deteriorate rapidly.

Q. What three things are necessary in order to induce a voltage and current in a wire?
A. A magnetic field, a conductor in a closed circuit, and relative motion between the two.

Q. What are the hazardous locations on vessels in which "explosion-proof" electrical equipment only should normally be installed?
A. In pump rooms on tankers.

Q. How would you change the field and armature connections in order to give opposite rotation of a D. C. motor?
A. Reverse <u>either</u> the field leads <u>or</u> the armature leads; not both.

Q. Why are interpoles used on D.C. motors? How are they connected in the machine?

A. They are used to offset armature reaction and give better commutation.

Q. State the formula for determining the frequency of an AC generator.

A. $f = \frac{P \times N}{120}$

f = frequency
P = number of poles
N = R. P. M.

Q. What is the purpose of the heaters which are placed within the inner shields of most large electric motors and generators?

A. To prevent accumulation of moisture on the windings during periods when machine is idle.

Q. How would you cut two generators in parallel?

A. Three things must be true. Both machines must have same voltage and same frequency, and their voltage must be in phase (going through the cycle at the same time).

1. Bring #2 machine up to speed, governor not operating.
2. Close #2 circuit breaker.
3. Cut in excitation.
4. Adjust voltage #2 to equal #1.
5. Synchronize the two machines (put them in phase with each other).
6. Close main switch and open throttle wide.

Q. Why are motor-driven blowers usually installed on main propulsion motors rather than shaft-mounted fans?

A. The speed of the propulsion motor is too slow for such purposes. A forced fan circulation through coolers permits greater power for a given size.

Q. List 4 D.C. loads which may be found on vessels with A.C. ship's service systems.

A. General alarm system, emergency lighting, emergency radio power, motor operated watertight doors, rectifying for battery charging.

Q. What is a battery?

A. A series of two or more cells that are capable of producing electricity by electro-chemical means.

Q. Describe a lead-acid storage-battery cell?

A. It has plates made up of lead sponge (—) and lead peroxide (+) insulated from each other by wood. These plates are immersed in a solution of H_2SO_4 (sulphuric acid) and H_2O (water) mixed in the proper proportions. The chemical action between the plates of different material creates a voltage.

Q. Describe a delta-connection.
A. In a delta-connected A. C. generator, the start end of 1 phase winding is connected to the finish end of the third, and the start end of the third is connected to the finish end of the first. The three junction points are connected to the line wires leading to the load Phase voltage is equal to line voltage because the phases are connected directly across the line wires.

Q. What is a phase balance relay?
A. The purpose of this relay is to open the field circuit of the main generator and propulsion motor by opening contactors. A phase unbalance would indicate that something is wrong in one or more of the phases such as a short circuit from phase to phase, or to ground, or to open circuit in one phase. The relay compares the current in each phase and its contacts close if any phase is carrying more than any other phase. Any increase in current will not cause the relay to operate as long as the current in each phase is equal.

Q. Describe batteries connected in series.
A. The negative terminal of the first battery is connected to the positive terminal of the second battery. The negative terminal of the second battery is connected to the positive terminal of the third battery, etc. The positive terminal of the first battery and the negative terminal of the last battery serve as the power take-off terminals.

The total voltage is equal to the sum of the voltages in all the batteries. The total current is equal to the current (amps) in one battery.

Q. How could you determine spacing intervals of brushes?
A. Lift all brushes. Place piece of heavy paper around commutator. Mark the edge of each brush on the paper. Lift brushes and remove paper. Measure distances between markings.

Q. Describe the routine steps to be taken whenever a new generating set is to be started for the first time.
A. Check direction of rotation. Check for loose parts and possible obstructions. Check lube oil system. Check hand operated shut-off controls. Check all valves.

Q. How are "eddy currents" reduced?
A. By laminating the armature core.

Q. What are some causes of short contact life on motor starters?
A.
1. **Copper oxide coating on contacts.**
2. **Loose contacts.**
3. **Vibration.**
4. **Overload.**
5. **Grease or dirt on contact surfaces.**

Q. What are the principal characteristics of the alkaline battery?

A. The alkaline battery generates energy due to the reaction between nickel-cadmium or nickel-iron and the electrolyte, caustic potash. This battery is not damaged by freezing and there is no change in the specific gravity of the electrolyte as the battery discharges as does the lead-acid battery. Since a hydrometer is useless, the battery charge is determined by checking the individual cell voltage. The cell voltage between discharged and charged condition ranges from 1.1 to about 1.36 volts under normal operating conditions. The caustic potash (KOH) can burn. Handle with care. To neutralize use boric acid solution.

Q. Explain the operation of the "no-field release" starting box shown in Fig. 45.

A. The operating handle is moved slowly from left to right across the contact points of the starting resistance. As each contact point is passed, less resistance is allowed to remain in the armature circuit. When the handle reaches the final contact point, the resistance is cut out entirely and the full line voltage is run through the armature. (The full line voltage passes to the field coil when the operating handle comes into contact with the first contact point of the starting resistance.) i.e., The magnetic is connected in series with the field.

Since the holding magnet has full line voltage connected in series with the field, its magnetic action holds the contact arm in its final running position.

A break in the field circuit such as caused by a voltage failure or a break in the field coil opens the circuit and de-energizes the magnet releasing the contact arm. A spring helps to pull the arm back and the motor stops.

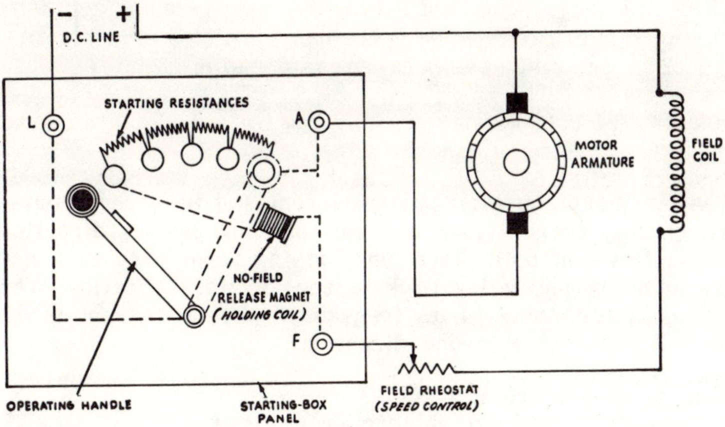

Figure 45.

Q. How are batteries rated for capacity?

A. In ampere-hours.

Q. When one of the line leads in a three-phase supply to an induction motor opens up after starting what may result?

A. A three-phase motor may run on single phase causing an overload of that phase and excessive heating of the motor if one of the line leads opens after starting.

Q. Explain the operation of the "no-voltage release" starting box shown in Fig. 46.

A. This operates about the same as the no-field release type except that the holding magnet is connected across the line. Since it does not have to carry the full flow of the field current, it draws only a small amount from the line. If the line voltage fails, it releases the contact arm and the motor stops.

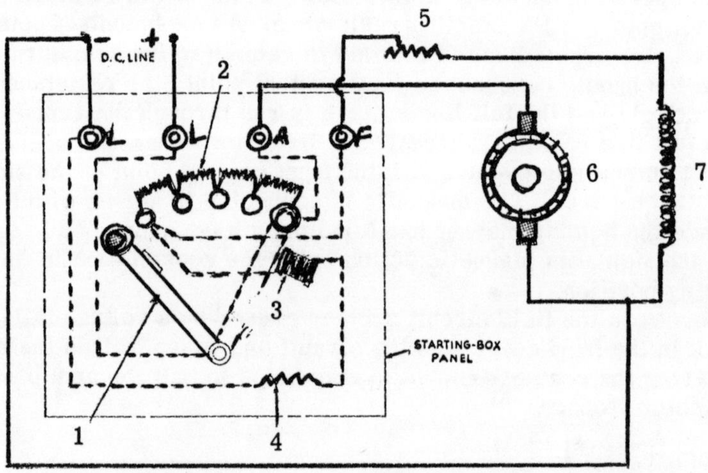

1. Operating handle
2. Starting resistances
3. No-voltage release magnet (holding coil)
4. Solenoid protective resistance
5. Field rheostat (speed control)
6. Motor armature
7. Field coil

Figure 46.

Q. What is meant by alternating current and by direct current?

A. Alternating current is a current that changes its direction of flow. It flows in both directions, rising from zero to a maximum intensity and going back to zero again, repeating over and over again according to its frequency. Direct current is a current that flows only in one direction.

Q. What is an electrical circuit?

A. A closed path through which current flows.

Q. What is meant by the "external circuit?"

A. It is that part of the electrical circuit leading from the source of supply back to the source of supply.

Q. How is a battery charged?
A. By reversing the current through the battery, using D. C. to restore the materials deposited in the electrolyte back on the plates in the proper position.

Q. What is the main difference between a primary and secondary battery?
A. A secondary battery can be recharged. A primary battery cannot be recharged after its chemical action on the plates has ceased.

Q. What kind of water would be used to replenish the water in a storage cell?
A. Distilled.

Q. What general care does a storage battery in operation require?
A. 1. Keep cool and well ventilated.
2. Maintain proper level of electrolyte (1/2" over the plates).
3. Maintain proper specific gravity of electrolyte.
4. Keep terminals clean.

Q. How would you test a battery for a dead cell?
A. Test with a hydrometer. The dead cell will not come up in specific gravity even after an unusually heavy charge. Full charge is approximately 1. 300 and a low battery has a specific gravity of approximately 1. 100.

Q. What is the function of a motor starter?
A. It limits the starting current to the armature to prevent destruction of same.

Q. What is the usual cause of grounds in a storage battery?
A. Grounds may be formed by dirty or acid wet cell tops and sides; the path of the ground in this case being from the terminals, along the sides of the casing, to the battery racks and finally to the ship's hull. A loss of charge by direct leakage from terminal to terminal may exist due to the presence of dirt and moisture without an actual path to the ground existing.

Q. In your opinion, what are the most important things for engine room personnel to watch in a turboelective drive ship?
A. Temperature and pressure of lubricating oil. Excitation and motor temperature.

Q. How would an open armature coil be detected?
A. Use a low voltage tester such as a flashlight tester. Place the ends of the test wires on each pair of adjacent segments. Normal coils will give a dim light; open coils will give no light. A mili-voltmeter can be used if available.

Q. Explain the operation of the "overload release" device shown in Fig. 47.

A. An overload release device automatically breaks the current if an excessive current is drawn by the motor. The contact points X and Y are normally open due to the fact that the plunger Z is held in its lower position by a spring and the weight of the plunger. If the motor draws an excess current the plunger is drawn up by the heavier magnetic field and shorts out the contacts X and Y, causing the holding coil to lose its magnetism and thereby releasing the contact arm shutting down the motor.

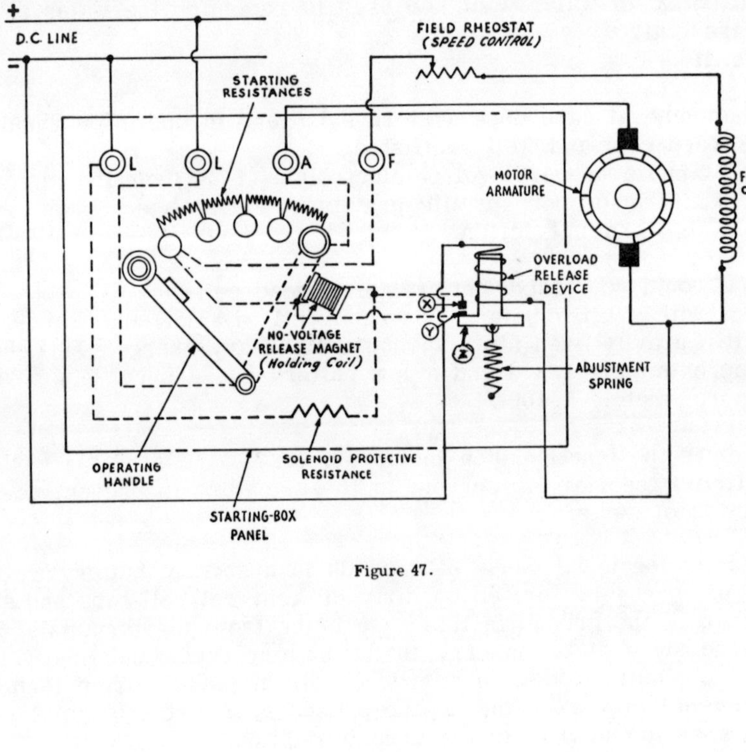

Figure 47.

MOTORS AND GENERATORS

PROBLEMS

Q. What is the H. P. of an 1800 Kw. motor?

A. $\frac{1800 \times 1000}{746} = 2412$ H. P.

Q. How many 30-watt lamps can be operated from a generator delivering 9 amps. at 110 volts?

A. 9 X 110 = 990 watts. 990 ÷ 30 = 33 lamps.

Q. How many Kw. will a generator put out if it is driven by a 75-H. P. engine?

A. 75 X 746 = 55950 watts. 55950 ÷ 1000 = 55.95. Kw.

Q. What factor would be used if we had horsepower and wished to find Kw?

A. As horsepower is smaller than kilowatt, we must use 1.34 and divide the horsepower by this factor. Or, we can use 746. By multiplying horsepower by 746, we get watts. By dividing watts by 1000, we get Kw.

Q. There are 150 40-watt lamps in an installation. The voltage is 220. What is the horsepower required to drive the generator if its efficiency is 92.5%?

A. Total watts = 150 X 40 = 6000

746 watts = 1 horsepower

$\frac{6000}{740} = 8.04$ horsepower

$$\text{Efficiency} = \frac{\text{electrical horsepower}}{\text{shaft horsepower}} \quad \frac{\text{Output}}{\text{Input}}$$

$$\text{Shaft horsepower} = \frac{\text{electrical horsepower}}{\text{efficiency}}$$

$$= \frac{8.04 \times 100}{92.5}$$

$$= 8.69$$

Q. An input of 32 H. P. is delivered to the shaft of a generator. If its efficiency at that input is 79.5 percent, find its output in watts.

A. 32 X 746 X .795 = 18978.24 watts

Q. A generator is 85% efficient. It is rated at 500 H. P. What is the output in Kw. ?

A. Output = Input X Efficiency
= 500 X .85
= 425

Since 746 watts = 1 HP
425 X 746 = 317050 watts
= 317.05 Kw.

Q. What is the relation of total E, I and R of a series circuit with that in each part of the circuit? Of a parallel circuit?

A. In Series:

Total volts = E1 + E2 + E3, etc.
Total amps. = I1 = I2 = I3, etc.
Total resistance = R1 + R2 + R3, etc.

In Parallel:

Total volts = E1 = E2 = E3, etc.
Total amps. = I1 + I2 + I3, etc.
Total resistance = $\dfrac{1}{\frac{1}{R_1}\ \frac{1}{R_2}\ \frac{1}{R_3}}$, etc.

Note: The total resistance in a parallel circuit is less than the resistance in any part.

Q. What is the electrical equivalent of 1 horsepower?

A. 746 watts = 1 horsepower.

Q. What is the relation between volts, amps., and watts?

A. 1 volt × 1 amp. = 1 watt.

Q. An ammeter reads 230 amps and the voltmeter 115. What power in Kw. is being used at the time of reading?

A. W = V X A
W = 115 X 230
W = 26,450 = 26.45 Kw.

Q. If resistances of 2, 4 and 6 ohms are connected in parallel, what is the joint resistance?

A.

$$T_R = \frac{1}{\frac{1}{2}\ \frac{1}{4}\ \frac{1}{6}}$$

$$T_R = \frac{1}{\frac{6}{12}\ \frac{3}{12}\ \frac{2}{12}} = \frac{1}{\frac{11}{12}} = \frac{12}{11} = 1.09 \text{ ohms}$$

Q. The output of an electric generator is 60,000 watts and its efficiency at this output is 90%. What is the input to the armature? Express in horsepower.

A. $$\text{Eff.} = \frac{\text{output}}{\text{input}}$$

$$\text{Input} = \frac{\text{output}}{\text{Eff.}}$$

$$= \frac{60,000}{.90}$$

$$= 66,666$$

Since 1 horsepower = 746 watts, $\frac{66,666}{746} = 89.3$ horsepower.

Q. What frequency would be developed by a turbo- alternator having 2 poles and turning at 3000 R. P. M.? If such a supply were connected to a synchronous motor having 60 poles, what would be the resulting speed?

A. $$f = \frac{PN}{120}$$

$$= \frac{(2)(3000)}{120}$$

$$= \frac{6000}{120}$$

$$= 50 \text{ cycles}$$

$$(2)(3000) = (60)(X)$$

$$60X = 6000$$

$$X = 100 \text{ R. P. M.}$$

Q. A D. C. generator delivers 22 Kw. at 110 volts. What is the current and resistance of the load?

A. W = V X A

$$A = \frac{W}{V} = \frac{22000}{110} = 200 \text{ amps}$$

E = I X R

$$R = \frac{E}{I} = \frac{110}{200} = .55 \text{ ohm}$$

REFRIGERATION

AMMONIA SYSTEM

Q. How would you detect leaks in an ammonia system?

A. Leaks in exposed piping, pipe connections, valves, valve stems, etc.

1. A soapy lather spread over joints or pipes will bubble at the leak.
2. Dampen a piece of litmus paper and run it along piping, joints, valve stems, etc. An ammonia leak will turn the litmus paper blue.
3. A sulphur stick passed along the piping, etc., will give off a dense white smoke on coming in contact with a leak.
4. Leaks in the condenser may be detected by applying litmus paper in the circulating water discharge. The litmus paper turning blue will indicate a leak.

Q. How would you shut down an ammonia refrigerating machine?

A.
1. Close liquid valve.
2. Allow the machine to continue to run until the low-pressure gage registers between 0 and 5 lbs.
3. Close suction valve.
4. Turn off power.
5. Close the discharge valve. (Never close the valve until the machine has stopped.)
6. Close water-supply valves.
7. If in a cold climate, drain all lines of water that might freeze.

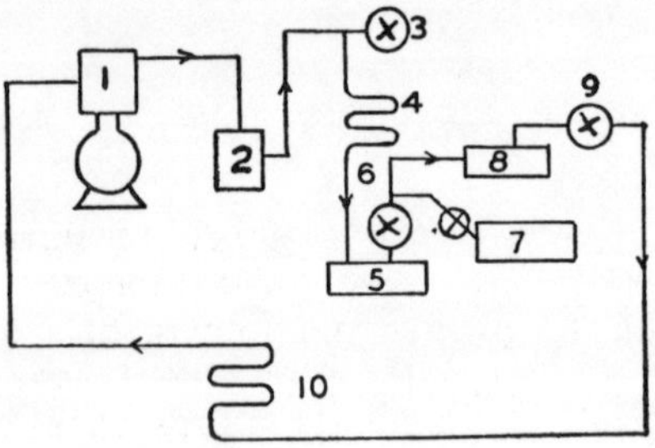

Fig. 48. Ammonia plant.

List of Parts

1. Compressor	6. Liquid valve
2. Oil separator	7. Charging cylinder
3. Purging valve	8. Scale trap
4. Condenser	9. Expansion valve
5. Receiver	10. Evaporator

Q. What are the characteristics of ammonia gas?

A. Colorless; pungent odor; very soluble in water; is combustible or explosive when mixed with air; even when breathed in small quantities, it attacks the lung tissues.

Q. Sketch an ammonia plant. Name all parts.

A. Figure 48.

Q. Why does vapor turn to liquid under pressure, and the liquid turn back to vapor after passing through the expansion valve?

A. The boiling point rises with the increase in pressure and vice versa.

Q. How would you shut down the plant if there were a bad leak, such as a gasket or a leaky pipe, on the high pressure side of the system?

A. Close compressor suction and discharge valves and open by-pass valves, Fig. 49. By doing this, the condenser becomes the suction side of the system and the expansion coils become the high pressure side. Keep the compressor running until the high pressure gage shows a vacuum. Stop compressor and secure.

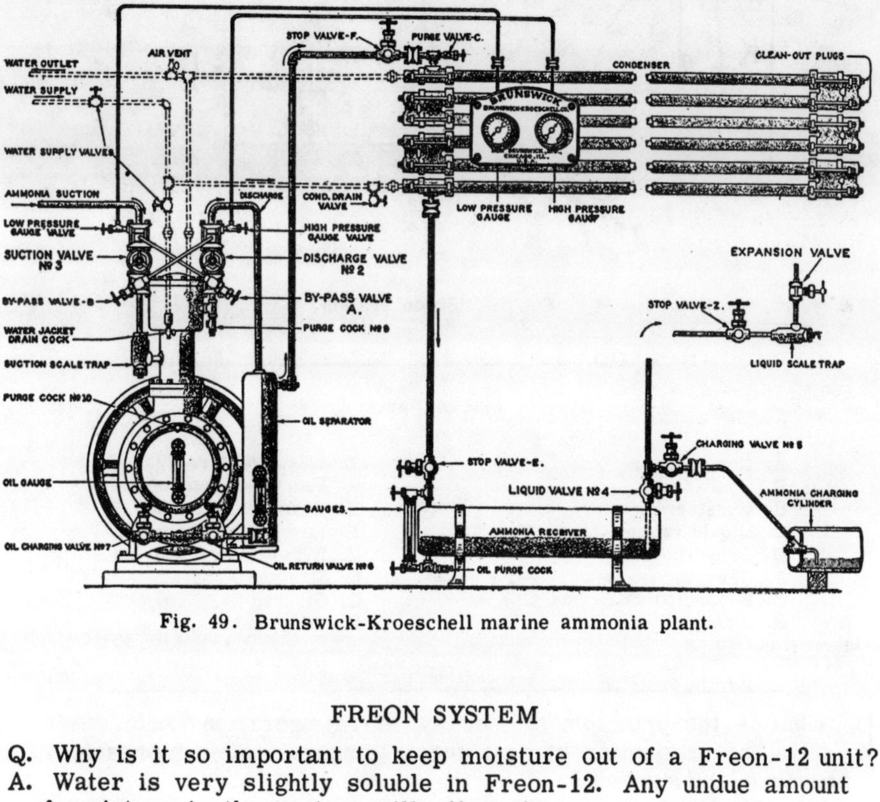

Fig. 49. Brunswick-Kroeschell marine ammonia plant.

FREON SYSTEM

Q. Why is it so important to keep moisture out of a Freon-12 unit?

A. Water is very slightly soluble in Freon-12. Any undue amount of moisture in the system will affect the proper functioning of the automatic controls. Since Freon-12 keeps the pipe lines clear of corrosion, any moisture present will immediately attack these exposed surfaces. There is also the usual seriousness of water mixing with the lubricating oil in the compressor crankcase.

Q. Explain the purpose of the water failure switch in a Freon automatic control system.

A. This switch should be set so that it remains closed as long as there is sufficient water flowing to the condenser. It should open the switch immediately in the event of water failure.

Q. Sketch a Freon system. Name all parts.
A. Figure 50.

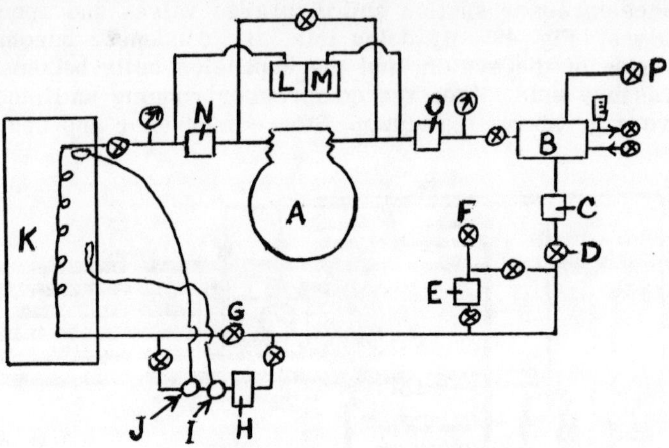

Fig. 50. Freon system.

List of Parts

A. Compressor
B. Condenser
C. Receiver
D. Liquid valve
E. Dehydrator
F. Charging valve
G. Expansion valve (manual)
H. Strainer
I. Solenoid valve
J. Thermo-expansion valve
K. Evaporator (icebox)
L. Low pressure cut-out
M. High pressure cut-out
N. Strainer
O. Oil trap
P. Purge valve

Q. What is the principle of mechanical refrigeration?
A. It is the absorption of heat under temperature, compression, pressure and expansion.

Q. How would you test for leaks in a Freon-12 unit?
A. By use of a Halide torch. Its normal blue flame will turn green in the presence of Freon. A bad feature of the torch is the fact that, if the leaks are large, the torch is almost useless, as it will burn green if there is the slightest bit of Freon in the atmosphere.

In case of large leaks, a solution of soapsuds works much better. This solution, applied to a joint where there is a leak, will bubble. A few drops of glycerin added to the solution helps it keep its wetness longer.

Q. Why is it difficult to accurately determine the oil level in the crankcase of a Freon-12 compressor? When should the oil level be checked?
A. The oil may contain quite an amount of the refrigerant. The greater the amount of Freon contained in the oil, the higher the oil level will be (not a true reading). It should be checked after a prolonged operating period. The least amount of refrigerant will be mixed with the oil at that time.

Q. What precaution should be taken after recharging the system?
A. Watch the oil level carefully. The new refrigerant will absorb a great deal of the oil.

Q. Why is the high pressure cut-out switch installed on a Freon-12 unit? At what pressure is this switch usually set?
A. To stop the machine when the head pressure gets too high. Usually cuts out at 135-150 psi and cuts in at 110-120 psi.

Q. Where are zinc rods usually found in the system? What is their purpose?
A. They are installed in the water boxes of the condenser. Where sea water is used for cooling, they reduce electrolytic action.

Q. Where would you find scale traps on a Freon system?
A. On the suction side of the compressor to prevent scale particles from getting into the cylinder. On inlet lines to regulating valves (solenoid and thermal expansion) to prevent clogging of these sensitive valves.

Q. What is the boiling point of Freon-12 at atmospheric pressure?
A. −21.7 °F.

GENERAL

Q. What is the function of the expansion valve?
A. It regulates the volume of liquid refrigerant to the expansion coils.

Q. What special care should be given to the expansion valve?
A. Keep gland tight and valve in perfect working condition.

Q. How is the expansion-valve adjustment controlled?
A. By thermostatic control or by hand-setting.

Q. What would happen if the expansion valve were closed?
A. No liquid would reach the expansion coils, and the back pressure would drop. Suction side would pump to a vacuum.

Q. What would happen if the expansion valve were open too wide?
A. Back pressure would rise and frost would form on the cylinders.

Q. What is meant by a "ton of refrigeration?"
A. The cooling effect of one ton or 2000 lbs. of ice melting in twenty-four hours. This is equal to 2000 lbs. multiplied by 144 B. T. U. or 288,000 B. T. U. per twenty-four hours, 12,000 B. T. U. per hour, or 200 B. T. U. per minute.

Q. Why are by-pass crossover valves installed on the ammonia compressor?
A. These valves are installed between the main suction and discharge lines. They permit the flow of ammonia to be reversed so that the compressor can pump the ammonia from the condenser and store it in the expansion coils in case of breakdown or leaks in the system.

Q. Explain how the valves in the previous question are connected to permit crossover operation.
A. One of the by-pass lines connects to the discharge line below the discharge valve and to the suction line above the suction valve. The other line connects to the suction line below the suction valve and to the discharge line above the discharge valve, Fig. 49.

Q. How is the head pressure in an ammonia machine controlled?
A. The pressure in the condenser is governed by the temperature and quantity of the cooling water used.

Q. Explain the purpose of the water regulating valve in a Freon automatic control system.
A. The water regulating valve should be adjusted so that it maintains a constant condensing pressure regardless of the circulating water temperature. It must also be adjusted so that it will shut off the flow of water when the compressor stops.

Q. How could you probably recognize air in the system?
A. Unusually high head pressure.

Q. Can condenser be purged when machine is running?
A. No. Machine must have been idle several hours before purging.

Q. What are the difficulties of removing oil from the liquid receiver and how may they be corrected?
A. Poor oil which forms gummy deposits and turbulence in the receiver, causing oil to remain suspended in the liquid. Corrections: use better oil; install baffles or bent pipes which provide still areas where the oil will settle and can be drained.

Q. What effect has liquid ammonia on petroleum lubricating oil?
A. Very little, unless water is present.

Q. Which is the warm side of a compressor, and which is the cold side?
A. Warm side: the high-pressure side from which the refrigerant is discharged. Cold side: the low-pressure side into which the refrigerant is drawn.

Q. Why are the valves and piping in an ammonia plant made of ferrous metals?
A. Because ammonia attacks non-ferrous metals such as brass or copper.

Q. What is the principle of the brine or indirect system of refrigeration?
A. The evaporator or cooling coils are placed in a brine tank which acts as a secondary agent. The brine is cooled and pumped in a continuous cycle through the refrigerating boxes and back to the brine tank.

Q. What is another name for the brine-circulating system of refrigeration?
A. Indirect system.

Q. Is frosting of the compressor suction a sure sign of efficient refrigeration?
A. No. It is usually a sign that the gas has circulated and returned after doing only a fraction of its intended work.

Q. Give the chemical symbols for the following:
A. Ammonia-NH_3
Carbon Dioxide-CO_2
Freon-12-CCl_2F_2

Q. What is the function of the compressor?
A. To pull the gas through the system, compress it to a higher pressure and temperature and discharge it to the condenser.

FIRE, EMERGENCY, ORGANIZATION, ETC.

Q. Before men are allowed to enter empty oil tanks, what precautions are taken?

A. Tanks are steamed and washed out. Flooding and pumping out helps to clean them. In a U. S. port, the inspection shall be made by a gas chemist certified by the American Bureau of Shipping. If services of such a chemist are not available, a competent person may be recommended by the Officer-in-Charge of Marine Inspection at the request of the vessel's owner.

When not in a U. S. port and a gas chemist is not available, the inspection may be made by the senior officer present and a proper entry made in the Log.

Make sure that no naked lights are near the tank. Have gas masks, air-breathing apparatus or oxygen-breathing apparatus handy. Secure a safety line around the man entering the tank and station a permanent watch until the man comes out of the tank.

Q. Where are relief valves required to be placed?

A. Relief valves must be fitted to the discharge side of fire pumps, reciprocating engine cylinders, air compressors, refrigerating machinery, reduced pressure sanitary mains, and all places where it is possible to have pressure in excess of safe limits.

Q. Where will you find listed the lifesaving and fire-fighting equipment required on your vessel?

A. In U. S. C. G. regulations.

Q. What are the regulations regarding the inspection of steering gear and communications between wheelhouse and engine room.

A. On all vessels making a voyage of more than 48 hours duration, the entire steering gear, whistle, and the means of communication between the bridge or pilothouse and the engine room shall be examined and tested by an officer of the vessel within a period of not more than 12 hours prior to departure. On all other vessels, similar examinations and tests shall be made at least once every week.

The date of the test and the condition of the equipment shall be noted in the official logbook.

Q. How often must the tail shaft be drawn?

A. Vessels operating in salt water having the tail shaft fitted with a continuous liner or where pressure lubrication is used, shall have the tail shaft drawn at least once every 3 years; in all other cases, they shall be drawn 2 years or whenever deemed necessary by Inspector.

Time may be extended to next dry-docking but not to exceed 4 months.

Where the propelling machinery is located aft, the maximum clearance shall be one grade (1/16") less than the above clearances.

Q. What are the regulations regarding the bilge suction on the main circulating pump?
A. Main circulating pumps must be fitted with a direct-suction connection and provided with a non-return valve in the machinery space. The diameter of the suction must be not less than 2/3 of the diameter of the inlet.

Q. What are the requirements for bilge pumps?
A. All vessels must be provided with an efficient pumping plant capable of drawing water from any watertight compartment under all possible conditions.

The number of pumps required is determined by the length of the vessels and the type of voyage it makes. The number can be found in G. R. &R. In general, the number is 2 or 3.

One of the power bilge pumps shall be available at all times for use under emergency conditions in which a vessel may be flooded at sea. Any of the following may be considered as meeting the above requirements:

a. A submersible type pump the source of power for which shall be located above the bulkhead deck.

b. A bilge pump attached to the propelling engine.

c. Sanitary, ballast and general service pumps having the required capacity may be accepted as independent power bilge pumps if fitted with the necessary connections to the bilge pumping system.

Q. How often must sea chests and valves be inspected?
A. Sea chests, sea valves, sea strainers, and bilge injection valves shall be opened up for examination at the time of drydocking if deemed necessary by an inspector.

Q. What type of power is used for the emergency lighting system? How often is this system tested?
A. Diesel or semi-diesel drive, equipped with means for quick starting. Located in a self-contained compartment, preferably on the deck above weather deck. Operated and checked once a week. Tested under load for at least 2 hours once a month. Storage batteries for emergency lighting and power systems shall be tested at least once every 6 months that the vessel is navigated.

The date of the tests and condition and performance are to be noted in the official logbook.

Q. Explain in detail how you would test an oxygen breathing apparatus?
A. 1. Before wearing, place apparatus in upright position so that plate H in Fig. 51 does not touch admission valve G.

2. Open pressure gage valve 1/2 turn.

3. Open main valve 3 turns. If oxygen cylinder is full, gage will show 135 atmospheres or approximately 2000 pounds.

4. Close main valve and if gage does not recede in about 10 seconds, the apparatus is tight up through the admission valve.

5. Check breathing bag. Hold hand over mouthpiece. Hold in on bumper plate. Crack main valve, allowing pressure to build up in apparatus. Close main valve, let out bumper plate and watch gage and bag. If gage pressure drops, there is a leak before the admission valve. If bag deflates, the leak is beyond the admission valve.

6. Test nose clip. Try to inhale and exhale deeply through nose. If unable to do so, it is in working order.

Q. What is an inflammable liquid?

A. Any liquid which gives off inflammable vapors at or below 80°F., as determined by the open cut test.

Q. What precautions would you take before using a flame safety lamp?

A. Check lamp completely. See that gauze is clean and that lamp does not smoke. Lock the lamp and leave the key outside the space that is to be inspected. This will prevent the operator from opening the lamp in the tank in the event that it is not working properly. Test for 5 minutes.

Q. What fuel should be used in a flame safety lamp?

A. Special fuel designated as safety lamp fuel, and no other.

Q. Sketch an oxygen breathing apparatus, and name all parts. Explain operation.

A.

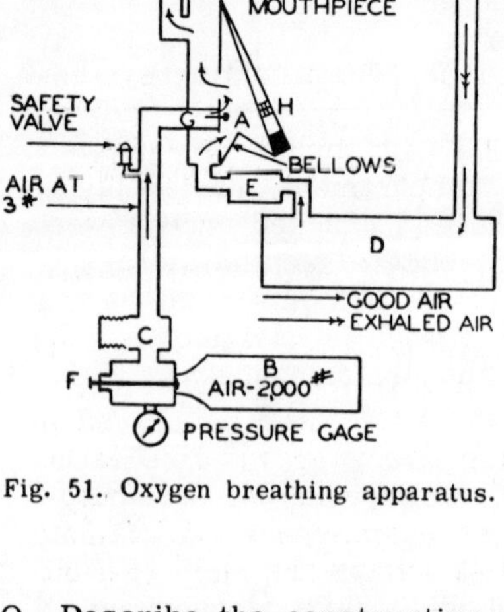

Fig. 51. Oxygen breathing apparatus.

List of Parts

A, breathing bag; B, air cylinder; C, automatic reducing valve; D, regenerator; E, cooler; F, main closing valve; G, admission valve; H, plate.

As wearer inhales, he draws air out of the breathing bag A. This permits the bag to collapse (bellows construction), and the plate H strikes the plunger of the admission valve G, allowing air from the tank, which has been reduced in pressure, to enter the breathing bag. The bag fills and the admission valve closes. The exhaled air goes through the regenerator where the CO_2 is absorbed. The resulting good air passes on and mixes with the air from the container.

Q. Describe the construction of a gas mask canister.

A. The canister usually consists of layers of absorbent materials such as charcoal, cotton, pumice, silica gel, etc. Different absorbents are used for different purposes inasmuch as they will each filter out certain gases. Each canister is painted with a definite color to designate the particular type of gas for which it is best suited. The mouthpiece and headgear are connected to a hose attached to the top of the canister. The air enters a hole in the bottom of the canister and the harmful gases are filtered out. The exhaled air is discharged through a flutter valve in the face piece.

Q. What are the rules regarding the number of oxygen breathing apparatus, gas masks and flame safety lamps carried aboard ship?

Service	Number of passenger staterooms	Self contained breathing apparatus	Additional gas masks or self-contained breathing apparatus	Special refrigeration gas masks[1]	Flame safety lamps	Portable electric drill
International voyage.	0 to 49	2		1	2	1
	50 to 100	2	2	1	2	1
	over 100	2	4	1	2	1
Ocean and coastwise, not on an international voyage.	0 to 49			1		
	50 to 100	1	3	1		
	over 100	1	5	1	1	
Great Lakes, and lakes, bays, and sounds.	0 to 49			1		
	50 to 100	1	1	1		
	over 100	1	3	1	1	
Rivers.	0 to 49			1		
	50 to 100	1		1		
	over 100	1	1	1	1	

[1]Required only on vessels equipped with refrigeration, small unit type refrigerators of not more than 20 cubic feet capacity excluded. A gas mask suitable for protection against each refrigerant used shall be provided.

Q. Where is the above equipment stored? What spare charges are required?

A. The refrigeration gas masks shall be stowed convenient to, but outside of the spaces containing the refrigeration equipment.

Half of the remaining equipment in the table shall be stowed in or near the pilothouse together with the fire axes and fire extinguishers required for that location. The other half of the equipment shall be stowed in a convenient accessible location, remote from the pilothouse, and preferably adjacent to the main entrance to the machinery space. Where only 1 of an item is required, it shall be stowed in the pilothouse.

A complete recharge shall be carried for each gas mask and self-contained breathing apparatus. The spare charge shall be stowed in the same location as the equipment it is to reactivate.

Q. What precautions must always be taken in connection with gas masks?
A. Test atmosphere with a flame safety lamp first. If flame safety lamp goes out, it is unsafe to use a gas mask. Check mask. Check log on canister. Check type of canister.

Q. What color is an all-purpose gas mask?
A. Red.

Q. What precautions should be taken with gas mask canisters?
A. Keep in dry place. Record on canister the date the seal was broken and do not use a canister that has been in service more than 1 year after breaking the seal. Discard the canister after the timing device registers 2 full hours of use. Mark all such canisters plainly so that they will not be used.

Q. Describe a CO_2 type fire extinguisher.
A. This extinguisher usually contains 15 lbs. of carbon dioxide gas, charged in the container ashore under high pressure. Used in fires of all types, as this gas is heavier than air and will smother fire.

The extinguisher is weighed at every inspection and is recharged if found to be more than 10% below the required amount.

At annual inspection, a tag is attached, giving date and findings at inspection.

Q. Describe a foam type fire extinguisher.
A. This extinguisher is similar in appearance to the soda-acid type but has a safety valve. The charge for this extinguisher comes in two separate packages which are placed in an inner and outer chamber of the extinguisher and dissolved in water at a specified temperature. When the extinguisher is turned bottom-up, the stopper falls out of the inner chamber, causing the two chemicals to mix and form gas and a heavy liquid which retains these gas bubbles. This smothers the flame.

Foam is used on oil fires. Must not be used in electrical fires due to the water content.

This extinguisher is discharged at annual inspection, recharged, and a tag attached showing date of recharge.

Q. Explain in general the operation of the automatic sprinkling system as prescribed by Regulations.
A. This system consists of piping fitted with sprinkler heads at suitable distances that will operate automatically in the event of a fire and spray water on the surrounding area. The system shall be supplied by a pressure tank or tanks and maintained at a required pressure, and by an automatically controlled pump so

arranged that when the pressure in the tanks falls to a predetermined point, the pump will cut in.

When the sprinkler pump is motor driven, it must be capable of being operated from the emergency electrical circuit in case of failure of the main power.

The water standing in the tank and piping system must be fresh. If the supply to the pump is salt water, check valves must be installed to prevent salt entering the tank.

Q. What fire-fighting equipment is required in the machinery space of electric propulsion motors?

A. One (1) 15 lb. CO_2 extinguisher for each motor.

Q. What are the rules regarding fixed fire extinguishing systems on passenger vessels burning oil for fuel?

A. All steam-propelled passenger vessels burning oil as fuel must be fitted with an approved, fixed carbon-dioxide, foam or water-spray system. This system must be capable of being operated from a convenient and accessible point outside of the space protected.

Q. What are the requirements for the fixed waterspray systems?

A. When fitted, the capacity of this system must be such as to blanket the entire area of the bilges and tank tops of largest boiler room with an adequate supply of water. The water supply shall be from outside the space protected and shall in no way be dependent upon power from the space protected. The pump shall either be reserved exclusively for the system or it may be one of the fire pumps.

Q. When are portable fire extinguishers recharged and how many charges are carried as spares?

A. Foam, soda-acid, and carbon tetrachloride extinguishers are recharged annually.

Spare charges shall be carried for at least 50% of each type of extinguisher carried.

If the unit is such that it cannot be readily recharged by the ship's personnel, one spare unit of the same classification shall be carried in lieu of spare charges for all such units of the same type.

Q. What are the regulations on spare safety valves for foam type extinguishers?

A. Extra safety valve units shall be carried for at least 50% of the extinguishers.

Q. Describe a carbon-tetrachloride type extinguisher.

A. This extinguisher is usually of one-quart size operated by a pump handle that is pulled out and pushed in. The liquid evaporates due to the heat of the fire and forms a blanket of gas.

There is a filling plug in the handle end.

This gas is somewhat poisonous and it is not advisable to remain in an enclosed space after discharging the extinguisher.

The pumping efficiency of the extinguisher is tested at annual inspection.

Q. What safety device is attached to the CO_2 system?
A. It is fitted with an alarm that operates automatically when the CO_2 is admitted to the space.

Q. What are the CO_2 fire extinguishing regulations for electrical propulsion equipment?
A. The number for pounds of CO_2 should be computed from tables in G. R. &R. The discharge of the required amount shall be completed within 2 minutes.

In addition to above, there shall be sufficient CO_2 available to permit delayed discharges of such quantity as to maintain at least a 25% concentration until the equipment can be stopped.

Q. What are the regulations on hand portable and semiportable fire extinguishers?
A. They shall be classified by a combination letter and number symbol. The letter indicates the type of fire which the unit could be expected to extinguish, and the number indicates the relative size of the unit.

The types of fires will be.

"A" For fires in ordinary combustible materials where the quenching and cooling effects of quantities of water, or solutions containing large percentages of water, are of first importance.

"B" For fire in flammable liquids, greases, etc., where a blanketing effect is essential.

"C" For fires in electrical equipment where the use of a nonconducting extinguishing agent is of first importance.

The number designations for size will start with "I" for smallest to "V" for the largest.

Sizes I and II are considered hand portable fire extinguishers and III, IV, and V considered semiportable systems which shall be fitted with suitable hose and nozzle or other practical means so that all portions of the space concerned may be covered.

Fire extinguishers of the above types shall be installed according to tables in G. R. &R.

Q. What provisions are made to prevent steam smothering system from freezing?
A. Shall be provided with proper drainage.

Q. What are the requirements for CO_2 extinguishing system?
A. The total available supply shall be at least sufficient for the space requiring the greatest amount.

The number of pounds of CO_2 required for each space shall be equal to the gross volume of the space in cubic feet divided by 30.

Branch lines to cargo holds and tween decks shall not be less than 3/4" standard pipe size.

For machinery spaces, paint lockers, etc., the number of pounds required for each space shall be computed from tables in G. R. &R. The size of pipe to these compartments shall conform to G. R. &R.

The controls and supply of CO_2 shall be outside the spaces protected and shall not be located in such space as might be cut off or made inaccessible in the event of fire in any of the spaces protected.

Distribution piping to cargo spaces shall be controlled from not more than 2 stations. The station controlling the main machinery spaces shall be located as convenient as practicable to one of the main escapes from such spaces.

The controls shall be protected by an enclosed cabinet or casing which shall be marked in red letters 2" high.

A set of instructions for operation of the system shall be kept in a conspicuous place at or near the releasing control device.

If the CO_2 supply is in a space that is locked, a key to the space shall be kept in a break-glass type box adjacent to the opening.

The CO_2 storage supply shall be properly ventilated. Cylinders shall be properly fastened and supported. They shall be readily accessible and capable of easy removal for recharging and inspection. Provisions shall be available for weighing the cylinders.

Regulations for systems installed prior to Nov. 1952: Mostly the same as above except that in boiler rooms, the bilges shall be protected by a system discharging below the floor plates. The amount of CO_2 and size of necessary piping to conform with G. R. &R.

Q. What are the requirements for a steam smothering system?

A. Steam shall be available from main or auxiliary boilers to provide at least 1 pound of steam per hour for each 12 cubic feet of the gross volume of the largest compartment to be protected. The size of the pipe to each compartment must conform to the area served according to regulations.

(Installations contracted for prior to Nov. 1952 shall have following requirements: The main pipes and branches to cargo compartments and similar spaces shall not be less than 1-1/2". Branches to paint lockers shall not be less than 3/4".)

Q. How and where must valves and manifolds be located and marked on a steam smothering unit?

A. Control cabinets or spaces containing valves or manifolds shall be distinctly marked in conspicuous red letters at least 2" high noting the type of system enclosed.

The distribution piping shall be located in not more than 3

stations in easily accessible locations on the weather deck. The distribution line to each compartment shall be fitted with a shut-off valve. Each valve shall be plainly and permanently marked indicating the spaces served.

The manifold steam supply line shall be fitted with a master valve at the manifold.

Q. What pressure is maintained on steam smothering lines?
A. A minimum of 100 p. s. i.

Q. When may fire hose be detached from hydrants?
A. When hose interferes with loading or discharge of cargo; where no protection is afforded in heavy weather. It shall be stored in an accessible nearby location.

Q. Can fire hoses be used for any other purposes?
A. No. It shall be used for fire extinguishing and fire drills only.

Q. What are the regulations on combination nozzles?
A. They shall be capable of producing a solid stream, a high velocity fog spray, or being shut off. Each nozzle shall be provided with an approved applicator with a low velocity water fog tip which can be attached directly to the nozzle or the hose.

Q. What is the regulation on nozzles in machinery spaces?
A. All vessels over 1000 G. T. The hose attached to the hydrants in propulsion machinery spaces containing oil fired boilers, internal combustion machinery, or oil fuel units shall be fitted with an approved combination nozzle. The applicator shall be not more than 6 feet in length.

Q. What are the requirements of the fixed foam system?
A. Installations contracted for prior to Nov. 1952.

The capacity of this system should be such as to rapidly discharge over the entire area of bilges and tank tops of largest boiler room a volume of foam 6" deep. To be completely discharged within a period of not more than 5 minutes for tanks and 3 minutes for other spaces.

Installations contracted for after Nov. 1952. For spaces other than tanks, a total of 1.6 gallons per minute shall be discharged for each 10 square feet of area protected. For tanks, the discharge rate shall be 1 gallon per minute for each 10 square feet of area protected.

The discharge rate specified shall be for at least 5 minutes for tanks and 3 minutes for all other spaces.

Q. Where must the controls and operating instructions for the foam system be located?
A. The foam agent container and all controls and valves shall be located outside the space protected. They shall not be located in

such space as might be cut off or made inaccessible in the event of fire in any of the spaces protected.

Complete and simple instructions for operating the system shall be located in a conspicuous place at or near the controls.

Valves to the various spaces shall be marked.

Q. Can fire pumps be used for other purposes?

A. Yes, provided at least 1 is kept available for use on the fire system at all times. In no case shall a pump having connection to an oil line be used as a fire pump.

Q. Can branch lines be cut into the fire main?

A. No branch lines shall be connected to the fire mains for other than fire and deck wash purposes.

Q. Where shall fire pumps be located on vessels with oil fired boilers and internal combustion propulsion machinery?

A. They shall be located in separate spaces and the arrangement, sea connections, and sources of power shall be such as to insure that a fire in any one space will not put all of the fire pumps out of operation.

Q. What is the regulation on shore connections to fire mains?

A. On vessels over 1000 G.T., these shall be one shore connection to the fire main on each side of the vessel in an accessible location.

Q. How many hose outlets are required on a vessel?

A. They shall be of sufficient number and so located that any part of the vessel, other than main machinery spaces, accessible to passengers or crew and cargo holds may be reached with at least 2 streams of water from separate outlets, at least one of which shall be from a single length of hose.

In machinery spaces, all sections shall be reached by at least 2 streams of water each of which shall be from a single length of hose from separate outlets.

Q. What is the regulation on fire mains on exposed decks?

A. They shall be protected against freezing or fitted with cut out valves and drain valves so that all such exposed piping may be shut off and drained in freezing weather. Except when closed to prevent freezing, such valves shall be sealed open.

Q. With what adjuncts are fire pumps equipped?

A. Each pump must be equipped with a pressure gage and a relief valve on the discharge side. The relief shall be set to relieve at 25 psi in excess of the pressure necessary to maintain the requirements of the G. R. &R. on outlet pressures, or 125 psi, whichever is greater.

Q. To what pressure is fire hose tested?
A. All fire hose shall be subjected to a test pressure equivalent to the maximum pressure to which they may be subjected in service, but in no case less than 100 psi.

Q. What fire detecting equipment is required on vessels? What type vessels require this equipment?
A. An approved system shall be installed on any vessel on an international voyage. Also any vessel over 150' in length having sleeping accommodations for passengers.

The types of systems are— automatic electric, pneumatic or smoke detecting systems as prescribed in G. R. &R.

Q. What vessels require fire smothering apparatus?
A. All mechanically propelled vessels carrying combustible cargo in holds should be equipped with a steam smothering system or CO_2 gas smothering system.

Q. How many fire pumps are required on cargo and passenger vessels? What pump capacity is required?
A. All vessels must be equipped with independently driven fire pumps in accordance with following:

1. Passenger vessels over 4000 gross tons: 3 pumps
2. Passenger vessels under 4000 gross tons: 2 pumps
3. Cargo vessels over 1000 gross tons: 2 pumps
4. Cargo vessels under 1000 gross tons: 1 pump

Each pump must be capable of delivering water simultaneously from the two highest outlets at a pressure of 100 p. s. i.

On vessels with oil-fired boilers, or with internal combustion propulsion machinery, where 2 fire pumps are required, they shall be located in separate spaces and the arrangement of pumps, sea connections, and sources of power shall be such as to insure that a fire in any one space will not put all of the fire pumps out of operation.

Q. What other fire equipment, besides pumps, is required on cargo vessels?
A. General

1. Manual alarm system on vessels carrying passengers or on an international voyage

Fireroom

1. Portable fire extinguishers: 2-15 lb. CO_2 or 2-2-1/2 gallon foam type in each boiler room
2. On vessels over 1000 G. T., 10 cu. ft. sand in container, with scoop; on vessels under 1000 G. T., 5 cu. ft. sand
3. Fixed system of CO_2, foam, or water spray
4. Drip pans under each oil burner, fuel oil pump and strainer, to prevent oil from reaching bilges

Cargo spaces

1. Inaccessible during voyage: A smoke detecting system and a

fixed CO_2 or steam installation (steam at 100 p. s. i. to be available at all times)

2. Accessible during voyage: A smoke detecting, electric, pneumatic or automatic sprinkling system.

Q. What is the required flash point for fuel oil?
A. 150°F. closed-cup test, minimum.

Q. What means are used for remote control of fuel oil service pumps?
A. They are equipped with means of control from an accessible place outside of the boiler room.

Q. How shall electric motors be installed in pump rooms on tank vessels?
A. Wiring is to be leaded and armored and shall run through approved gastight fittings having stuffing glands at inlets and outlets.

Joints in wiring shall be made only in wiring appliances, such as junction boxes, outlet boxes, etc., and such boxes shall be completely metallic and shall be gastight.

Electric motors shall be of approved type either totally enclosed or ventilated to the atmosphere by suction and discharge air ducts. Separately ventilated motors are to have pressure type ventilation and shall be arranged with an automatic shutoff to open the circuit, when the ventilating fan motor stops. This system is to be interlocked so that the pump motor cannot be started prior to a circulation of air.

Lighting fixtures shall be of approved type.

Q. When shall a fusible plug not be used?
A. When the steam temperature exceeds 425°F.

Q. What is the normal fusing temperature of a fusible plug?
A. 450°F-500°F.

Q. How far through the sheet shall a fusible plug extend?
A. Fusible plugs shall be so fitted that the smaller end of the filling is in direct contact with the radiant heat of the fire, and shall be at least 1" higher on the water side than the plate or flue in which they are fitted. In no case shall it be more than 1" below the lowest permissible water level.

Q. Where must the suction valves on fuel oil supply tanks be located?
A. Piping subject to head pressure from oil in the tank shall be fitted with shutoff valves at the tank which shall be remotely controlled from a readily accessible and safe location outside of the compartment in which the valves are located.

Q Why should watertight doors be closed during vessel operation?
A. To be sure that they are in working order at all times in the event of a grounding or collision.

Q. What purpose is served by the reducing valve on the oxygen breathing apparatus?
A. This valve reduces the high pressure in the tank to the lower pressure required for proper breathing pressure. The valve is designed to maintain a pressure of 3 to 4 pounds at the valve outlet. It is a very sensitive mechanism and should not be tampered with. Spare valves should be on hand for emergency use.

Q. Explain in detail how you would prepare a safety lamp for use and how you would proceed to use it.
A. The flame is lighted with a flint. The control handle on the bottom of the lamp is attached to a small rod which passes through a tube in the tank. The rod moves the igniter and a spark is produced which ignites the wick. The height of the flame is adjusted with a special wrench which turns the adjusting screw.

Q. Name the various pieces of equipment that may be used to enter a gaseous tank.
A. Flame safety lamp, oxygen breathing apparatus (OBA), gas mask, fresh air hose mask, Chemox.

Q. How is a CO_2 fire extinguisher used?
A. Remove locking pin, aim horn at fire and squeeze trigger handle.

Q. When is a fire under control?
A. When it stops spreading.

Q. State three methods by which fire spreads and the means you would take to prevent such spreading of the fire.
A. Radiation: Wet down opposite side of bulkheads or partitions
Conduction: Use proper extinguishing agent: foam, CO_2, etc.
Convection: Use fog spray.

Q. What should be done when a vessel grounds? When a collision occurs?
A. Grounding: The owner, agent, master, or person in charge of a vessel shall notify the nearest U.S.C.G. inspection office by letter, including full details, and report in person if in the vicinity of an inspection office.
Collision: Take soundings. Start bilge pumps. Check all compartments. Break out emergency fire and safety equipment.

Q. What provision is made by boiler manufacturers to check for furnace plate thickness?
A. Furnace is drilled at crest of corrugation, four or less corrugations from either end, and fitted with a 3/8" screw plug.

Q. Describe a fixed foam system for engine room spaces.
A. The container of foam and all controls and valves shall be located outside the protected area. They shall not be located in such space as might be cut off or made inaccessible in the event of fire in any of the spaces protected. Complete and simple instructions for the operation of the system shall be located in a conspicuous place at or near the controls. All valves to various areas shall be marked. For spaces other than tanks, a total of 1.6 gallons per minute shall be discharged for each 10 square feet of protected area. For tanks, the discharge rate shall be 1 gallon per minute for each 10 square feet of protected area. The discharge rate shall be for at least 5 minutes for tanks and 3 minutes for all other spaces.

Q. When should the steering gear be tested?
A. On vessels making a voyage of over 48 hours, the entire steering gear must be tested by an officer within a period of not more than 12 hours prior to departure. On all other vessels similar examinations shall be made at least once a week.

Q. During the vessel's annual inspection, what are the duties of all engineering officers in regards to defects or imperfections in the machinery?
A. Inform the Coast Guard inspector and assist in every way possible.

Q. What are the various emergency signals used aboard ship?
A. Fire & Emergency: Rapid ringing of ship's bell and continuous ringing of general alarm for at least 10 seconds.
Boat Drill: More than 6 short blasts and 1 long blast on ship's whistle.
Abandon Ship: More than 6 short blasts and 1 long blast on ship's whistle and the same signal on the general alarm bell.
Man Overboard: Pass the word to the bridge.
Dismissal from Boat Stations: 3 short blasts on ship's whistle and 3 short rings on general alarm.

Q. What markings are required with respect to portable fire apparatus?
A. They are classified by a combination of letter and number symbols. The letter indicates the type of fire which the unit could be expected to extinguish. The number indicates the size of the unit. The letters are A, B and C. The numbers range from small units to large units by the Roman numeral numbers from I to V. All hand-portable and semiportable types shall have a metallic name plate attached, giving the number of the item, name and address of person or firm for whom approved and the identifying mark of the actual manufacturer.

Q. What determines the number of fire hoses in machinery spaces?
A. The gross tonnage of the ship.

Q. Discuss factors which will aid in the prevention of engine room fires. What duties would you have during a fire?

A. Wipe up all oil spills, empty trash cans each day, keep oily rags in covered containers, fix all leaky joints. In general, practice Good Housekeeping. Duties: Start fire pumps, hook up hoses, have portable extinguishers ready, have someone stand by master CO_2 valve.

Q. What are the requirements for normal operation of safety valves?

A. To insure safety valves being free, each safety valve shall have a substantial lifting device by which the valve disk may be positively lifted from its seat when there is at least 75% of the maximum allowable pressure on the boiler.

Such mechanism shall be connected by suitable relieving gear so arranged that controls may be operated from the fireroom or engine room floor.

When the pressure is so high as to make it impracticable to fit a relieving gear, suitable piping and a valve having a relieving capacity of not less than the safety valve(s), may be connected from the boiler to the escape pipe or condenser. The valve shall be arranged to permit operation from the fireroom or engine room floor.

Q. Who sets safety valves and adjusts them to keep the boiler within safe working pressure?

A. Duly authorized inspectors only, except in emergency.

Q. What is the hydrostatic test pressure applied on a new boiler?

A. When a new boiler is installed on a vessel, a hydrostatic test pressure equal to 1-1/2 times the maximum allowable working pressure is applied before the boiler is lagged.

After installation of lagging, brickwork, etc., on boilers to be operated over 400 psi, or steam temperature over 700° F., at the request of builder or owner, an additional test at 1-1/4 times the maximum allowable working pressure may be given.

Q. What hydrostatic test pressure is applied at annual inspection?

A. 1-1/4 times the maximum allowable working pressure if the inspector is able to make an internal inspection; otherwise, 1-1/2 times the allowable working pressure.

Q. How much may a furnace in a Scotch boiler be out of round?

A. A furnace may have a deviation of 1-1/2" from a true circle before being jacked back to form, if the pressure is reduced 1-1/2% for each 1/10" out of round.

If the maximum distortion does not exceed 1" and the length of the distorted area is not more than three corrugations, or if the maximum distortion does not exceed 3/4" for a length greater than three corrugations of distorted area, the repairs or reduction in pressure will not be required unless considered necessary by the inspector.

Q. Why do G. R. &R. prefer that the suction valve on all tanks supplying the burners be placed inside the tank?
A. Because of the danger of the valve being broken off and because of damage in case of fire. Many installations have two valves, one inside and one outside of the tank, the flanges of both valves being attached directly to the tank.

Q. Why are extension rods required on suction valves installed in tanks, and where must these rods lead?
A. The rods can be controlled from a readily accessible and safe location outside of the compartment in which the valves are located in case of a fire in that compartment.

Q. What are the requirements for clamps or gags for safety valves?
A. Each valve must be provided with clamps or gags to be kept on board at all times. (It is not permissible to gag safety valves by compressing the springs.)

Q. What are the requirements for water indicators on a boiler?
A. Each boiler shall have two independent means of indicating the water level in the boiler, one of which shall be a gage glass. The secondary indicator may consist of a gage glass, or other device approved by the commandant. If allowable steam pressure does not exceed 250 psi, three test cocks may be used attached directly to head or shell of boiler. If allowable steam pressure exceeds 250 psi, the gage glasses shall be of the flat type.

Q. Where is the water glass located in relation to the water level in the boiler?
A. The lowest visible part of the water glass must not be lower than the lowest permissible water level.

Q. What is the location of the lowest gage cock on a shell or water column?
A. The lowest gage cock is located 2" above the lowest visible part of the gage glass.

Q. What are the regulations regarding salinometer cocks on boilers?
A. In vessels operating in salt water, each boiler shall be equipped with a salinometer cock or valve which shall be fitted directly to the boiler in a convenient position. They shall not be attached to the water gage or water column.

Q. How many safety valves are required on a boiler?
A. Each boiler and its superheater shall be provided with one or more approved safety valves. Boilers of more than 500 square feet of water heating surface, without superheaters, shall be fitted with at least two safety valves. Boilers with integral superheaters shall have at least two safety valves attached to the drum and one safety valve fitted to the superheater outlet.

Q. (a) On passenger vessels fitted with magnet controls on doors, is it permissible to install holdback hooks or other devices to keep the door permanently open? (b) On firescreen doors and other doors which are not normally locked on passenger vessels, what is the maximum list that the door is required to close against? If you were obliged to adjust the spring tension on such a door what other considerations should be borne in mind?

A. (a) No Holdback hooks, or other means of permanently holding the door open, not subject to control station release, will not be permitted.

(b) Doors, others than those which are normally locked, such as from staterooms, fan rooms, lockers, etc., shall be of the self-closing type capable of closing against a 3-1/2° list. In adjusting tension on springs for doors such as those described it is wise not to greatly exceed the requirements as this might impede exit of women, children, etc., and unless a regulating device is fitted on the door causing it to slam shut, rather than close gently.

Q. What are the requirements governing the number and location of fire hydrants serving main machinery spaces aboard ocean passenger vessels?

A. Fire hydrants shall be of sufficient number and so located that all portions of main machinery spaces shall be capable of being reached by at least two streams of water, each of which shall be from a single length of hose from separate outlets.

Q. Will the mere washing of a gasoline tank free it of gases?

A. No. It should be well ventilated also

Q. What is a class "A" bulkhead?

A. A bulkhead composed of steel or equivalent metal construction, suitably stiffened and made intact with the main structure of the vessel.

It should be capable of preventing the passage of flame or smoke for 1 hour. It should have sufficient insulation so that the average temperature on the unexposed side will not rise over 250°F. above the original temperature, nor would the temperature at any point rise more than 325°F above the original temperature within the time limit alloted by G. R. & R.

GENERAL QUESTIONS

Q. What is atmospheric pressure?
A. It is the pressure of the air, at sea level, pressing down on the earth, usually considered as 14.7 lbs. per sq. in. (roughly, 15 lbs.).

Q. What is absolute pressure?
A. It is gauge pressure plus atmospheric pressure.

Q. What is gauge pressure?
A. It is that pressure, in lbs. per sq. in., above atmospheric pressure.

Q. What is one degree Fahrenheit temperature, and one degree Centigrade temperature?
A. A degree of Fahrenheit temperature is 1/180 of the interval between the melting point of ice and the boiling point of water.
A degree of Centigrade temperature is 1/100 of the interval between the melting point of ice and the boiling point of water.

Q. What is specific heat; sensible heat; latent heat; total heat?
A. 1. Specific heat: the number of B. T. U.'s that must be added or subtracted to change the temperature of a substance 1°F.
2. Sensible heat changes the temperature of a substance without changing the form.
3. Latent heat changes the form but not the temperature, as when water turns to steam or ice to water.
4. Total heat: the sum of the sensible and latent heat when both temperature and form are changed.

Q. What is a vacuum?
A. Vacuum is the absence of pressure. Since no perfect vacuum has ever been created, the word "vacuum" is generally used to designate a space in which a partial vacuum exists. Since atmospheric pressure equals approximately 15 lbs. per sq. in. and a perfect vacuum is represented by 30" Hg, we generally say that 2" of vacuum = 1 lb. pressure.
Example: 26" vacuum = 2 lbs. absolute pressure.

Q. In steam-engine and turbine operation, how is the vacuum formed?
A. Exhaust steam is condensed and is relieved of air; the condensate occupies much less space than did the steam from which it was formed, and the space which it occupied is an area of lower pressure than that at the exhaust. This forms the partial vacuum, which is maintained by the continuous condensing process.
The ratio of steam to water by volume at atmospheric pressure is approximately 1600:1. Thus, 1600 cubic inches of exhaust

steam, in coming into contact with the cooled surface of the condenser tubes, condenses to 1 cubic inch of water. Therefore, we have no steam or pressure or, in other words, a vacuum, this in turn being maintained by air pumps or ejectors.

Q. What is a condenser?
A. An apparatus for converting exhaust steam to water by means of heat transfer.

Q. What are the advantages of a condenser?
A. 1. Economy of converting the exhaust steam back to water to be returned to the boiler.
2. Reduces the back pressure on the engine.

Q. What is a 2-pass condenser?
A. One in which the circulating water enters at the bottom of one end of the condenser, passes through the lower half-section of tubes, is reversed, comes back through the upper half-section, and is discharged at the top of the end it entered, **Figs. 52 & 53.**

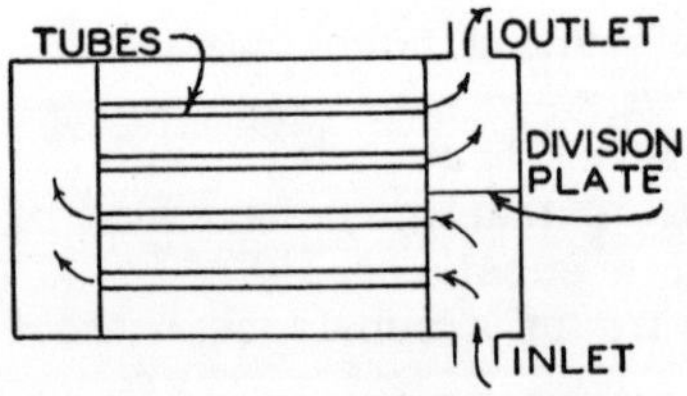

Fig. **52**. Two-pass condenser.

Q. In a two-pass condenser, what might cause a drop in vacuum and a drop in the temperature of the overboard discharge at the same time?
A. This would show that the cooling water is not getting through the tubes; therefore, the division plate must have carried away.

Q. Sketch a 2-pass surface condenser, showing internal construction. Name all parts.
A.

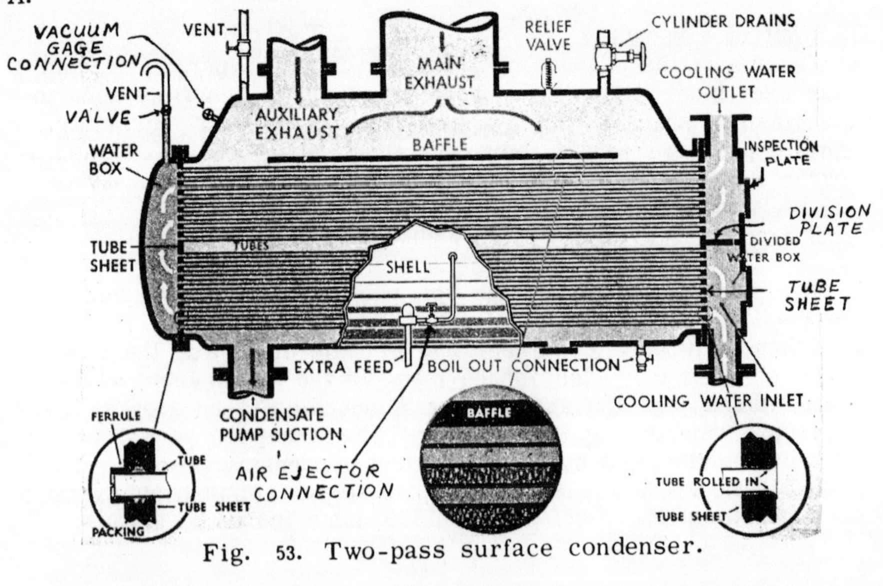

Fig. 53. Two-pass surface condenser.

Q. What is a scoop condenser?
A. One in which the cooling water flows in one end and out the other, the sea water running through due to the motion of the ship through the water, Fig. 54.

Fig. 54. Scoop condenser.

Q. What precaution must be taken with scoop circulation condensers when expecting or receiving slow, stop or astern bells?
A. Start independent circulating pump on condenser before ship loses headway or starts to move astern.

Q. What are the signs of too much and too little injection?
A. Too much—condensate too cool; overboard discharge temperature too low. Too little—condenser heats up; vacuum decreases; overboard discharge temperature too high.

Check the temperature of the overboard discharge. It should be approximately the same temperature as the steam in the exhaust trunk.

Q. What is a jet condenser and how does it operate?
A. Consists of a chamber in which the exhaust steam and injection water enter at the top. The steam condenses on contact with the injection water and falls to the bottom of the chamber to be removed by the air pump.

Q. List the most common causes of inadequate condenser vacuum.
A. Excessive air leakage into the vacuum system; Improper functioning of air-removal equipment; Improper drainage of condensate from condenser; Insufficient flow of circulating water; High injection temperature; Dirty condenser.

Q. Explain why the pressure differential between the main condenser and air ejector intercondenser does not equalize during normal operation.
A. The slight difference in pressure between the main condenser and the air ejector intercondenser is maintained by the loop seal in the interconnecting drain line from the intercondenser. The unequal heights of the water in the two legs of the loop seal will equal the difference in pressure between the two units.

Q. What immediate action should be taken if at any time loss of vacuum is accompanied by a hot or flooded condenser?
A. If at any time loss of vacuum is accompanied by a hot or flooded condenser, the units exhausting into it must be slowed down or stopped until the condensing plant is again put into working order.

Q. Explain latent heat of evaporation and latent heat of fusion.

A. Latent heat of evaporation: amount of heat required to convert 1 lb. of water at 212° F. into steam at the same temperature.
Latent heat of fusion: amount of heat required to melt 1 lb. of ice at 32° F. without changing the temperature.

Q. What is the unit of measurement of energy?
A. British Thermal Unit (B. T. U.). A B. T. U. is the quantity of heat (energy) necessary to raise 1 lb. of water 1° F. under ordinary atmospheric conditions.

Q. About how many B. T. U.'s are contained in a pound of fuel oil, and how does it compare with coal?
A. Fuel oil contains about 18,300 to 19,300 B. T. U.'s per pound. Coal contains about 13,300 to 14,300 B. T. U.'s per pound. Fuel oil contains about 20% more B. T. U.'s per pound than coal.

Q. What is the definition of work?
A. The result of a force overcoming a resistance through space. (Motion is necessary—pushing on an immovable wall is not work.) Work = Force acting X Distance through which it acts.

$$W = F \times D$$

Q. What is the unit of measurement of work?
A. One foot-pound (1 ft.-lb.), which is the work done by a force of one pound acting through a distance of one foot.
Illustration:
1. In lifting a weight of 15 lbs. through a height of 6 feet, how much work is done?

Work = Force X Distance
= 15 X 6
= 90 ft.-lbs.

2. A ship's propeller exerts a push of 10,000 lbs. in driving a ship at a certain speed. How much work is done per nautical mile? (One nautical mile = 6080 feet.)

Work = Force X Distance
= 10,000 X 6080
= 60,800,000 ft.-lbs.

Q. What is potential energy?
A. Stored energy or energy of position. For example, the weight in a pile driver is 2000 lbs., and it is lifted to a height of 35 feet. While the weight is at the top, it has potential or stored energy. On falling, this potential energy is changed to kinetic energy or energy of motion. The work it does in falling is equal to the weight times the distance, or 2000 X 35 = 70,000 foot-pounds of work.

Q. What is kinetic energy?
A. Energy of motion. A body has kinetic energy due to its velocity. Examples: running or falling water; a projectile in flight; winds in motion, etc.

Q. What is the unit of measurement of power?

A. Horsepower (H. P.). One horsepower = 33,000 ft.-lbs. of work per minute.

$$H.P. = \frac{\text{Work done per minute}}{33,000}$$

Q. What is shaft horsepower (S. H. P.)?

A. The power delivered to the shaft. (S. H. P. is always less than I. H. P. for a given engine, because some of the work developed by the cylinders is used to overcome the friction of running the engine.)

Q. What is indicated horsepower (I. H. P.)?

A. The power developed by the action of a piston within a cylinder, so named because it is measured by use of an indicator.

Q. What is brake horsepower (B. H. P.)?

A. The power at the shaft as it might be determined by a brake. (Same as S. H. P.)

Q. What is a horsepower (H. P.)?

A. One horsepower is 33,000 ft.-lbs. of work per minute.

Q. Define horsepower hour.

A. Work done at the rate of 33,000 ft.-lbs. per minute for a period of one hour.

Q. What heating equipment do the condensate and feedwater normally pass through from the condenser to the boiler on a modern turbine-driven vessel?

A. It would depend on the type of ship. In general, it would be:

1. Inter- and after-condenser of air ejector
2. Gland seal condenser
3. Drain cooler
4. First-stage heater
5. Second-stage heater (D. C. heater)
6. Third-stage heater, or economizer

Q. Why are separators used in steam lines? What precautions must be used in installing and operating them?

A. They remove entrained water, oil, dirt, and other impurities. Must be installed in a level position and kept clean.

Q. How would you check the alignment of shafting attached by couplings?

A. Remove coupling bolts. Insert feeler gage between coupling faces at various points around the circumference. Record readings.

Q. Would any damage be likely to occur if the discharge valve were closed just before securing the power to the centrifugal condensate pump?

A. Not as long as the stand-by pump was started to remove additional condensate building up in the condenser.

Q. If a bearing were exceedingly hot, would you stop the engine? Why?

A. No. If possible, keep the engine in operation, but remove the load from the bearing and apply extra oil. If the engine is suddenly stopped, there is danger of the bearing freezing on the shaft due to excess heat.

Q. What should be done when a bearing starts to heat?

A. The flow of oil should be increased and, if necessary, a lighter oil used to wash out any particles of dirt or grit in the bearing. Oil being used should be examined to make sure it is free from dirt.

(Do not pour water on a bearing to cool it, as bending, warping, or cracking may result.)

Q. What are the causes of hot bearings?

A. 1. Bearing is out of line.
2. Insufficient or improper lubrication.
3. Dirt or grit in the oil.
4. Bearing is set up too tightly.
5. Uneven surface on the bearing or journal.

Q. What is the purpose of the stern tube?

A. To provide a support for the tail shaft and allow for a watertight fitting to prevent water from coming into the ship.

Q. How is a stern tube lubricated?

A. By allowing a small leakage of sea water from the outside to the inside by adjusting the stern gland. Some stern tubes are provided with means for oil lubrication.

Q. Explain the construction of the stern-tube bulkhead stuffing box.

A. The gland is made in two halves and fitted around the shaft. Flax packing is fitted to the stuffing box and the two halves pulled up evenly.

Q. What is "pitch" as referred to a propeller?

A. If the pitch of a propeller is 18', each revolution theoretically would push the vessel through the water 18'.

Q. How is a propeller secured to a shaft?

A. A tapered fit, with a key fitted to the shaft and a left-handed nut on the end of the shaft (for a right-handed propeller).

Q. What is a built-up propeller? What are its advantages over the solid type?

A. One that is built up in sections as compared to the solid one-piece type. Consists of a hub to which a designed number of blades are bolted on. Its advantage is that if one blade is damaged it may be replaced. With the solid type, if a blade is badly damaged it is often necessary to install a whole new propeller or spend a great deal of time in dry dock waiting for a repair.

Q. What is apparent slip?
A. The difference between the speed of the propeller and the speed of the ship.

Q. What are the main points of a forced feed lubricating system through which the oil may become contaminated by moisture?
A. Excessive blowing at steam seals. Condensed water vapor in gear boxes and sumps.

Q. Why is the tail shaft of most single screw vessels drawn inboard when being inspected in dry dock?
A. Tail shafts of single screw vessels are usually constructed with a flange on the inboard end. (The flange will not fit through the stern tube.)

Q. What preparations would you make prior to renewal of a gasket in a steam line?
A. 1. Close off pressure valves on the line so they cannot be opened accidentally.

2. Close valve on outlet end of line to prevent exhaust from backing up.

3. Drain the line if possible.

4. Leave at least two nuts in place and ready to tighten instantly if an extreme flow develops when the flange is loosened. (Valves may be leaking.)

5. Scrape flange faces absolutely clean and dress out rough spots.

Q. A centrifugal pump fails to deliver water. What troubles would you look for?
A. Pump not primed; insufficient speed; air leaks in suction line; discharge head too high; suction lift too high; impeller plugged or broken; wrong direction of rotation.

Q. What is the purpose of the air pump?
A. To remove the water and air from the condenser and help maintain the vacuum in the condenser.

Q. What is meant by displacement of a pump?
A. In a reciprocating pump, it would be the volume of the liquid cylinder times the number of strokes per minute. In a rotary pump, the manufacturer indicates the capacity in gallons per minute (gpm) on the name plate.

Q. What is the difference between displacement and the actual discharge of a pump?
A. The displacement is actual measurement as in the previous question. The actual displacement would be less due to losses.

Q. During constant flow conditions, when water flows horizontally through a pipe that has a narrow constriction, how is the speed (velocity) and pressure of the fluid affected in passing through the constriction?
A. Pressure drops; velocity increases.

Q. Do automatic control devices, such as pump pressure governors, relieve the engineer from worry about control?
A. No. They increase the accuracy of control and reduce the labor devoted to control, but they must be watched constantly to make sure that they are functioning accurately.

Q. Why are the valves on steering engines, deck machinery, etc., built without lap?
A. As lap cuts off the steam to allow it to expand in the cylinder, full load is not obtained. In the type of machinery mentioned, it is necessary to have full load power at all times.

Q. If there are two coolers for the lubricating oil so one can function while the other is being serviced, under what circumstances should both be turned on?
A. Whenever one is failing to carry the cooling load, or whenever enough extra equipment is going to be cut in so that one cooler will probably not carry the load.

Q. In which side of an oil cooler should the pressure be kept heavier, the oil side or the water side, and why?
A. The oil side, so if there is any leakage, it will be from oil into water and not water into oil.

Q. What is the surest sign that the oil should be withdrawn from the system for batch cleaning and that sumps should be cleaned out?
A. The strainers begin to show unusual quantities of foreign matter.

Q. How can you settle a mixture of oil and water without running it through a purifier?
A. Open steam coils in a settling tank until the mixture reaches a temperature of approximately 180°. Hold this temperature and keep the mixture as stationary as possible. Open bottom drain and allow a flow until all water is out and oil appears. Close drain valve. After a time, more water will settle in the bottom and can be drained from the tank. Repeat this until as much water as possible has been removed.

Q. Trace the flow path of a typical evaporator.

A. The evaporator usually has three sources of water supply.

1. Salt water, usually obtained from a branch leading from the sanitary line. This line should be blocked off when not in use to prevent accidental opening.

2. Once the plant is in operation, the salt water discharge from the evaporator distiller is usually fed back to the evaporator as feed. Thus, instead of using cold sea water, warm sea water is fed to the evaporator and less BTU's have to be added to bring it to the boiling point.

3. Water source from inner-bottom tanks. This is the most commonly used source, as it is considered advisable to distill all shore water entering high-pressure watertube boilers. A pump with pressure governor forces water to the evaporator regulating valve.

Vapor produced in the evaporator may pass through a distiller where it is condensed due to cooling effect of sanitary-line water running through coils in the distiller. From here it is piped to an inspection tank, and from there it branches to the distilled-water tank, or the drinking-water tanks. A branch line may lead from the distilled-water tank line to the double bottoms, where distilled water may be stored in specified tanks of the double bottom.

Vapor may also be conducted to the drain cooler where it joins the condensate, or it may be led to the exhaust steam line where it runs through the condenser.

Q. Explain the "cracking-off" method of removing scale from evaporator coils.

A. Close vapor valve. Allow vapor pressure to build up to approximately 25 pounds. Close steam valve. Open bottom blow valve wide. The vapor pressure will blow the water out of the shell overboard. When the pressure decreases, the sea water will run back into the shell by gravity.

Allow the sea water to flow into the shell until it reaches the top of the gage glass. Open steam valve and allow vapor pressure to again build up to approximately 25 or 30 pounds. Close steam valve. Open bottom blow and let all water blow out of shell and run back in by gravity. Shut off bottom blow when water level reaches normal operating level.

Continue to operate plant in usual manner. The sudden cooling of the coils by the cold sea water and then the sudden increase of heat on opening the steam valve causes the coils to expand and contract rapidly thus cracking off the scale from the coils.

The scale falls to the bottom of the shell and much of it is blown out through the bottom valve on the next operation.

Q. Explain the operation of an injector.

A. Steam is admitted to a steam chamber and passes through a nozzle or a series of nozzles creating a high velocity. On leaving the nozzles, the steam passes over the open end of the feed suction pipe from the tank, creating a vacuum in the pipe. The atmospheric pressure on the surface of the water in the tank forces the water up the pipe, where the pressure is less than atmospheric (partial vacuum), into the chamber, and the high steam velocity forces it into the boiler.

Q. What is the "bilge injection?"
A. A non-return valve in the sea suction line to the main circulator, to pump out the bilges in case of emergency.

Q. Explain the principle of the hydraulic telemotor.
A. Two closed cylinders A and B are connected by pipes E and F. Both cylinders and pipes are filled with a mixture of glycerine and water (gives low freezing point). The liquid being practically incompressible, a movement of piston C results in a corresponding displacement of piston D, Fig. 55. On shipboard, one of these cylinders is located on the bridge and the other aft at the steering engine, with its piston connected to the crank of the steering-engine control valve.

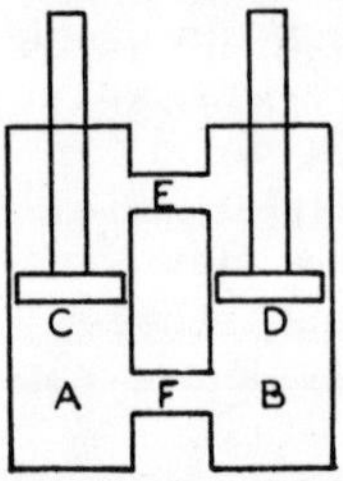

Fig. 55. Hydraulic telemotor cylinders and pistons.

Q. What are some of the most important things an engineer should do when going aboard a ship for the first time?
A. 1. Familiarize himself with all piping and engine-room equipment through personal inspection and with the aid of blueprints.
2. Learn to start and stop all auxiliary equipment and know exactly how he would proceed for all routine or emergency situations.
3. He should study boiler feed lines and boiler valves and be able to cut a boiler in or out.

Q. What are some of the most important duties of an engineer going on watch aboard a steamship?
A. 1. Check steering engine for oil level, bearing temperatures, loose nuts, unusual sounds, etc.
2. Note temperatures of refrigeration equipment, level of refrigerant, over-heating of pump; check automatic control, smoothness of operation.
3. Examine main engine for unusual sounds, bearing temperatures, oil flow; make sure all gauge readings are normal, inspection tank is oil-free, etc.
4. Look for sparking or overheating of electric motors, and examine all auxiliary machinery. Check bilges, as oil or excessive water may indicate leaks.
5. Boiler-room water gauge glass should be blown down, and all instruments and controls checked.
6. Read the log and learn from the engineer you relieve of any conditions that will affect your watch.

Q. Should you erase any figures made in the bell book?
A. No. They should be crossed out and the new figures entered below. Erasing is not only forbidden but, in case of an investigation, the engineer would be suspected of changing figures to his own advantage. Time at which speed changes were ordered at the bridge and the time at which the order was executed are vitally important. In case of collision or running aground, they may be used as a factor in fixing responsibility.

Q. What do the letters - WOG - stand for on valve bodies?
A. Water, Oil, Gas. They refer to cold service.

Q. What effect does "linking-in" on any of the valve gear on a reciprocating engine have when the links are in the astern position?
A. In the astern position, the slot in the expansion block is in a vertical position. Turning the adjusting screw will have no effect on the drag links and, therefore, no effect on valve travel.

Q. Describe the usual connections for venting and sealing centrifugal condensate pumps.
A. To prevent the entrance of air to the pump, the shaft packing gland or stuffing box is fitted with a watersealed lantern ring in order to maintain a pressure slightly above atmospheric, on the packing. Thus a slight outward leak of water is allowed rather than an inward leak of air to the pump. Water for sealing is taken from the discharge side of the pump or any other outside source (latter preferable). The vents are usually located at the highest point in the pump casing.

Q. What is the main difference between safety valves and relief valves?
A. Relief valves are constructed on the general design of safety valves to relieve excess pressure. They may be used for service where an approved safety valve is not required by U. S. Coast Guard Regulations.

Q. What type of valve is used in the machinery space for controlling bilge suction from various compartments?
A. Stop-check valves.

Q. What are the three methods of fighting fire?
A. 1. By removing the fuel or letting it burn itself out
2. By removing the oxygen from fire with smothering agents
3. By removing the ignition temperature by cooling or quenching

Q. Why are vent pipes required to be covered with corrosion-resistant wire screens and what care should they be given?
A. They act as flame screens in the event of an open fire on the deck. They must be kept clean at all times and never painted.

Q. What precautions should be taken before changing an electric light globe in the pump room of a tank vessel?

A. Shut off power.

Q. Describe a diffusion ring of a centrifugal pump. What is its purpose?

A. A section of a turbine type centrifugal pump which helps to convert velocity energy to pressure, Fig. 56.

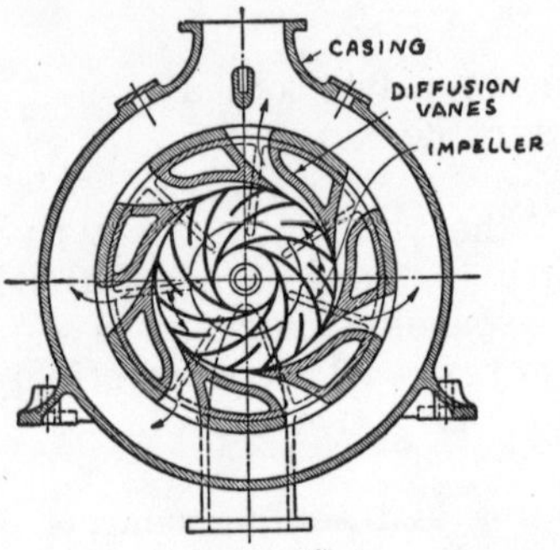

Figure 56.

Q. Why will most pumps not handle hot water from below the level of the pump suction?

A. When the hot water comes in contact with the cool suction pipe and enters this area of lower pressure, it flashes into steam.

Q. How should a steam gage be connected in a steam line to protect it from high temperature?

A. Make a loop in the steam line near the gage. Condensation will form in the loop and water will fill the Bourdon tube and prevent live steam from entering and possibly melting the soldered ends of the tube.

Q. Why must air be removed from both cooling water side and condensate side of a condenser?

A. To prevent air from getting into condensate water to boiler and keeping it out of condenser itself. Oxygen is very corrosive.

Q. Assuming an unlimited supply of cooling water and a clean tight condenser, what will be the limiting factor on the vacuum that can be obtained?

A. In a reciprocating cycle, it will be the condition of the air pump valves. In a turbine cycle, it will be the operation of the air ejector.

Q. Explain how the water side plunger of a vertical simplex pump is repacked.

A. The installation of the pump will often determine the method to be used. If the pump is high enough to remove the bottom head and the follower plate, chances are that the plunger can be packed from the bottom. In the case where the bottom head is too close to the tank tops for removal, the upper head will have to be removed. By removing the steam cylinder head, the complete assembly of steam piston, piston rod, crosshead, liquid rod and liquid plunger may be lifted high enough to work on the liquid plunger. This can be done because the plunger is usually constructed in three pieces, a center barrel and two ends or follower plates. TUX packing is usually used and should be soaked before installing.

Q. Describe the emergency bilge suction valve and explain how high water in the bilge can be disposed of.

A. This is a non-return type valve located in the suction line to the main circulator pump. If all bilge pumps are operating and no headway is being made on the water in the bilge, this bilge suction can be opened and the sea suction (main injection) closed. This will pump the bilge water through the main condenser and overboard. The only objection to this procedure is the fact that the condenser tubes may be contaminated with oily bilge water and will have to be cleaned out at the first opportunity.

Q. What uses are made of the auxiliary exhaust in a reciprocating plant?

A. Most of it is used in the feedwater heater; any excess is sometimes used for heating.

Q. What are the advantages gained by using feedwater heaters?

A. An increase in economy. The hotter the water entering the boiler, the less fuel needed to convert it to steam. An approximate rule is an increase of 1% efficiency for each 11° increase in feedwater temperature.

Q. Describe a "spring bearing." Why are they used?

A. The bearings in the shaft alley, with the exception of the last one at the stern tube, are spring bearings. They consist of a babbitted bottom half and a shell top half. Since there is no vertical pull on the shaft, no babbitt is needed in the top half.

Q. Where are rubber sheet packings used?

A. Flange or valve head gaskets on water or air lines.

Q. How should packing be installed in stuffing box of pump rods?

A. Provide about 1/8 to 3/16 inch clearance on circumference to allow for expansion. Stagger cuts to prevent excess leakage along the rod.

Q. How would you set the valves on a duplex pump?

A. Remove the steam chest bonnets and place the pistons in their midstroke position. To do this, open the drain cocks and move each piston by prying on the crosshead, never on the valve lever, until it comes in contact with the cylinder head. Make a mark on the piston rod at the steam end stuffing box gland. Move each piston back until it strikes the opposite head and make a second mark on the piston rod. Halfway between these two marks make a third mark. Move each piston until this last mark coincides with the face of the gland. The pistons will then be exactly at their mid-stroke position.

After setting pistons in their mid positions, proceed to set the valves centrally over their ports. Adjust the lock nuts to allow about 3/16" lost motion on each side. The best way to test for equal division of lost motion is to move each valve back and forth until it strikes the nuts and see if the port openings are equal. When the port openings have been equalized, the valves are set. The valve motion should not be changed when setting the valves. Too much lost motion will give a long stroke and may cause the piston to hit the head. Too little lost motion will give a short stroke. If only one valve of a duplex pump is to be set, remember that it is operated by the piston of the opposite pump.

Q. Describe the general procedure for starting the main air ejectors. Under what conditions will both sets of ejectors be used?

A. The general procedure for starting the main air ejectors should be accomplished as follows:

1. Drain the steam supply lines to the air ejector assembly.

2. Start circulating condensate cooling water through the air ejector inter-and-after condensers. Recirculation of condensate will probably be necessary to provide sufficient cooling water.

3. Open valves in inter-and-after condenser drain lines.

4. Open first-and-second stage suction and discharge valves of the air ejector elements to be started.

5. Open wide the second-stage ejector steam inlet valve, and check the steam supply line pressure.

6. When the condenser vacuum rises to 20-inch mercury or above, open wide the first-stage steam inlet valve.

7. The ejector should now be in full operation.

In order to raise vacuum more quickly or when large air leaks are present, both second-stage elements may be started simultaneously and both first-stage elements started when a vacuum of 20-inch mercury has been obtained. Adequate recirculation must be provided for condensing the additional steam being discharged to the air ejector condensers.

Q. What is the purpose of the auxiliary exhaust automatic unloading valve?

A. The automatic unloading valve is fitted into the auxiliary exhaust system for the purpose of unloading excessive pressures into one or more of the main or auxiliary condensers, thus relieving the excess pressure without losing the steam to the atmosphere while maintaining sufficient back pressure for the proper operation of the various heating systems, gland seal systems, etc., which are dependent upon the back pressure.

Q. (a) Where is the back-pressure valve installed? (b) Why is this valve so important?

A. (a) It is a spring-loaded valve fitted in the exhaust line.

(b) To provide constant pressure in the exhaust line, to provide a cushion to all reciprocating auxiliaries and to maintain a set pressure of steam on the feedwater heater

Q. (a) What provisions are made in the bilge system of a vessel to protect the pipes against being clogged or the pumps from becoming damaged? (b) How can the possibility of clogging the bilge suction be reduced when carrying granular cargo such as grain, coal, rock ballast, etc.?

A. (a) Each individual bilge suction should be fitted with a suitable bilge strainer having an open area of not less than three times that of the suction pipe. In addition, a mud box or basket strainer should be fitted in an accessible position between the bilge suction manifold and the pump.

(b) When carrying cargo of a granular nature such as grain, coal, rock ballasts, etc., the possibility of the bilge suction becoming clogged can be reduced by increasing the strainer area. This may be accomplished by constructing a box over the bilge strainer covered with burlap or wire mesh; or other devices may be employed with the same result.

Q. When a vessel which is fitted with heating coils in her tanks is carrying light fuel such as kerosene or gasoline, what precautions should be taken to eliminate any possibility of contaminating boiler feedwater, inspection tanks, heating pipelines, etc.?

A. When a vessel which is fitted with steam coils in her cargo tanks is carrying light oils such as kerosene or gasoline, the steam lines to the heating coils should have all valves tightly closed and blank flanges inserted where possible to eliminate any possibility of contaminating boiler feedwater, inspection tanks, steam lines, etc.

Q. What is usually required in order to have a centrifugal pump operate properly with a negative suction head?

A. When there is a negative suction head it is usually necessary to install a check valve in the suction line below the water level and to prime the suction pipe as well as the pump before starting.

Q. List the procedure to be followed if the steam reciprocating simplex pump fails to start.

A. 1. Secure the pump. Do not attempt to adjust the tappet collars.

2. Examine the discharge and the exhaust lines for closed valves or for a valve disc that has become detached from its stem.

3. Jack the pump with a bar to determine if there is excessive friction.

4. Disconnect the auxiliary valve stem from the operating gear without disturbing the adjustment of the tappet collars. Open the exhaust, suction and discharge valves and then crack the throttle. Work the auxiliary valve by hand. Should the pump still refuse to start, secure the pump. Remove the steam valve chest cover and examine the main valve to see if it has overridden or stuck.

5. If the pump cannot now be started, a complete overhaul of the working parts of the steam end will probably be necessary.

Q. Describe a constant-pressure pump governor for a steam-driven pump.

A. The pump governor consists mainly of a valve body containing the main valve and its piston and a controlling diaphragm and spring with connecting linkage to an auxiliary valve. The underside of the controlling diaphragm is exposed to the pump discharge pressure and the movement of the controlling diaphragm against the spring pressure is transmitted by linkage to the auxiliary valve which controls the opening of the main valve so as to admit a greater or lesser amount of steam to the driving mechanism, thereby maintaining a constant discharge pressure regardless of flow.

Q. Explain the operation of the direct-contact deaerating heater used in the closed-feed system.

A. The condensate enters the heater at the top through the tubes of a vent condenser. From the vent condenser the condensate is led to the center of the tank where it is sprayed through a bank of nozzles, upward and outward. Exhaust steam is also led to the tank and mixes with the sprayed condensate. The combined action of spraying and heating of the condensate releases the dissolved oxygen which rises to the top of the heater. The oxygen then passes through the vent condenser and out of the air vent. The heated condensate falls to the lower section of the tank and passes down to the feed pump. In the heater the condensate is heated to 230° to 240°F. with exhaust steam at 10-14 p.s.i. gage pressure.

Q. What heating equipment does the condensate and feedwater normally pass through from the condenser to the boiler on a modern turbine-driven vessel?

A. The heating equipment that the condensate will pass through consists of the inter, after, and gland exhauster condensers of the main or auxiliary air ejector units, and the deaerating feed heater including its vent condenser.

The feedwater heating equipment may consist of a high-pressure feed heater and/or an economizer

Some vessels have a drains cooler installed between the air ejector condensers and the deaerating feed heater.

Q. (a) What are the advantages of the centrifugal pump in comparison with the vertical reciprocating pump? (b) Would any damage be likely to occur if the discharge valve was closed just before securing the power to the centrifugal condensate pump?

A. (a) The centrifugal pump delivers an even pressure, is easily adaptable for turbine or motor drive, requires less headroom, and has fewer moving parts.

(b) The centrifugal pump is not a positive displacement pump and hence there will not be a dangerous rise in pressure and no damage should occur.

Q. What is the purpose of the wearing rings of a centrifugal pump?

A. The wearing rings, by their close clearances, restrict flow from the high pressure side to the low pressure side. These rings suffer erosion, due to water leakage, but are easily and economically replaced in comparison with the costly renewal of an eroded casing or impeller.

Q. Describe the mechanical arrangement by which the stroke is varied on the triplex-crankshaft type of variable-stroke feed pump.

A. The lower end of each crankshaft connecting rod is pin-connected to a link, the upper end of which is connected to the plunger or cross-head. The lower end of each of these links carries a curved pivoting guide shoe which slides in the curved track of the "stroke transformer." Control of the pump stroke is obtained by varying the angular position of the stroke transformer guide by means of the hydraulic servo piston which tilts this guide about its trunnion supports.

Q. What may be the causes of a direct-acting steam feed pump making short jumps of the plunger at the beginning of the stroke?

A. This jumping indicates that the liquid cylinder is not completely filled with liquid at the beginning of the discharge stroke. This may be a result of running the pump too fast; a restricted suction line; an air leak in the suction line; or the pumping of excessively heated liquids which vaporize under the lower pressures encountered during the suction stroke.

Q. Explain the operation of a hydraulic telemotor and tell what fluid is used in the system.

A. It consists of a transmitter, which is placed in the wheelhouse, or on the bridge in some cases. There is a receiver located near the steering engine at the stern of the vessel, and there is a copper pipeline which connects the transmitter to the receiver.

The entire telemotor system is completely filled with a fluid that is used to transmit pressure from the transmitter to the receiver. Pistons and in some installations, hydraulic rams, in the cylinders of the transmitter, are moved by gears and pinions on racks, connected with the hand steering wheel. The motion of the hydraulic rams is transmitted to similar rams in the receiving part of the mechanism at the steering engine, by the pressure of the fluid in the pipes that connect the transmitter to the receiver. The motion of the rams of the receiver is transmitted through a system of gears and levers to the valve which controls the operation of the engine.

A telemotor holds about 5 gallons of fluid. The fluid is filled into the system under pressure of a hand pump for that particular purpose, or by gravity from a tank placed slightly higher than the rams in the wheelhouse and which allow any loss due to leakage at the glands to be replaced through various types of equalizing valves and air contained in the system escapes through a pet cock which is located at the highest point in the system.

Some vessels use a mixture of glycerine and water for the telemotor system. It is not too slow in moving and will not freeze in the pipes. Others use a mineral oil which is light in body and quick in action.

Q. What are the two principal functions of the main and auxiliary condensers?

A. The primary function of the condensers is the production of the condensers is the production and maintenance of a low exhaust pressure, thereby increasing the work delivered by the engine. A secondary function is the conservation of the ship's supply of fresh water by returning the steam used in the engineering plant to the condensate system.

Q. (a) For what purposes are manometers usually used? (b) How do you read a U-tube manometer? (c) How is a manometer connected?

A. (a) Manometers are usually used to indicate small differences of pressure such as draft pressures in a furnace or uptake of a boiler.

(b) The difference in pressure between two points is registered by the difference in the water level of the two tubes and is read on the graduated scale in inches or fractions of an inch.

(c) One of the two upper ends of the U-tube is connected to the boiler stack, or to any other point where it is desired to measure draft pressure, while the other end of the tube is open to the atmosphere.

Q. What materials are usually used to insulate low temperature spaces; high temperature spaces?

A. For low temperatures cork or rock wool is usually used.

For high temperatures basic minerals are used, such as asbestos, carbonate of magnesia, diatomaceous earth, mica, aluminum foil, and fibrous glass.

Q. What are the usual causes of sticking stems on stop valves? Give the remedy for each.

A. Sticking of valve stem is caused by:

(a) Stuffing box set or packed too tightly. Slack up on the gland and relieve the packing pressure.

(b) Stuffing box gland cooked due to uneven setting up of the gland nuts. Correct the positions of the nuts.

(c) Paint or rust on valve stem which should be removed by cleaning.

(d) Valve jammed shut while hot, with the result that subsequent cooling causes contraction which binds the disc tightly to the seat. To relieve the strain, carefully slack the yoke nuts; if not a yoke valve, slack back slightly on the bonnet nuts. This may permit freeing the disc from the seat.

(e) Valve jammed open while cold, with the result that subsequent heating causes expansion which binds the valve open. Usually can be started with a wrench, care being taken not to spring the stem. After opening valve wide, close in a half turn so that the danger of binding will be eliminated.

(f) Burred threads in stem or bent valve stem. Straighten and clean, or renew the damaged stem.

Q. What is the main difference between safety valves and relief valves?

A. Safety valves function by popping wide open at the set pressure and remaining in that position until the pressure has dropped a predetermined amount after which the valve snaps shut. They cannot be used for liquids. Relief valves start to open at a set pressure but require about 20% over pressure to open wide. As the pressure drops they start to close gradually and seat at about the set pressure.

Q. During constant flow conditions, when water flows horizontally through a pipe that has a narrow constriction, how is the speed and pressure of the fluid affected in passing through the restriction?
A. During constant flow conditions, when water flows horizontally through a pipe that has a narrow constriction, the water speeds up as it approaches and flows through the constriction, and this increase in speed is accomplished by a decrease in pressure.

Q. Define "ratio of expansion."
A. It is the ratio of the total volume of the final stage to that of the first stage.

$$\text{Ratio of expansion} = \frac{\text{final volume}}{\text{initial volume}}$$

Q. What is the accepted standard atmospheric pressure at sea level? How is it registered on a mercury barometer?
A. 14.7 p.s.i.; 30" Hg.

Q. What is the difference between a Fahrenheit and a Centigrade thermometer?
A. Fahrenheit: Water freezes 32°; boils 212°. Centigrade: Water freezes 0°; boils 100°.

Q. What immediate action should be taken if at any time loss of vacuum is accompanied by a hot or flooded condenser?
A. Secure plant. Open atmospheric exhaust. Locate trouble.

Q. What means are provided to enable an engineer to be assured that the steam drains from the heaters in the fuel system are not returning to the boilers?
A. All such drains lead to the inspection tank before reentering the feedwater system. If a leak occurs the drains are led to the bilge until repairs are made.

Q. A cylinder contains 15 cu. ft. of steam at 40 lbs. per sq. in. absolute pressure and occupies 10.5 cu. ft. per lb. Find the weight of this volume of steam.
A. $15 \div 10.5 = 1.428$ lbs.

Q. When both welding and riveting are to be used in vessel repair, which shall be done first? Why?
A. Welding is done first so as not to put rivets in shear as a result of heat from the weld.

Q. What is the side view of a ship called?
A. Profile or elevation.

Q. What is a template?
A. A wire, heavy cardboard or piece of wood that can be shaped or cut to a specific design to be used as a pattern when making a new piece.

Q. Explain how steering control may be changed to various stations, to another drive unit, and how hand steering may be performed with a typical electro-hydraulic unit.
A. By utilizing a system of dog clutches, clevis pins and equalizing valves, the telemotor may be shifted over to direct control through the trick wheel. In some vessels a duplicate selsyn system replaces the telemotor and the steering systems may be changed by a switch control.

Q. What is a weld reinforcement?
A. A welded joint with a doubler plate or some other type of backing piece to add strength.

Q. State the proper procedure to be followed when loss of steam pressure occurs while the vessel is under way.
A. Cut out fires, take boilers off the line, secure feedwater to boilers, investigate trouble and correct.

Q. Would any damage be likely to occur if the discharge valve was closed just before securing the power to the centrifugal condensate pump?
A. No, as there is built-in recirculation through the vent back to the condenser.

Q. Why are high and low suctions fitted to settling tanks?
A. High suction should be used whenever possible to eliminate possibility of getting water or sludge into the burners which may happen if the low suction is used.

Q. Explain how the governor of the inertia type controls the speed of the engine.
A. This type of governor controls the speed of the engine by varying the point of cut-off of the valve. A bar is attached to the flywheel. This bar is weighted on the ends (heavy weight on one end and a lighter weight on the other end). One end of a spring is attached to the flywheel and one end of the bar. A small pin projects from the bar to which is attached the end of the valve stem. This pin changes its position thus changing the travel of the valve, as the weights move in and out against the spring pressure.

SKETCHES

Q. Sketch a globe valve; a gate valve; a two-way valve; an angle valve.

A. Figure **57**.

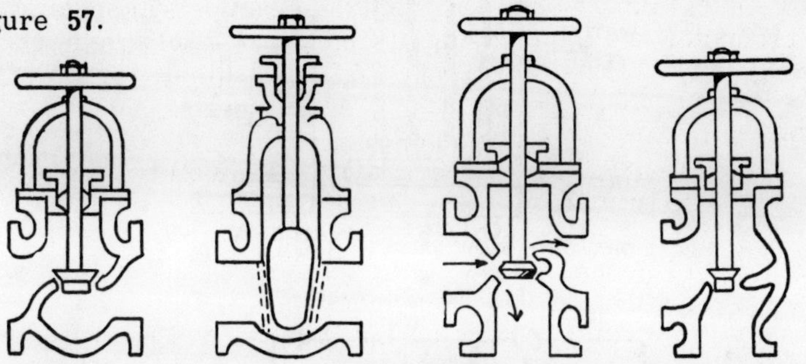

Fig. 57. A. Globe valve. B. Gate valve. C. Two-way valve. D. Angle valve.

Q. Sketch an expansion joint of the slip-joint or sliding type.

A. Figure **58**.

Q. Sketch a non-return valve.

A. Figure **59**.

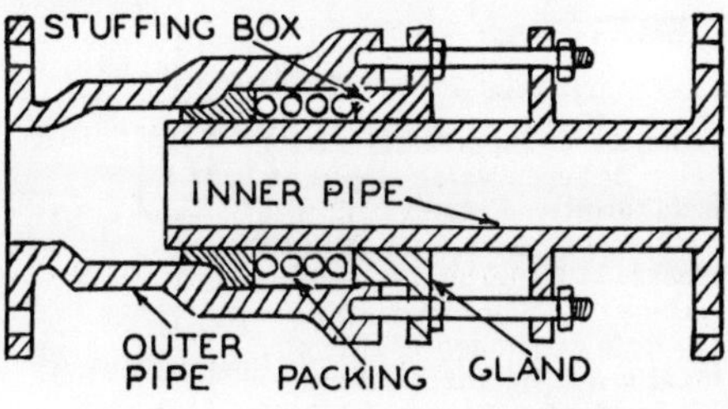

Fig. 58. Sliding type expansion joint.

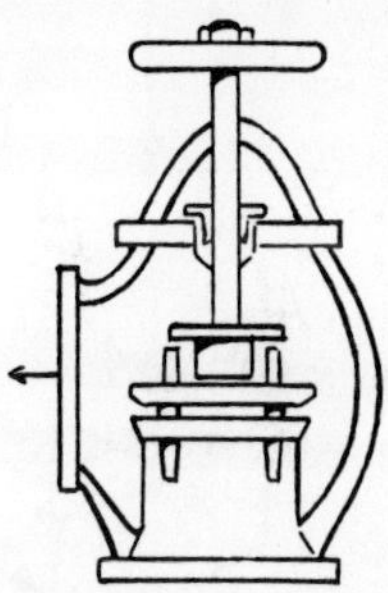

Fig. 59. Non-return valve.

Q. Sketch an evaporator. Name all parts.

A. Figure **60**.

Q. Draw a simple schematic sketch of a steam and water cycle of a modern marine steam turbine power plant with a closed feed system.

A. Figure **61**.

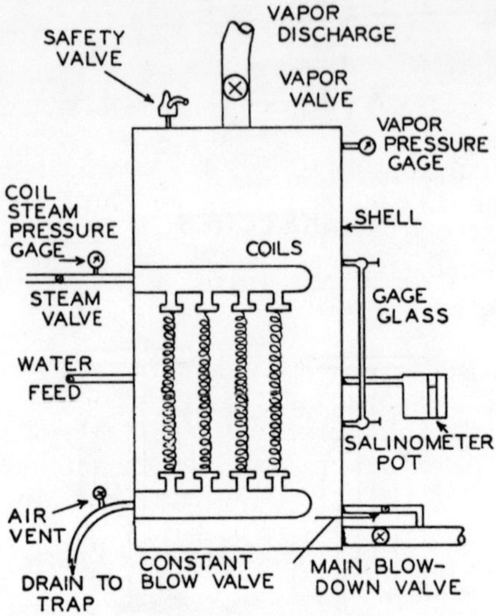

Fig. 60. Evaporator.

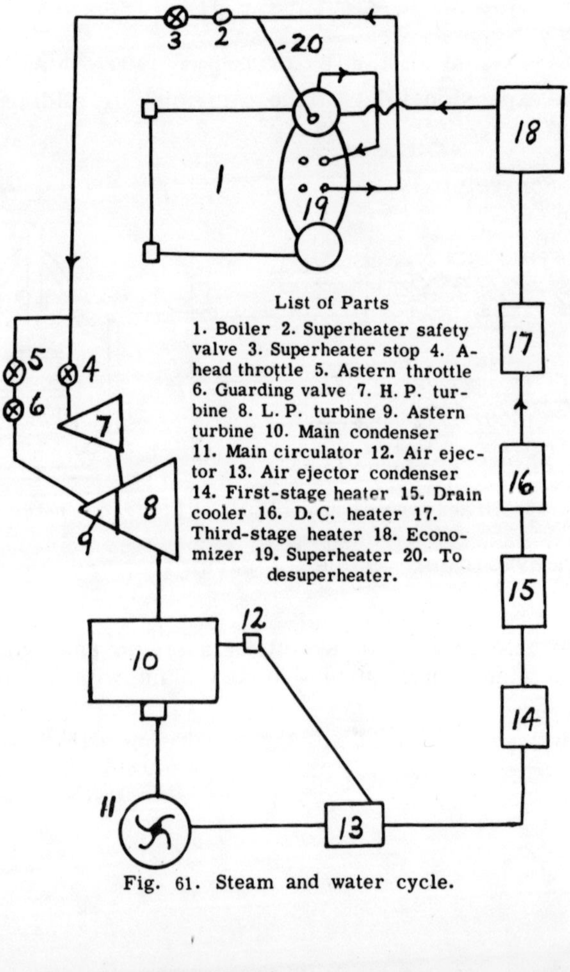

List of Parts

1. Boiler 2. Superheater safety valve 3. Superheater stop 4. Ahead throttle 5. Astern throttle 6. Guarding valve 7. H. P. turbine 8. L. P. turbine 9. Astern turbine 10. Main condenser 11. Main circulator 12. Air ejector 13. Air ejector condenser 14. First-stage heater 15. Drain cooler 16. D.C. heater 17. Third-stage heater 18. Economizer 19. Superheater 20. To desuperheater.

Fig. 61. Steam and water cycle.

MATHEMATICS

Q. Fuel oil containing 18,500 B. T. U.'s per lb. is fed to a boiler and it is found that 15,000 B. T. U.'s go into forming steam. What is the efficieney of the boiler?

A. $$\text{Efficiency} = \frac{\text{Output}}{\text{Input}} = \frac{15,000}{18,500} = 81\%$$

Q. A flat horizontal plate 9' square lies 12' below the surface in salt water. What is the hydrostatic pressure per sq. ft. on this plate and what is the total pressure on 1 side of the plate?

A. Pressure in lbs. per sq. ft. = weight of sea water in lbs. per cu. ft. X head of water on plate.
= 64 X 12 = 768 lbs.

Plate is 81' square. Therefore, 768 X 81 = 62,208 lbs. pressure on one side of plate.

Q. What is the weight of a steel plate 10' 4" long, 4' 8" wide, 1-1/4" thick? A square foot of plate 1/8" thick weighs 5 pounds.

A. Area of plate = 10' 4" X 4' 8"
Change to inches
= 124" X 56"
= 6944 sq. in.
Change to sq. ft.
6944 ÷ 144 = 48.22 sq. ft.
As 1/8" plate weighs 5 lbs. per sq. ft., change the 1-1/4 to 8ths = $\frac{10}{8}$.
Therefore, it will be 10 times 5 or 50 lbs. per sq. ft.
Weight of plate = 48.22 X 50
= 2411 lbs.

Q. A cast iron block has the dimensions 14" X 12" X 18". It has a 6" diameter hole cut through its largest dimension. What is the weight of the block, allowing 640 lbs. per cu. ft.?

A. Total volume = 14" X 12" X 18" = 3024 cu. in.
Volume of hole = .7854 X $(6)^2$ X 18 = 508.94 cu. in.
Actual volume = 3024 — 508.94 = 2515.06 cu. in.
There are 1728 cu. in. in 1 cu. ft. Therefore,
2515.06 ÷ 1728 = 1.45 cu. ft.
1.45 X 640 = 928 lbs.

Q. A steamer has covered 1200 miles at 10 knots using 1400 barrels of oil. She has 1300 miles to go. Only 1100 barrels of oil remain. What speed is required to complete the voyage?

A. Rule: Total consumption varies as square of speed times distance.

1. $1400 : (10)^2 \times 1200 :: 1100 : (x)^2 \times 1300$

$$1400 \times x^2 \times 1300 = 1100 \times (10)^2 \times 1200$$

2. $\frac{1400}{(10)^2 \times 1200} = \frac{1100}{x^2 \times 1300}$

Either method 1 or 2 can be used.
Carrying out No. 2, we have:

$$\frac{\overset{7}{\cancel{1400}}}{100 \times \underset{6}{\cancel{1200}}} = \frac{110\cancel{0}}{x^2 \times 13\cancel{00}}$$

Cross-multiplying, we get:

$$91x^2 = 6600$$
$$x^2 = 72.52$$
$$x = 8.5 \text{ — Required speed}$$

Q. Convert 100° Centrigrade (C.) to Fahrenheit (F.).

A.
$$F. = \frac{9}{5} C. + 32$$
$$= \frac{9}{\cancel{5}} (\overset{20}{\cancel{100}}) + 32$$
$$= 180 + 32$$
$$= 212°$$

Q. Convert 212° F. to C.

A.
$$C. = \frac{5}{9}(F. - 32)$$
$$= \frac{5}{9}(212 - 32)$$
$$= \frac{5}{\cancel{9}}(\overset{20}{\cancel{180}})$$
$$C. = 5\ (20)$$
$$= 100°$$

Q. How many B. T. U.'s are there in 1 H. P. hour?

A. 1 H. P. = 33,000 ft. lbs. of work per min.
1 B. T. U. = 778 ft. lbs. of work

$$1 \text{ H. P. hour} = \frac{33,000 \times 60}{778} = 2545 \text{ B. T. U.'s}$$

Q. If the distance by engine is 340 miles and the distance by observation is 333.2 miles, what is the slip in per cent?

A.
$$\% \text{ slip} = \frac{E - O}{E}$$

$$= \frac{340 - 333.2}{340}$$

$$= \frac{6.8}{340}$$

$$= .02 = 2\% \text{ slip}$$

E = Engine
O = Observation

Q. If the average R. P. M. for 24 hours and 18 minutes is 102, pitch of propeller is 16. 2', distance by observation is 360 miles, what is the slip in per cent?

A.

$$\text{Engine distance} = \frac{16.2 \times 102 \times (24 \times 60 + 18)}{6080}$$

$$= \frac{16.2 \times 102 \times (1440 + 18)}{6080}$$

$$= \frac{16.2 \times 102 \times 1458}{6080}$$

$$= 396.25 \text{ miles}$$

$$\%\ \text{slip} = \frac{\text{Engine} - \text{Observation}}{\text{Engine}} \qquad \frac{E - O}{E}$$

$$= \frac{396.25 - 360}{396.25}$$

$$= \frac{36.25}{396.25}$$

$$= .091$$

$$= 9.1\%$$

Q. The diameter of propeller is 24'. Width of blade 42". Forward edge of blade leads after edge by 19. 5". Pitch radius (distance from center of hub to point of measurement) is 5' 9". What is the pitch of the propeller?

A.

$$\text{Pitch} = \frac{\text{Circumference} \times \text{Piece of pitch}}{\text{Piece of circumference}}$$

1. $P = \frac{Cp}{c}$

2. $C = \pi D$

   a. Pitch radius = 5' 9"
      Pitch diameter = 2 X 5' 9"
      = 11' 6"

$C = 3.1416 \times 11\text{-}1/2$

$C = 36.1284$ or 36.13

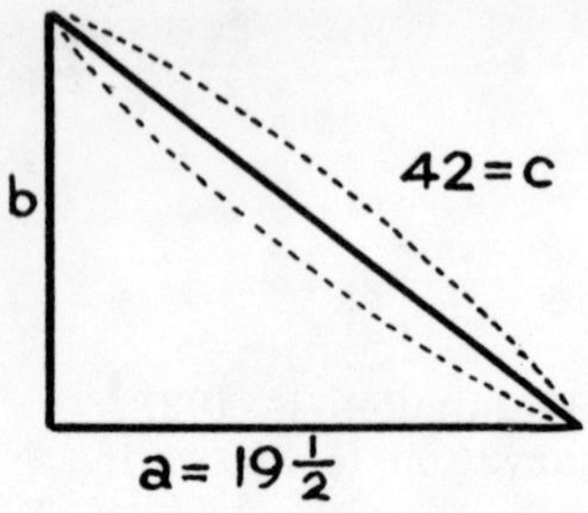

Figure 62.

3. Referring to Fig. 62:
   a = 19-1/2, or the amount by which the forward edge of the blade leads the after edge

   c = 42 or width of the blade

4. $c^2 = a^2 + b^2$

   $b^2 = c^2 - a^2$

   $b = \sqrt{c^2 - a^2}$

   $b = \sqrt{(42)^2 - (19.5)^2}$

   $b = \sqrt{1764 - 380.25}$

   $b = \sqrt{1383.75}$

   b = 37.19 = piece of circumference

5. Substituting in No. 1:

$$P = \frac{Cp}{c} \quad \frac{36.13 \times 19.5}{37.19}$$

$$P = 18.94'$$

Q. At midnight, the revolution counter shows 376510. At 4 A. M., it reads 397630.
   (a) What is average R. P. M. ?
   (b) Pitch of propeller 16', 13% slip. What distance has ship traveled?

A.
$$\begin{array}{r} 397630 \\ -376510 \\ \hline 21{,}120 \end{array}$$ Revs.

60 X 4 = 240 mins.

21,120 ÷ 240 = 88 R. P. M.

21,120 X 16 X .87 = 293,990 ft.

293,990 ÷ 6080 = 48.3 miles

Q. An alloy called "antifriction metal" is $\frac{37}{1000}$ copper, $\frac{111}{125}$ tin, and $\frac{3}{40}$ antimony. Find the weight of each metal in a mass of the alloy weighing 1250 pounds.

A. Find a common denominator from which to work.

$$\frac{37}{1000} = \frac{37}{1000}$$

$$\frac{111}{125} = \frac{888}{1000}$$

$$\frac{3}{40} = \frac{75}{1000}$$

$$\frac{1000}{1000}$$

This shows we have a 100% mixture with the new common denominator. Now figure each metal separately on the basis of the solid - 1250 pounds.

$$\frac{37}{1000} \text{ of } 1250 = 46.25 \text{ lbs. copper}$$

$$\frac{888}{1000} \text{ of } 1250 = 1110.00 \text{ lbs. tin}$$

$$\frac{75}{1000} \text{ of } 1250 = \underline{93.75 \text{ lbs. antimony}}$$

$$1250.00 \text{ lbs. total}$$

Q. A single-acting power pump making 100 R. P. M. has dimensions 4" X 4" X 4". Slip 3%. What is its actual discharge in gallons per minute (G. P. M.)?

A. G. P. M.

$$= \frac{\text{Vol. of cylinder X number working strokes X efficiency}}{231}$$

$$= \frac{.7854 \times (4)^2 \times 4 \times 100 \times .97}{231}$$

$$= 21.107 \text{ G. P. M.}$$

Q. A double-bottom tank holds 6530 gals. A duplex double-acting pump 8" X 6" X 10" makes 35 double strokes per minute. Leakage 10%. How long will it take to pump out the tank?

A. G. P. M.

$$= \frac{\text{Vol. of cylinder X number working strokes X efficiency}}{231}$$

$$= \frac{.7854 \times (6)^2 \times 10 \times (35 \times 4) \times .90}{231}$$

$$= 154.224 \text{ G. P. M.}$$

$$6530 \div 154.224 = 42.34 \text{ minutes.}$$

Q. A pump having the dimensions 10" X 8" X 12" makes 70 S. P. M. with 80 lbs. steam pressure. How many gallons per hour will it pump?

A.

*1. $\text{G. P. H.} = \frac{MANS}{231}$

M = Minutes in 1 hour
A = Area liquid cylinder
N = Number strokes
S = Length of stroke

or

2. $\text{G. P. M.} = \frac{ALNE}{231}$

$\text{G. P. H.} = \frac{ALNE \times 60}{231}$

A = Area liquid cylinder
L = Length of stroke
N = Number of strokes
E = Efficiency

---

*1 R. P. M. = 1 double stroke.

Using No. 1, we have:

$$G.P.H. = \frac{60 \times .7854\ (8)^2 \times 70 \times 12}{231}$$

$$= 10967.04$$

G. P. H. = Gallons per hour
G. P. M. = Gallons per minute

Q. A duplex double-acting pump 6" X 8" X 6" makes 45 S. P. M. Slip 4%. What is its actual discharge in gallons per minute (G. P. M.)?

A. G. P. M.

$$= \frac{\text{Vol. of cylinder} \times \text{number working strokes} \times \text{efficiency}}{231}$$

$$= \frac{.7854 \times (8)^2 \times 6 \times (45 \times 2) \times .96}{231}$$

$$= 112.80 \text{ G. P. M.}$$

Q. A duplex double-acting pump 4" X 6" X 6" makes 25 R. P. M. Slip 4%. What is its actual discharge in G. P. M.?

A. G. P. M.

$$= \frac{\text{Vol. of cylinder} \times \text{number working strokes} \times \text{efficiency}}{231}$$

$$= \frac{.7854 \times (6)^2 \times 6 \times (25 \times 4) \times .96}{231} = 70.502 \text{ G. P. M.}$$

Q. A duplex double-acting pump 6" X 4" X 6" makes 50 double strokes per minute. What is its theoretical discharge in gallons per minute?

A. G. P. M.

$$= \frac{\text{Vol. of cylinder} \times \text{number working strokes}}{231}$$

$$= \frac{.7854 \times (4)^2 \times 6 \times (50 \times 4)}{231}$$

$$= 65.28$$

Q. A simplex double-acting pump 12" X 8" X 8" makes 100 S. P. M. Slip 6%. What is its actual discharge in G. P. M.?

A. G. P. M.

$$= \frac{\text{Vol. of cylinder} \times \text{number working strokes} \times \text{efficiency}}{231}$$

$$= \frac{.7854 \times (8)^2 \times 8 \times 100 \times .94}{231}$$

$$= 163.635 \text{ G. P. M.}$$

Q. A simplex single-acting pump 6" X 5" X 5" makes 50 S. P. M.

Slip 5%. What is its actual discharge in gallons per minute (G. P. M.)?

A. G. P. M.

$$= \frac{\text{Vol. of cylinder X number working strokes X efficiency}}{231}$$

$$= \frac{.7854 \times (5)^2 \times 5 \times \left(\frac{50}{2}\right) \times .95}{231}$$

= 10.094 G. P. M.

Q. A tank 14' high and 28" in diameter is filled with water. What is the total pressure on the bottom of the tank?

A. Area of bottom of tank = .7854 X 28 X 28
= 615.75 sq. in.

Pressure in lbs. per sq. in. = .433 X height of column in feet
= .433 X 14
= 6.06

Total pressure = 615.75 X 6.06
= 3731.5 lbs.

PUMP PROBLEMS

The following information on pump problems should be noted carefully:

1. A stroke should be considered as a movement of one piston or plunger over its maximum travel in one direction.
2. A double stroke or revolution should be considered as a movement of one piston or plunger through a maximum travel in both directions, or the equivalent of two single strokes.
3. Regardless of the type of pump, a stroke refers to the movement of only one piston or plunger over its maximum travel in one direction.
4. The strokes, as given in the problem involving the duplex double-acting pump, should be multiplied by 2 in order to obtain N. Where double strokes of a duplex-acting pump are given, they should be multiplied by 4 in order to obtain the total strokes.
5. In a simplex double-acting pump where strokes are given, they should be considered the total power strokes to be used as N. Where double strokes are given, they should be multiplied by 2 in order to obtain N.
6. In regard to simplex single-acting pumps, the strokes given should not be considered as the strokes used by N, but should be divided by 2 in order to obtain the effective working strokes. Where double strokes are given, the given number should be used as N.

The following information regarding the figures on pump name plates should be memorized:

10" X 6" X 8"

1. The first figure represents the diameter of the steam piston —10".

2. The second figure represents the diameter of the water or liquid piston—6".

3. The third figure represents the stroke—8".

Q. How many bbls. of oil will a wing tank of the following dimensions hold? 30' long, 12' high, 10' wide at top, 6' wide at middle, 4' wide at bottom.

A. 1. Volume = Width X Height X Length

The average width is found by the following rule:

$$\frac{\text{top} + (4 \text{ X middle}) + \text{bottom}}{6}$$

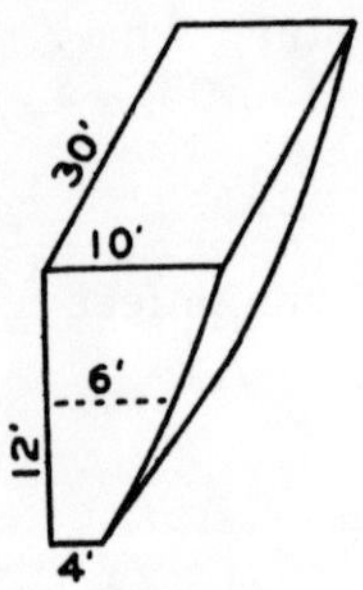

Figure 63.

Substituting in (1)

$$\text{Volume} = \frac{10 + 4(6) + 4}{6} \text{ X } 12 \text{ X } 30$$

$$\text{Volume} = \frac{10 + 24 + 4}{6} \text{ X } 12 \text{ X } 30$$

$$= \frac{38}{\not{6}} \text{ X } \overset{2}{\not{12}} \text{ X } 30$$

$$= 38 \text{ X } 2 \text{ X } 30$$

$$= 2280 \text{ cu. ft.}$$

2. There are 7.5 gals. per cu. ft.
3. 2280 X 7.5 = 17,100 gals.
4. There are 42 gals. per bbl.
5. 17,100 ÷ 42 = 407.1 bbls.

Answer: 407.1 bbls. (See Fig. 63.)

Q. How much must a valve be raised from its seat to obtain a full opening equal to the size of the valve?

A. 1/4 of its diameter.

<u>Example</u>: How much will a 5" valve have to be lifted from its seat to obtain full opening?

<u>Answer</u>: 1/4 X 5 = 1.25 or 1-1/4"

<u>Solution or Proof</u> (this is often asked in exams):

Area of opening = $2\pi rh$

Area of valve = πr^2

We want the two to be equal; therefore,

$$2\pi r h = \pi r^2$$

$$h = \frac{\pi r^2}{2\pi r}$$

$$h = \frac{r}{2}$$

$$h = \frac{2r}{4}$$

$$h = \frac{D}{4}$$

Q. A cylindrical tank is 8' 4" high and 3' 7" in diameter. How many gallons will it hold?

A. Volume = $\pi r^2 h$ or $.7854\ D^2 h$

Changing all dimensions to inches, we have:

$$= .7854 \times (43)^2 \times 100$$

$$= .7854 \times 1849 \times 100$$

$$= 145220.46 \text{ cu. in.}$$

There are 231 cu. in. to the gallon.

$145220.46 \div 231 = 628.66$ gals.

Q. A double-bottom tank is 18' wide, 24' long and 4' deep. How many tons of fresh water will it hold?

A. Volume of tank = 18 X 24 X 4

= 1728 cu. ft.

There are 36 cu. ft. in 1 ton of fresh water.

$1728 \div 36 = 48$ tons of fresh water.

Q. Find the maximum allowable working pressure on a plate stayed 7" X 8", plate 9/16" thick. Constant = 340.

A.

$$W = \frac{CT^2}{P^2 + p^2}$$

$$= \frac{340 \times 9 \times 9}{7 \times 7 \quad 8 \times 8}$$

$$= \frac{340 \times 81}{49 + 64}$$

$$= \frac{27540}{113}$$

$$= 243.71 \text{ lbs. per sq. in.}$$

W = Maximum allowable working pressure in lbs. per sq. in.

C = Constant

T = Thickness of plate in sixteenths of an inch

P = Pitch of stays in one direction, in inches

p = Pitch of stays in other direction, in inches

Q. What is the maximum allowable working pressure in psi for seamless carbon steel pipe 4" in diameter, 1/2" thick, allowable fibre stress 3500 psi. M = .8; A = .1; D = 4 1/2 1/2 = 5.

A.

$$P = \frac{2S(T - A)}{D - M(T - A)}$$

$$P = \frac{(2)(3500)(.5 - .1)}{5 - .8(.5 - .1)}$$

$$P = \frac{(7000)(.4)}{5 - .8\,(.4)}$$

$$P = \frac{2800}{5 - .32}$$

$$P = \frac{2800}{4.68}$$

$$P = 598.2$$

P = maximum allowable pressure psi

T = minimum wall thickness of pipe in inches

A = allowance for threading or grooving

D = external diameter of pipe in inches

M = multiplier from table in G. R. & R.

Q. Find the necessary weight to be placed 30" from the fulcrum of a lever-type safety valve of the following dimensions:

Length of lever (L_1) = 46"
Distance of weight from fulcrum (L_2) = 30"
Weight of lever (W_L) = 10 lbs.
Weight of valve and stem (W_V) = 5 lbs.
Distance of valve from fulcrum (L_3) = 4"
Diameter of valve = 3"
Steam pressure (P) = 150 lbs.
Area of valve = A

A.

$$PAL_3 = W_VL_3 + W_L\left(\frac{L_1}{2}\right) + (W)L_2$$

(The (W) is the weight we are looking for.)

$$PAL_3 - W_VL_3 - W_2\left(\frac{L_1}{2}\right) = (W)L_2$$

$$W = \frac{PAL_3 - W_VL_3 - W_L\left(\frac{L_1}{2}\right)}{L_2}$$

$$= \frac{150 \text{ X } .7854 \text{ X } (3)^2 \text{ X } 4 - (5 \text{ X } 4) - \left(10 \text{ X } \frac{46}{2}\right)}{30}$$

$$= \frac{4241.16 - 20 - 230}{30}$$

$$W = \frac{3991.16}{30}$$

Weight (W) = 133.038 lbs.

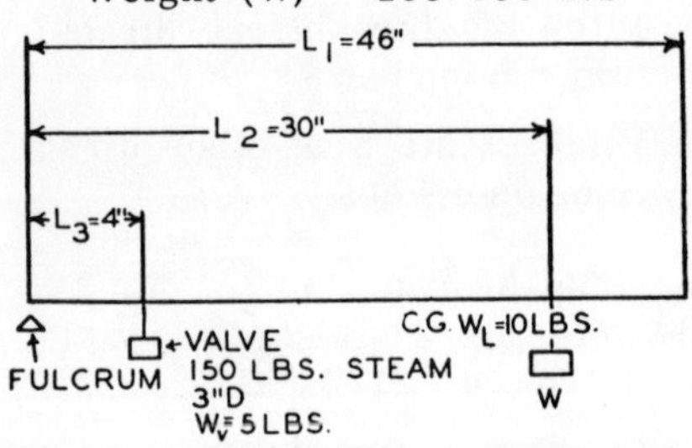

Figure 64.

In figuring the safety valve problem, many men cannot remember the formula. In this case, set the problem up by diagram in the manner shown in Fig. 64, in which all the "up" forces must equal all the "down" forces.

(The center of gravity (C. G.) or weight of a uniform bar is always considered at its center point.)

Then

1. $\uparrow = \downarrow$
2. $PAL_3 = W_VL_3 + W_L\left(\frac{L_1}{2}\right) + WL_2$

a. The only up force is the steam pressure on the valve P X A, and as the force applied is a certain distance from the fulcrum, we multiply this by that distance, PAL_3.

b. There are 3 down forces:

1) Weight of valve and stem times distance to fulcrum, W_VL_3.

2) Weight of lever times 1/2 length of lever,

$$W_L \frac{L_1}{2}$$

3) The unknown weight times its distance from the fulcrum: WL_2.

Any unknown can be figured from the formula as set up in step No. 2.

Q. With stays placed 6" X 6", what diameter of stay would be required for a steam pressure of 160 lbs. per sq. in.? The strain allowed per sq. in. cross-sectional area = 6000 lbs.

A. Strain on 1 stay = 6 X 6 X 160 = 5760 lbs.
Required area of stay = 5760 ÷ 6000 = .96 sq. in.

$$\text{Area} = .7854D^2$$

$$.7854D^2 = \text{Area}$$

$$D^2 = \frac{\text{Area}}{.7854}$$

$$D = \sqrt{\frac{\text{Area}}{.7854}}$$

$$= \sqrt{\frac{.96}{.7854}}$$

$$= \sqrt{1.222}$$

$$D = 1.105''$$

Q. What is the maximum allowable working pressure on a stay rod 2-1/2" in diameter supporting a surface 14" X 16"? Use constant of 7,500

A.

$$W = \frac{aC}{A}$$

$$W = \frac{\overset{\overset{.0561}{\cancel{.1122}}}{\cancel{.7854}} \times 2.5 \times 2.5 \times \overset{1875}{\cancel{7500}}}{\underset{4}{\cancel{16}} \times \underset{\cancel{2}}{\cancel{14}}}$$

$$= \frac{.0561 \times 6.25 \times 1875}{4}$$

$$= \frac{657.422}{4}$$

W = 164.35 lbs. per sq. in.

W = Allowable maximum working pressure in lbs. per sq. in.
a = Cross-sectional area of stay in inches
C = Constant
A = Area supported by stay in sq. in.

Q. What is the maximum allowable working pressure on a Morison furnace of the following dimensions:

Diameter of furnace = 34"
Thickness of plate = 3/4"

A.

$$P = \frac{CT}{D} = \frac{14,700 \times 3/4}{34} = \frac{11025}{34} = 324.26 \text{ lbs. per sq. in.}$$

P = Maximum allowable working pressure in lbs. per sq. in.

T = Thickness of plate in inches

D = External diameter of furnace at bottom of corrugations, in inches

C = Constant

C = 15,700 for Leeds furnace
= 14,700 for Morison or Deighton furnace
= 12,500 for Fox, Purves or Brown furnace

Q. Find the allowable working pressure on an unstayed vessel with a spherically dished head, pressure on the convex side. Head dished to 46" radius, T.S. 60,000 psi, plate 1.3" thick, efficiency 80%. (The allowable working pressure on a vessel with pressure on the convex side is 60% of that allowed on vessel with pressure on concave side of head.)

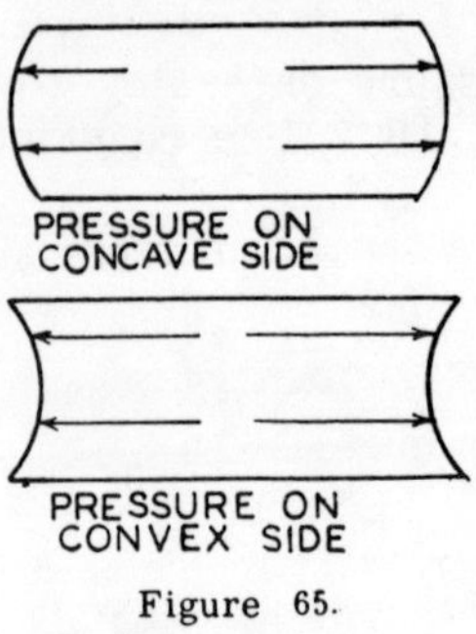

Figure 65.

A. $$P = \frac{60,000 \times 1.3 \times .80}{.833 \times 46} \times 60\%$$

$$P = 1630 \times 60\%$$

$$P = 978 \text{ psi}$$

Formula for hemispherical heads is (with pressure on concave side):

$$P = \frac{2\ STE}{R}$$

(With pressure on convex side, take 60% of answer from above formula.) (See Fig. 65.)

Q. Determine the pitch of stays to be equally spaced in both directions where plate is 7/16" thick. Steam pressure 250 psi. Stays screwed with riveted heads.
C = 170.

A.

$$p_1 \text{ or } p_2 = \sqrt{\frac{CT^2}{2P}} = \sqrt{\frac{(170)(7 \times 7)}{(2)(250)}} = \sqrt{\frac{8330}{500}}$$

P = maximum allowable pressure in psi

T = thickness of plate in sixteenths of an inch

p_1 = pitch of stays in one direction, in inches

$= \sqrt{16.66}$

$= 4.08''$

p_2 = pitch of stays in other direction, in inches

C = constant

(Where pitch in one direction is greater than that in the other direction)

$$p_1 = 2\sqrt{\frac{CT^2}{5P}}$$

With the greater pitch thus determined for one direction, the maximum pitch in the other direction shall be computed by the formula

$$p_2 = \sqrt{\frac{CT^2}{P} - p_1{}^2}$$

Q. If the length of a D slide valve is 25-5/8" and the distance between the outside edges of the steam ports is 24-1/2", what is the steam lap of the valve?

A. 25-5/8"
−24-1/2"
1-1/8" ÷ 2 = 9/16"

Q. Find the allowable working pressure on an unstayed vessel with a spherically dished head, pressure on the concave side. Head dished to 36" radius, tensile strength 30,000 psi, plate 2" thick, efficiency 90%.

A. $P = \frac{STE}{.833\ R}$

$P = \frac{30,000 \times 2 \times .90}{.833 \times 36}$

P = 1800

P = maximum allowable working pressure in psi

S = maximum allowance stress in psi

T = thickness of head in inches

E = efficiency of weakest joint

R = radius to which head is dished, in inches

Q. A cylinder has 37 — 1-1/4" studs. Diameter of cylinder 37". Diameter of head 45". Studs set in 2" from outer edge. Steam pressure 175 lbs. Find pitch of studs and stress per square inch of cross-section area of each stud.

A. Pitch diameter of studs = 45 — 4 = 41"
Circumference of pitch circle = 41 X 3.1416 = 128.8"
Pitch of studs = 128.8 ÷ 37 = 3.48
Area of cylinder = .7854 $(37)^2$ = 1075.2 sq. in.
Total load under head = 1075.2 X 175 = 188160 lbs.
Load each stud supports = 188162 ÷ 37 = 5085.4 lbs.
Cross-section area of 1 stud = .7854 $(1.25)^2$ = 1.227
Stress per square inch of cross-section area of each stud, = 5085.4 ÷ 1.227 = 4144.58

a. Pitch of studs = 3.48"

b. Stress per square inch of cross-section area of each stud = 4144.58 lbs.

Q. The stroke of an engine is 28". If the engine makes 40 S.P.M. (strokes per minute), what is the piston travel in feet per minute?

A. 28 X 40 = 1120 in. per min.

converting to feet

1120 ÷ 12 = 93.33 ft. per min.

Q. An engine with a 3' stroke makes 80 R.P.M. Find the mean speed of the piston.

A. 3 X 80 X 2 = 480 ft. per min.

Q. What is the horsepower of a reciprocating engine, M.E.P. 200 lbs., stroke 4', diameter of piston 24", 90 R.P.M.?

A.
$$\text{H.P.} = \frac{\text{PLAN}}{33{,}000}$$
$$= \frac{200 \times 4 \times .7854 \times 24 \times 24 \times 90 \times 2}{33{,}000}$$
$$= 1974.0672$$

Q. By use of a constant, determine the I.H.P. of a cylinder with a piston diameter of 48", stroke 42", M.E.P. 38 lbs., and making 75 R.P.M.

A.
$$\text{Constant} = \frac{\text{cylinder area} \times \text{stroke (2)}}{33000}$$
$$= \frac{.7854\ (48)^2 \times (3.5 \times 2)}{33000}$$
$$= .3838 \text{ or } .384$$
$$\text{I.H.P.} = \text{constant} \times \text{M.E.P.} \times \text{R.P.M.}$$
$$= .384 \times 38 \times 75$$
$$= 1094.4$$

Q. A gear wheel making 156 R.P.M. has 56 teeth. It drives another gear at 91 R.P.M. How many teeth are there on the second gear?

A. The speed of the gear times the number of teeth on the first gear equals the speed times the number of teeth on the second gear.

$$(91)\ (X) = (156)\ (56)$$
$$91X = 8736$$
$$X = 96 \text{ teeth}$$

Q. What is the minimum diameter of round stock necessary to make a square key 3/4" on a side?

A. From the Pythagorean (right triangle) formula, **Fig. 66** we have:

$$X^2 = (3/4)^2 + (3/4)^2$$
$$= 9/16 + 9/16$$
$$= 18/16$$
$$X^2 = 1.12$$
$$X = 1.058 \text{ (square root of 1.12)}$$

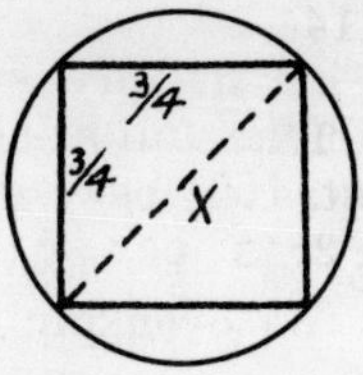

Figure 66.

Q. The outside diameter (O. D.) of a pipe, including the lagging, is 12". The lagging is 2-1/2" thick. The pipe thickness is 1/2". What is the size of the pipe?

A. Two thicknesses of lagging = 5"
Two thicknesses of pipe = 1"
Total thickness of pipe and lagging = 6"

12"–6" = 6" pipe

Q. A piece of metal 4 feet long is tapered 7/8" per foot. If the piece is 10" in diameter at the large end, what is the diameter of the small end?

A. 7/8 X 4 = 3-1/2" less on small end

10" — 3-1/2" = 6-1/2" diameter of small end

Q. What is the weight of a 3/4" bronze liner for a 12" shaft 10 feet long. Specific Gravity (S. G.) of bronze is 8.

A. The dotted line (Fig. 67A) represents the middle of the liner. The diameter of this circle is 12-3/4".

The circumference of this circle is

$$C = \pi D$$
$$= (3.1416)(12.75)$$
$$= 40.05'' \qquad (40'')$$

If we split the liner from end to end and opened it up, we would have a piece of metal with dimensions as shown in Fig. 67B.

The volume of this liner now becomes

$$v = 120 \text{ X } 40 \text{ X } .75$$
$$= 3600 \text{ cu. ins.}$$
$$3600 \div 1728 = 2.08 \text{ cu. ft.}$$

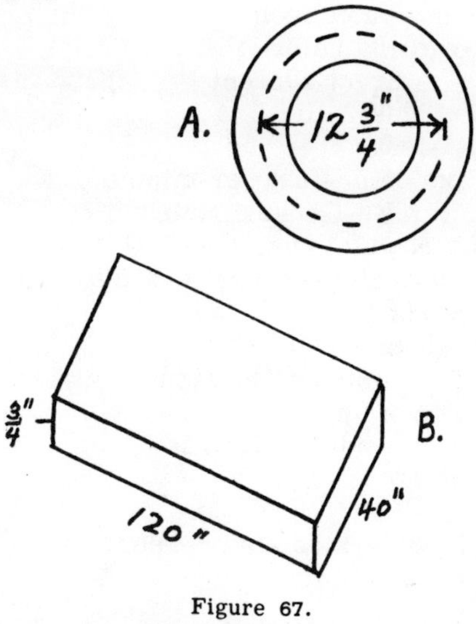

Figure 67.

Since S. G. of water is 62.5 lbs. per cu. ft. and bronze is 8 times heavier, we have

62.5 X 8 = 500 lbs., wt. per. cu. ft. of bronze

Since there are 2.08 cu. ft. in the liner, 2.08 X 500 = 1040 lbs., wt. of liner

Q. Find the pressure per square inch on the crankpin of an engine with the following dimensions: piston diameter 22", steam pressure 150 lbs., diameter of pin 12", length of pin 18".

A. Press. per sq. in. on crankpin $= \dfrac{\text{Area piston X Steam pressure}}{\text{Diameter pin X Length pin}}$

$$P = \frac{A \times P}{P \times L}$$

$$= \frac{.7854\,(22)^2 \times \overset{25}{\cancel{\overset{50}{\cancel{150}}}}}{\underset{4}{\cancel{12}} \times \underset{9}{\cancel{18}}}$$

= 263.98 lbs. per sq. in.

ABBREVIATIONS

A-area
A. C. -alternating current
amp. -ampere
B. H. P. -brake horsepower
B. T. U. -British Thermal Unit
C. -Centrigrade; circumference
cu. ft. -cubic foot
cu. in. -cubic inch
D or d-diameter
D. C. -direct current
F. -Fahrenheit
ft. ;'-foot
ft. per min. -feet per minute
G. R. & R. -General Rules & Regulations, U. S. C. G.
H. P. -high pressure; horsepower
hr. -hour
I. C. E. -Internal Combustion Engine
I. H. P. -indicated horsepower
in. ;"-inch
I. P. -intermediate pressure
Kw. -kilowatts
lbs. -pounds
lbs. per sq. ft. -pounds per square foot
lbs. per sq. in. -pounds per square inch
L. P. -low pressure
M. E. P. -mean effective pressure
min. -minute
p. f. -power factor
p. s. i. -pounds per square inch
R or r-radius
R. P. M. -revolutions per minute
sec. -second
S. H. P. -shaft horsepower
S. P. M. -strokes per minute
sq. ft. -square feet
sq. in. -square inches
T. D. C. -top dead center
T. S. -tensile strength
v. -volt
vol. or V-volume
W. or W. P. -working pressure
°-degrees

MEASURES

2 pts. = 1 qt. (quart)
4 qts. = 1 gal. (gallon)
1 gal. fresh water = 8.3 lbs.
1 gal. = 231 cu. in.
1 cu. ft. fresh water = 62.5 lbs.
1 cu. ft. salt water = 64 lbs.
1 B.T.U. = 778 ft. lbs. work
1 lb. coal = 13,000-15,000 B.T.U.
1 lb. oil = 18,000-20,000 B.T.U.
1 ton fresh water = 36 cu. ft.
1 ton salt water = 35 cu. ft.
1 ton fresh water = 268 gals.
1 gal. salt water = 8.5 lbs.
7.5 gals. oil = 1 cu. ft.
1 gal. fuel oil = 8 lbs.
1 bbl. oil = 42 gals.
Atmospheric pressure = 14.7 lbs.
12 in. = 1 ft.
144 sq. in. = 1 sq. ft.
1728 cu. in. = 1 cu. ft.
1 H.P. = 33,000 ft. lbs.
π = 3.1416
1 nautical mile or 1 knot = 6080 ft.
1 short ton = 2000 lbs.
1 long ton = 2240 lbs.
Pressure in lbs. per sq. ft. = 62.5 X height of column in ft.
Pressure in lbs. per sq. in. = .434 X height of column in ft.

FORMULAS

1. Area of rectangle = Length X Width

   A = L X W

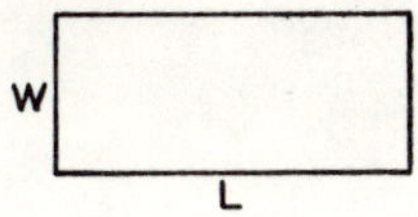

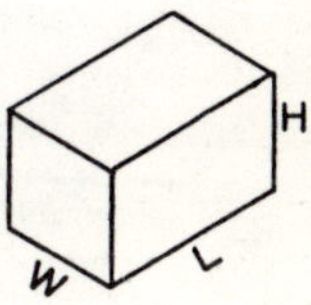

2. Volume of rectangular solid = Length X Width X Height

   V = L X W X H

3. Area of circle = .7854 X $(D)^2$

   A = $.7854D^2$

   or

   A = πR^2

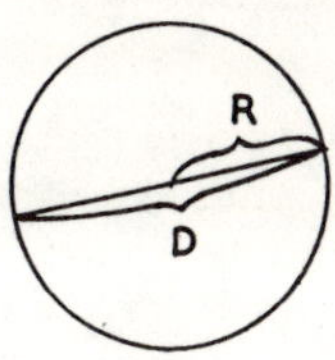

4. Circumference of circle = π X D

   C = πD

5. Volume of cylindrical vessel = Area of base X height

   V = $.7854D^2h$

   or

   V = πR^2h

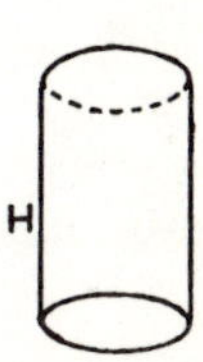

6. The following rule will be found useful for figuring problems with right-angle triangles.

$$c^2 = a^2 + b^2 \qquad c^2 = 25$$
$$c^2 = (4)^2 + (3)^2 \qquad c = \sqrt{25}$$
$$c^2 = 16 + 9 \qquad c = 5$$

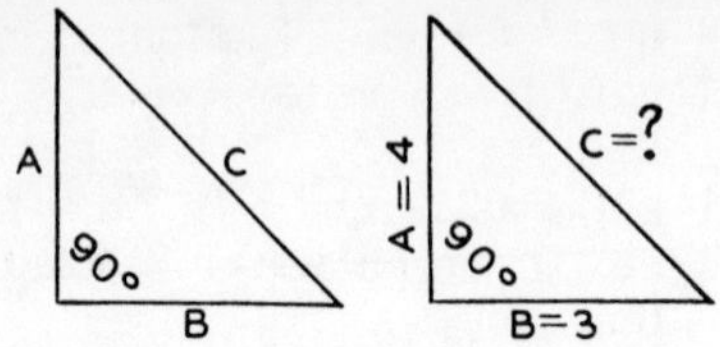

7. $°F. = \frac{9}{5}C. + 32$

$°C. = \frac{5}{9}(F. - 32)$

° Absolute = °C. + 273

° Absolute = °F. + 460

PREFACE

An applicant sitting for examination on MOTOR will be asked the following groups of questions:

1. For a Motor license over 2000 HP, the applicant will be asked all the usual Steam questions plus the Motor questions.

2. For a Motor license under 2000 HP, the applicant will be asked all the usual Steam questions <u>except</u> the sections on Boilers, Turbines and Reciprocating Engines, plus the Motor questions.

MOTOR

CONSTRUCTION

Q. What is a dry liner?
A. Usually a very thin liner (insert) which does not come in direct contact with the cooling water.

Q. What is a wet liner?
A. A cylinder which comes in direct contact with the cooling water and must have some means of sealing at the bottom in order to prevent cooling water from leaking into the base of the engine and contaminating the lube oil.

Q. How are valves actuated on a Diesel engine?
A. Mechanically, hydraulically, pneumatically or electrically. If mechanically operated, they are actuated by means of cams, push-rods, rocker arms and rollers.

Q. Trace the path of cooling water through a Diesel engine.
A. The water passes through strainers and into the suction side of the circulating pump which is usually of the centrifugal type. It is discharged from the circulating pump through the oil cooler and thence to the engine. The water enters the engine at the coolest part, which is usually the bottom of each cylinder water jacket in a single-acting engine, and circulates upward toward the cylinder head. There may be either water jumpers or passages to carry the water from the cylinder to the cylinder head. From the cylinder head the water passes through the jackets around the exhaust header and it may then go to an exhaust gas boiler if so provided.

Q. Describe three types of slides common to Diesel piston and rod construction.
A. 1. A crosshead with a single slipper and guide, such as the common single-slipper crosshead found in steam practice.
2. A crosshead which has a double slipper and guide, thus being secured on two opposite sides.
3. A crosshead which has double sets of guides. The crosshead pin is extended, and double sets of guides are added to each end of the pin. Thus there is alignment fore and aft and also athwartships.

Q. Explain a secured wristpin.
A. The secured-type wristpin has the wristpin secured tightly in the bosses of the piston casting. A screwed dowel is used to secure the pin.

Q. Explain floating wristpin.

A. The floating-type wristpin is free to move in both the eye of the connecting rod and the bosses of the piston casting. A spring-clip retainer is placed in grooves in each end of the piston boss in order to prevent the pin from scraping the cylinder walls.

Q. What does the term "port scavenging" mean?

A. It means that the exhaust gases are forced out through ports (openings) in the cylinder wall rather than through exhaust valves.

Q. Describe the various methods used to introduce water or oil into Diesel-engine pistons for cooling purposes.

A. Methods of piston cooling consist of two types of telescopic pipes, the packed and the unpacked type, and the swing-joint type. (Rubber hoses are also used in some cases.)

In the first type, the moving member of the telescope assembly is attached directly to the underside of the piston and the stationary member to the engine housing or frame.

In the second type, used with crosshead engines, the inlet and outlet pipes from the piston are fixed and lead to the crosshead, thence to the outer end of a bracket bolted to the crosshead, where connection is made to the moving member of the telescope assembly.

The details of the telescope assembly vary with different engine designs, but in every case the entire assembly or the internal pipe is given a small amount of freedom to float to allow for side movement of piston and crosshead.

Q. Where do we find the fuel valves placed on an opposed-piston Diesel engine?

A. They are placed on opposite sides in the middle of the cylinder. They are usually tangent to the cylinder in order to give turbulence and better distribution of the fuel.

Q. Describe the following: 1. piston displacement; 2. compression ratio.

A. 1. Piston displacement is the volume swept by the piston from top dead center to bottom dead center.

2. Compression ratio is the ratio of the piston displacement plus the clearance volume to the clearance volume.

Q. What is the difference between gasoline and Diesel engines?

A. The Diesel engine may use low-grade oil and ignite it directly in the cylinder by the heat of compression, while the gasoline engine requires high-grade fuel (gasoline) and ignites it by an electric spark after the gasoline has been mixed with air in the carburetor, and compressed.

Q. Describe the general construction of a Diesel installation in a ship.

A. The hull of the ship which supports the engine consists essentially of a series of frames, shaped to form the outlines of the hull, on the outside of which are riveted the plates forming the outer skin of the ship.

Between the outer skin and the tank top in double-bottom ships are vertical plates, called floors, one such floor being riveted to each frame. Longitudinal girders between the floors furnish the necessary stiffening and support for the mat or heavy plating upon which the bedplate of the engine is secured.

The proper alignment of the engine is accomplished by the fitting of steel chocks or shoes between the foundation mat and bedplate. The engine is finally secured with bodybound bolts through bedplate and mat.

Q. What styles of engine frames are common in Diesel-engine construction?

A. In the early days of Diesel development, the supporting structure for the cylinders took the familiar form of the A-frames, characteristic of steam-engine construction.

This type of frame is still used for large engines, but has been supplanted mainly by the box frame in all but these large sizes.

The box frames of large engines may be recognized as an evolution of the early A-frames.

The side flanges of the A-frames were simply extended and joined, with a door between each pair of frames for access to the crankcase. The true box frame is found on smaller engines, being cast as a complete box in one piece. From the original box frame were developed a number of variations. In one of these the frame and bedplate are cast in one piece, with the cylinders resting on top of the frame. In another, the separate cylinders are dispensed with and the frame and cylinder block is simply a water box with bores to receive the cylinder liners.

The box frame is particularly suited for use with trunk piston engines, because there is no need for any crosshead guide in such engines.

It has the additional advantage of cheapness and lower height, and is universally used for small engines, and for the large sizes, in some cases, up to about 3000 H. P.

Q. Describe a trunk-type piston.

A. The trunk-type piston has a long skirt and has the connecting rod attached directly to it by means of an oscillating wristpin.

Q. Describe a crosshead-style piston.

A. The crosshead-type piston usually has a short skirt and has a piston rod either screwed or bolted to the bottom of the piston. The piston rod is connected to the connecting rod at the crosshead.

Q. What is the difference between the trunk-type and crosshead-style piston?

A. The difference between them is the means by which each is attached to the connecting rod. The trunk type is attached by a wristpin and the crosshead type is attached by a piston rod and crosshead.

Q. Describe several methods of driving camshafts.

A. 1. By a train of gears, which drive the camshaft gear by means of an intermediate gear, and a driven gear on the crankshaft.

2. By a chain which is driven from a gear on the crankshaft and drives the camshaft gear.

3. By gears and a vertical shaft. There is a bevel gear on both the camshaft and crankshaft and a shaft which has pinion gears on each end is placed between the camshaft and crankshaft gears.

Q. Describe at least two methods of drilling oil passages in crankshaft to convey lube oil from main bearings to crankpin bearings.

A. To provide for circulation of the oil, when forced-feed lubrication is used, the crankshaft has a radial hole drilled at the center of each main bearing connecting with an axial hole, which in turn connects with holes in the two adjacent crank webs, through which the oil is conveyed to an axial hole in the crankpin. A radial hole in the center of each crankpin makes a connection with the hole leading up through the connecting rod. This is shown in Fig. 68A, and an alternative method of drilling sometimes used is shown in Fig. 68B. When the engine housing is not closed in and oil-throwing by the cranks must be prevented, a ring is turned on the shaft at each end of each main bearing for the purpose of throwing off any oil draining out of the bearing before it reaches the cranks.

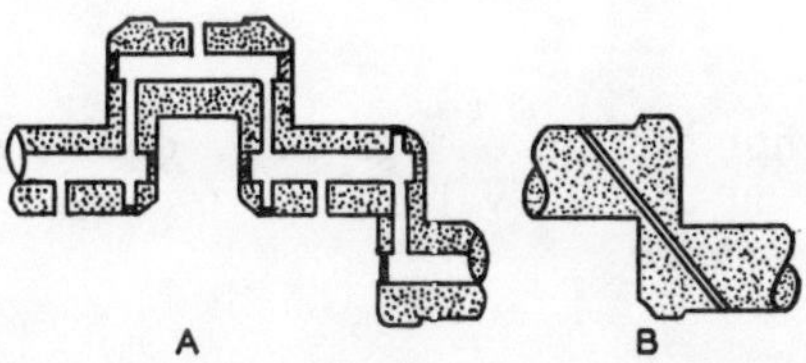

Fig. 68. Oil passages in crankshafts.

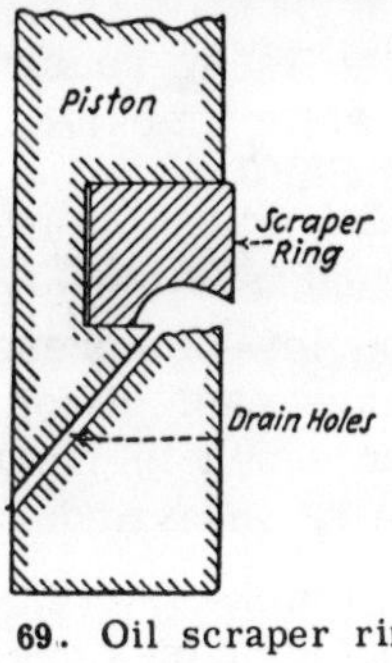

Fig. 69. Oil scraper ring.

Q. What is the purpose of the scraper ring on Diesel engine pistons?

A. The scraper ring is usually the lowest ring on the piston. The ring is fitted with a lip to scrape off the lubricating oil from the cylinder walls on the downstroke so that the oil will not pass to the top of the piston and cause improper combustion. The piston is drilled with a drain hole so that the oil scraped off will drain to the inside of the piston and drop to the crank case, Fig. 69.

Q. (a) Describe a ported liner in 2-cycle engines. (b) What is the relative position of the exhaust and scavenging ports?

A. (a) The liner used in 2-cycle engine construction usually has two

rows of ports which are at slightly different positions. The top row of ports are exhaust ports. At a slightly lower level are the scavenging ports.

(b) The position of the exhaust ports is about 75 percent down the stroke, and the scavenging ports are about 80 percent down the stroke.

OPERATION

Q. How is lubrication of the cylinder liners of Diesel engines possible in view of the intense heat caused by the combustion of the fuel?

A. Oil is forced into the cylinder by means of a mechanical force feed lubricator. It is fed through openings in the cylinder wall. In most large engines, it is timed to enter between ring segments when the piston is at or near the end of its stroke. On many large engines with piston and liner cooling, no oil is used for lubrication.

Q. What preparations should be made to ensure easy starting of a large marine Diesel engine during cold weather?

A. Heat the circulating water and lube oil to about 110°F. Circulate through the engine. If there are no heating facilities, it will be necessary to prime the fuel pump with kerosene and start with this mixture. Run with this mixture for a few minutes to warm up. Some plants use the cooling water discharge from the auxiliaries.

Q. Give as many reasons as possible for smoky exhaust (blue, black, white smoke).

A. White Smoke:

1. One or more cylinders not getting enough fuel, due to plugged fuel valves or improper setting of fuel valves.
2. Too low compression pressure.
3. Leaks past piston rings.
4. Too much lubricating oil.
5. Water in cylinders and exhausts.

Black Smoke:

1. Too high fuel pressure.
2. Fuel valve open too long.
3. Too low compression pressure.
4. Fuel valve opening worn and too large.
5. Carbon in exhaust pipes.
6. Piston rings sticking.
7. Cylinders scored.
8. Dirt in air filters.
9. Overload on engine.

Blue Smoke:

1. Excess lubrication and the burning of lube oil.

Q. Why is the compression pressure comparatively higher on air-injection engines than on solid-injection engines?

A. The compression pressure is slightly higher on the air-injection engine because the high-pressure injection air tends to chill the heated compressed air in the cylinder. Thus the compression pressure must be slightly higher to compensate for this cooling effect.

Note: Diesel engines depend entirely on heat of compression for fuel ignition.

Q. Describe in detail the working strokes of a 2-cycle Diesel engine.

A. The piston on the first stroke up compresses pure air alone in the engine cylinder.

On the second stroke down, fuel is introduced, burned and expanded. Exhaust and scavenging take place at the bottom of the stroke, thus completing the cycle in two strokes or one revolution.

Q. What are the relative advantages and disadvantages of a 4-cycle Diesel engine?

A. Advantages:

1. Better volumetric efficiency.
2. Lower fuel consumption.
3. No air scavenging pump needed.
4. Less cylinder liner trouble, since there are no ports.
5. Higher engine speeds.

Disadvantages:

1. More weight and space than the 2-cycle.
2. More complex cylinder-head casting.
3. More valves and moving parts than the 2-cycle engine.

Q. What are the relative advantages and disadvantages of a 2-cycle Diesel engine?

A. Advantages:

1. Less weight and space per H. P.
2. Greater H. P. per cylinder.
3. More uniform turning effect.
4. Less complicated cylinder head.

Disadvantages:

1. Less volumetric efficiency.
2. Greater fuel consumption than the 4-cycle.
3. A scavenging air pump is required.
4. Trouble with cylinder liners due to ports.

Q. Describe in detail the working strokes of a 4-cycle Diesel engine.

A. 1. Stroke 1 (down): pure air alone is taken into the cylinder.
2. Stroke 2 (up): the pure air alone is compressed.

3. Stroke 3 (down): fuel is introduced, burned, and expansion takes place.

4. Stroke 4 (up): the burned gases are exhausted from the cylinder.

Q. Describe in detail the working strokes of a gasoline engine.

A. 1. Stroke 1: a mixture of fuel and air is taken in the cylinder.

2. Stroke 2: the fuel and air mixture is compressed in the cylinder.

3. Stroke 3: the fuel and air mixture is ignited by an electrical spark and expansion takes place.

4. Stroke 4: the burned gases are exhausted.

Q. Show by sketch the valve and crank positions at each stroke.

A. In Fig. 70, E is exhaust, I is inlet and N is nozzle.

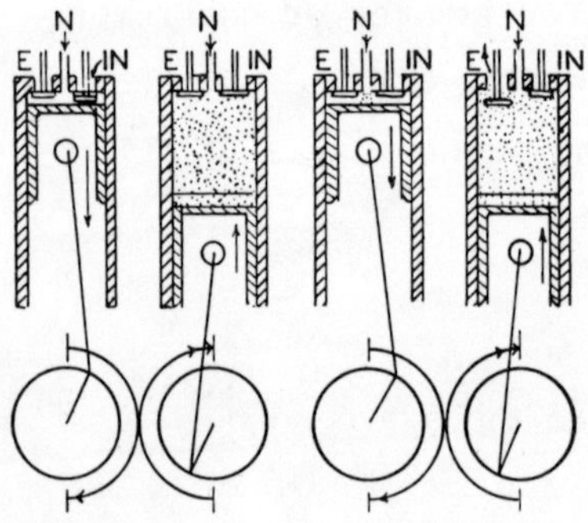

Fig. 70. Valve and crank positions in a 4-cycle gasoline engine.

Q. Name 5 reasons for a smoky Diesel engine exhaust.

A. 1. Overload.

2. Injector not working.

3. Choked exhaust pipe.

4. Dirty fuel or water in fuel.

5. Leaky rings.

Q. In a Diesel installation, what is the danger of admitting cold water into the cooling system?

A. 1. Cylinder heads may be cracked.

2. Cylinder liners may be warped.

3. Pistons may be cracked.

4. There is danger of piston seizure.

Q. (a) What might cause a cylinder liner to become overheated?

(b) What immediate precautions should be taken?

(c) What might happen if these precautions were neglected?

A. (a) Want of cooling water, want of lubrication, or broken rings may cause a cylinder liner to overheat.

(b) If the heating is considered serious, it may become advisable to cut out that particular cylinder altogether by shutting off the fuel before serious damage ensues.

(c) A hot liner may result in fracture below the flange, breakage of piston rings, seized piston, and similar defects.

Q. Why are fine-meshed strainers necessary on a Diesel in order that no impurities will reach the fuel pumps and injection valves?

A. Any impurities would cause damage to the pump plungers and needle valves in the fuel-injection valves.

Q. Discuss the operation maintenance of a Diesel engine using bunker "C" fuel oil.
A. When bunker "C" fuel is used, there must be some means for reducing the viscosity of the oil to permit its being handled from the bunkers to the engine, and to insure proper atomization for burning. A greater heat is required in the engine to burn the asphalt residue. This type of fuel may cause irregular combustion with its attendant troubles, due to water in the fuel, and there will be rapid wear of liners and rings, due to the abrasive solids and ash. There is a greater burden of responsibility placed on the engineers in keeping all operating conditions right in order to prevent shutdowns due to clogging of rings, corrosion of valves, etc.

Q. Since the compression pressure in the working cylinders is about 500 psi, how is it that a much lower starting air pressure acting on the same area is able to start a Diesel engine?
A. The high compression pressure is necessary to obtain the heat (from compression temperature) for igniting the fuel. Starting air is used only to get the engine rolling prior to fuel operation and is, therefore, much lower.

GENERAL

Q. What is an internal combustion engine?
A. One from which work is obtained by combustion of the fuel within the cylinders themselves. This combustion may take place either at constant volume at a very rapid rate, such as in the gasoline or Otto cycle; at constant pressure, such as in the true theoretical Diesel cycle; or at a combination of two, such as the mixed or dual cycle.

Q. What is a Diesel engine?
A. The Diesel engine is an internal combustion engine which operates on the constant-pressure or Diesel-cycle principle. Fuel is admitted directly into the cylinder and combustion takes place as a result of the heat of compression.

Q. What is a gas engine?
A. The gas engine is an internal combustion engine which operates on the constant-volume or Otto-cycle principle. The carburetor is used to mix the fuel (gasoline) with the proper amount of air, outside the cylinder. Upon admission to the cylinder, ignition is accomplished by using an electric spark.

Q. What is the rule concerning the flash point of fuel oil?
A. The flash point for Diesel fuel oil must be above 150° Fahrenheit.

Q. What are the desirable characteristics of a good fuel oil (Diesel)?

A. 1. A maximum heat value.

2. A minimum of water, sediment, sulphur, sand, and other impurities.

3. A fluidity such that it can be pumped without heating it too much.

4. The oil should not be too slow-burning, so as to cause delayed ignition in the cylinder.

5. The oil should burn completely without leaving any carbon residue in the form of ash or soot.

Q. Describe the path of the fuel oil from the storage tanks to the engine cylinders.

A. The fuel is taken from the storage tanks by an oil service pump, passed through a centrifugal purifier, and then to the fuel-oil service tank (day tank). Usually a low-pressure transfer pump takes suction from the day tank and pumps the fuel to the engine fuel pumps at a pressure of about 10 lbs. The engine fuel pumps send fuel to the fuel-injection valves, and the fuel passes through the fuel valves into the combustion chamber.

Q. What effect does dirty and gritty fuel oil have on Diesel engines?

A. It exerts a great amount of wear on the working parts of the fuel pumps and fuel-injection valves. In many cases, it causes fuel-injection valves to become inoperative, and will also cause stopped-up fuel lines and filters.

Q. How is Diesel fuel oil cleaned before it is burned in the engine?

A. By settling, filters and strainers, and in many cases by means of a centrifugal purifier before being used.

Q. What are the advantages of a closed cooling system? The disadvantages?

A. Advantages:

1. There is no danger of scale formation.
2. There is no danger of galvanic action due to sea water.
3. Better efficiency may be obtained since the engine can be kept at a higher operating temperature.
4. There is better control of the engine temperature.

Disadvantages:

1. Two circulating pumps are required.
2. There is danger of leaky tubes in the heat exchanger.
3. This system is more complicated.

Q. What relation exists between compression pressure and compression temperature? What is meant by the heat of compression?

A. 1. Compression pressure is the pressure to which air is compressed in the cylinder of the Diesel engine. Compression temperature is temperature caused by compressing air in the cylinder.

2. When air is compressed, there is a definite increase in temperature for every pound that the air is compressed. This increase is called the heat of compression.

Q. What is the average compression ratio of a Diesel engine?
A. From 14 to 1 to 16 to 1.

Q. In what way is heat related to work performed in a Diesel engine?
A. Each heat unit (B. T. U.) will produce 778 ft. lbs. of work, but since no engine is a perfect converter of heat to work, only about 36% of the 778 ft. lbs. is converted into work by the Diesel engine.

Q. What is the I. H. P. formula for a 2-cycle double-acting engine; for a 4-cycle double-acting engine?
A.

$$\text{I. H. P.} = \frac{(PLAN) + [PL(A - a)N]}{33,000} \times \text{Number of cylinders}$$

When P = Mean effective pressure per sq. in.
L = The length of stroke in feet.
A = The area of the piston in sq. in.
a = The area of the piston rod in sq. in.
N = The number of power impulses per minute.

Q. Explain the term "Cetane Number".
A. C.N. is an indication of the ignition quality of a fuel oil. The ignition quality of Diesel fuel is related to ignition lag. Cetane has an extremely small ignition lag which means it has a high ignition quality.

Q. Explain the difference in valve requirements between the 2-stroke cycle and the 4-stroke cycle of a Diesel engine.
A. In the 2-stroke cycle, it is necessary to get rid of the exhaust gases (scavenge) and admit a new supply of fresh air, all in the same stroke. During this time, it is also necessary to keep the back pressure as low as possible. For this reason, the exhaust valves must be as large as possible. In the 4-stroke cycle, the exhaust gases are discharged by the positive displacement of the piston on the exhaust stroke.

Q. At what part of the cylinder should the highest temperature be found in a (1) Single acting engine? (2) Double acting engine? (3) Opposed piston engine?
A. (1) Top. (2) Top and bottom. (3) Center.

Q. Where would the greatest wear be found in cylinder liners of each of the above types of engines?
A. (1) Top. (2) Top and bottom. (3) Center.

Q. State the differences between full and semi-Diesel engines.

A. The full Diesel ignites its fuel solely by the heat of compression, while the semi-Diesel must have some outside means of igniting its fuel, such as a glow plug, hot bulb, or some other type of heated surface. The full Diesel has a high compression (500 lbs. per sq. in.) and the semi-Diesel has a low-compression pressure of from 180 to 250 lbs. per sq. in.

Q. What is meant by a hot bulb engine? Describe the process of combustion in this type of engine.

A. A hot bulb engine has an uncooled chamber, or vaporizer, connected to the cylinder by a restricted passage. This chamber contains most of the clearance volume. The chamber has a hot surface on which the fuel is sprayed. For starting purposes, the surface of the chamber is heated by a blowtorch or by an electric plug. After starting, the heat of combustion keeps the surface of the chamber hot enough to ignite the fuel, except in cases of very low load.

Q. What causes ignition in a (1) Diesel engine? (2) Gasoline engine?

A. 1. Ignition in a Diesel engine is caused by the heat of compression.

2. Ignition in a gasoline engine is caused by a mechanical spark.

Q. What is meant by the following: (1) sulphur content; (2) ash content; (3) asphalt content; (4) acidity; (5) pour point; (6) congealing point?

A. 1. The per cent of sulphur contained is called sulphur content and should be less than 1-1/2%.

2. The per cent of hard ash contained in the fuel is called ash content and this should not be more than .05%.

3. The per cent of the original asphalt base which remains in the fuel is called the asphalt content.

4. The per cent of acid which the fuel contains is called the acid content; this should not be more than .05%.

5. The lowest temperature at which the oil will flow is the pour point.

6. The temperature (just above pour point) at which the fuel begins to solidify and the solid substances such as paraffin begin to crystallize is called the congealing point.

Q. Define the following: (1) flash point; (2) burning point; (3) viscosity; (4) heating (calorific) value; (5) specific gravity.

A. 1. The temperature to which oil must be heated before the oil vapor over the oil will ignite when a light is passed across the oil's surface.

2. The temperature at which the oil will ignite and continue to burn.

3. Viscosity is internal friction or resistance to flow.

4. The greatest amount of heat that can be obtained from a given quantity of fuel by complete combustion is called heat or calorific value.

5. The weight of a solid or liquid as compared with an equal volume of pure water at 62° Fahrenheit.

Q. What type of gasket material would you use on piping handling fuel oil, sea water, fresh water, lube oil, steam?

A. 1. Fuel piping: Soft brass or copper. Fiber gasket material may be used for low-pressure fuel piping.

2. Sea water: Rubber or asbestos sheet gasket material.

3. Fresh water: Rubber gaskets for low pressures and asbestos sheet packing for high pressures.

4. Lube oil: Use composition oil packing (Velomoid and Gaskoid).

5. Steam: Use metal gaskets or asbestos sheet packing. Ground joints are used without any gasket, in case of very high pressures.

Q. What approximate percentage of combustion heat in a Diesel engine is transferred to: useful work, cooling water, exhaust gases, friction, radiation?

A.

Useful work	32%
Cooling water	23%
Exhaust	25%
Friction	13%
Radiation	7%
Total	100%

Q. Why do some Diesel ships maneuver on a lighter fuel oil?

A. Because the temperature of the cylinders drops when the engine is stopped and run irregularly, therefore the heavy fuel oil will not burn completely and rings would gum up. This lighter fuel also insures that the fuel lines will not clog up and gives more positive starting and reversing.

Q. Where do we find temperature thermometers on a Diesel engine?

A. 1. Cooling-water inlet and outlet.

2. Lube-oil inlet and outlet, also lube-oil sump tank.

3. Cooling water to and from air compressor.

4. Cooling-water outlet to each individual cylinder.

5. Both water and lube oil to and from heat exchangers.

6. Scavenging air in some cases.

Q. Describe a high- and low-pressure alarm system found on air, lube-oil and water systems.

A. An electrical device by which a bell is rung in case of excess temperatures or pressures, and when the temperatures or pressures drop too low. The purpose is to warn the operator that one of these systems is at a critical temperature or pressure.

Q. What is the decompression or compression-release valve? How is it used?

A. It relieves the compression in the cylinders when the engines are stopped, and facilitates restarting in either direction. This is also used in connection with electric or hand starting. With the compression-release valve open, the torque necessary to turn the engine is reduced. After obtaining a brisk speed, the release valves are closed, and the flywheel continues to turn the engine for starting purposes.

Q. Name 3 methods of starting a Diesel engine.

A. Electric motor; requiring that electric energy be readily available. Auxiliary gas engine; may be attached directly to Diesel engine. Hand starting; for gasoline engine and small Diesel engine. Compressed air, previously stored for the purpose of starting.

Q. How does ignition take place in a low-compression or semi-Diesel engine?

A. It is caused by the fuel being sprayed against an uncooled surface, this surface being heated by some outside means, such as a heated plate or bulb (surface ignition).

Q. What is a precombustion chamber? Why is it used?

A. A chamber connected to the main combustion chamber by means of a small throat or passage. Fuel is sprayed into this chamber and, therefore, combustion begins in the precombustion chamber. The reason for using the precombustion chamber is that a great amount of turbulence is gained as the hot gases rush through the narrow throat of the precombustion chamber.

Q. What duties are performed by exhaust systems and how should they be installed on a Diesel-propelled ship?

A. They start at the exhaust ports or exhaust valves and run from here into a manifold; from the manifold the gases are piped to a silencer or muffler, then up the stack in which, in marine work, the heat remaining in the exhaust gases is absorbed in a waste-heat boiler. In the stack, in or near the muffler, a spark arrester is installed. An exhaust system's primary purposes are to carry off products of combustion such as CO or CO_2 with its subsequent heat and water vapor; to deaden the explosive noise of burning fuel; to extract as much heat as possible from the burned fuel.

Q. What are the causes of crankcase deposits?

A. 1. Partly burned fuel leaving carbon deposits.
2. Carbon caused by high-temperature oil.
3. Carbon caused by blow-by.
4. Carbon caused by excessive oil spray (oxydization).
5. Mineral matter from dust, rust, cylinder wear, or bearing wear.
6. Water from humid air (climate), lubricating system, or leaks in piston cooling or cylinder cooling.

Q. What are the causes of cylinder wear?

A. 1. Abrasive material in air, fuel, or lubricating oil.
2. Engine too hot or too cold.
3. Excessive piston clearances or cylinder distortion.
4. Not enough oil.
5. Water in fuel or lubricating oil.
6. Fuel corrosive.
7. Oil of unsuitable viscosity.
8. Blow-by.

Q. Show by sketch the valve and crank positions at each stroke in a 2-cycle engine.

A. See Fig. 71. 1. Scavenging; 2. Compression; 3. Fuel admission; 4. Exhaust.

Q. Sketch fuel arrangement operating with the Common-Rail system of fuel injection.

A. Figure 72.

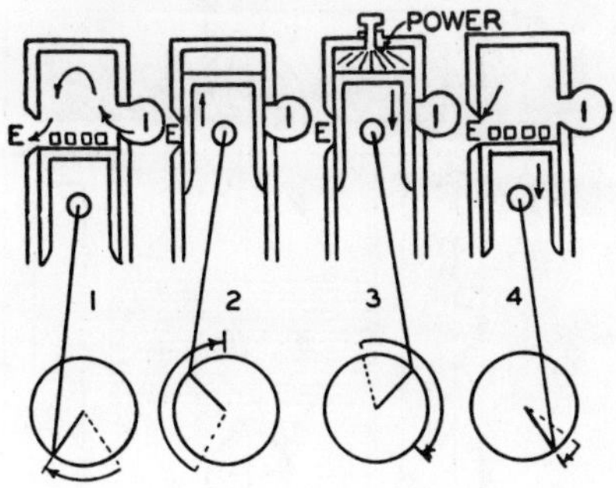

Fig. 71. Valve and crank positions in a 2-cycle engine.

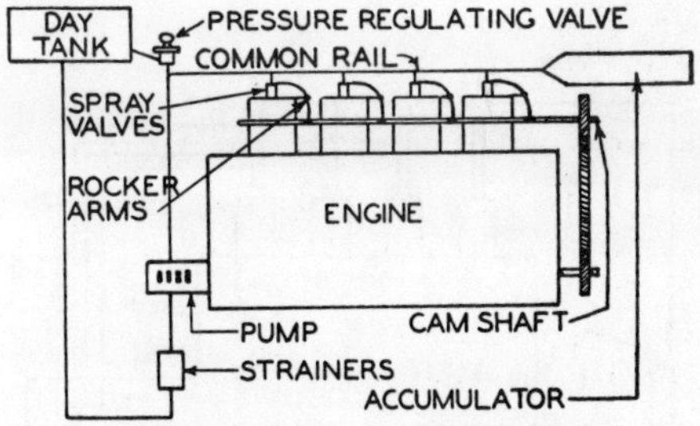

Fig. 72. Fuel arrangement—Common-Rail system.

Q. What are the regulations pertaining to the flashpoint of the fuels used to drive internal combustion engines on passenger vessels exceeding 100 gross tons? What type of emergency generator is allowed to be installed on new passenger vessels or replaced on existing passenger vessels?

A. All internal combustion engines installed on passenger vessels exceeding 100 gross tons shall be driven by a fuel having a flashpoint exceeding 110°F. Any emergency generator installed on a new or existing passenger vessel must be driven by a Diesel engine.

Q. What might cause a cylinder liner to become overheated (Diesel)? What immediate precautions should be taken? What might happen if these precautions were neglected?

A. Insufficient cooling water; check cooling water temperature and supply. Insufficient lubrication; check lube oil temperature and supply. Pistons may freeze and cause serious damage.

Q. Sketch the fuel arrangement used in an air-injection system operating with a closed fuel valve.
A. Figure 73.

Q. Sketch a pre-combustion chamber.
A. Figure 74.

Q. Sketch an air-type injection valve.
A. Figure 75.

Q. Sketch a silencer or spark arrester.
A. Figure 76. 1. Absorbent type. 2. Resistant type.

Q. Sketch an exhaust-valve mechanism from cam to valve.
A. Figure 77.

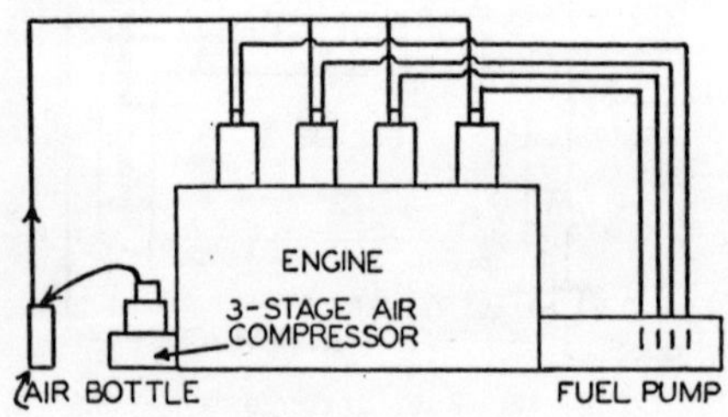

Fig. 73. Fuel arrangement for air-injection system with closed fuel valve.

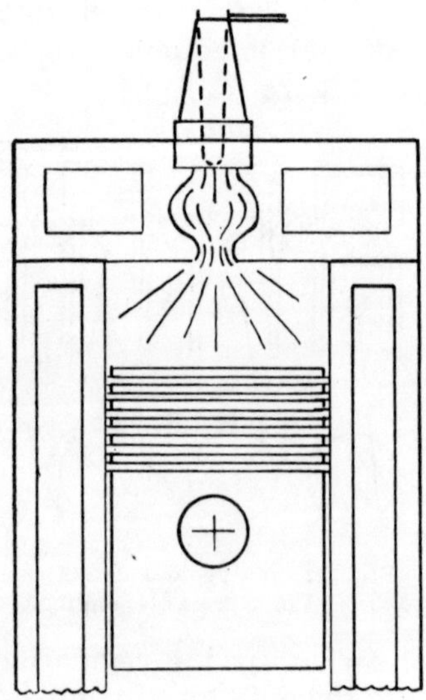

Fig. 74. Pre-combustion chamber.

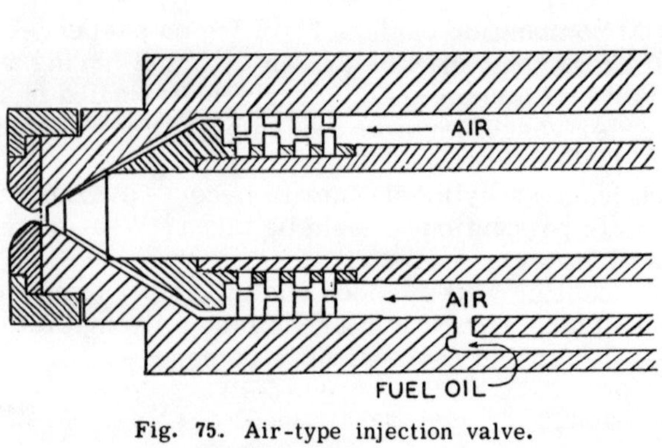

Fig. 75. Air-type injection valve.

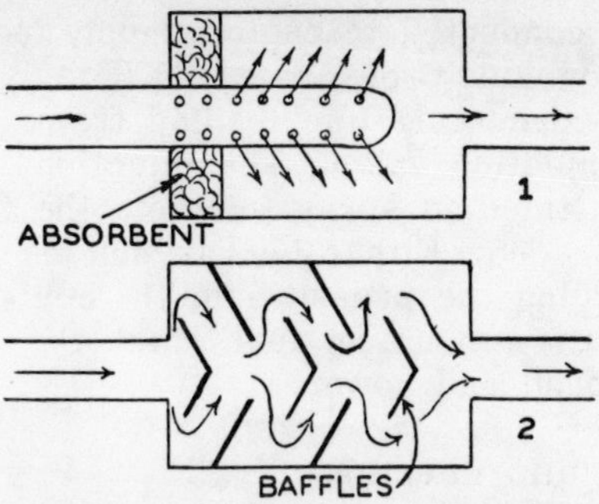

Fig. 76. Silencer or spark arrester.

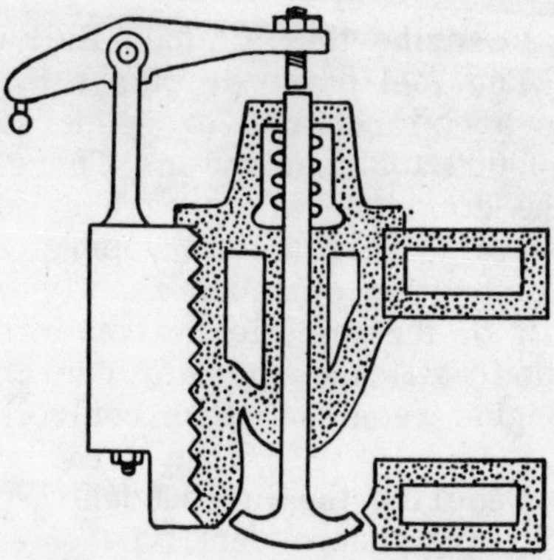

Fig. 77. Exhaust valve mechanism.

FUEL INJECTION

Q. If injection starts late and ends early, how is timing corrected?
A. This symptom indicates excessive valve tappet clearance. Check manufacturer's specifications and adjust valve tappet to minimum.

Q. What are the effects of early and late fuel injection?
A. Early fuel injection causes oil to ignite too soon causing a back pressure on the piston, and a resultant loss of power. Pre-ignition may cause damage to cylinder, cylinder head or even a bent rod. Late ignition causes a smoky exhaust, loss of power and incomplete combustion. The latter fault results in carbon accumulations which affect piston rings and valve stems. Will carbonize silencer, also, which may lead to a fire.

Q. Describe the process of combustion in a Diesel-engine cylinder.
A. Combustion in Diesel engines takes place in three stages:

First, a delay period during which the surface of each individual droplet of fuel is surrounding itself with an envelope of vapor, the outer surface of which must retain a temperature high enough to bring about self ignition;

Second, a period of very rapid burning during which the many droplets, which have accumulated and become surrounded by air during the delay period, burst into flame in rapid succession;

Third, a period of controlled burning when, because of the high temperatures within the cylinder, the last droplets of fuel burst into flame almost as they enter the combustion chamber.

Q. What factors would you say controlled combustion? Name at least four.
A. 1. The proper amount of air.
2. The proper amount of fuel.
3. The proper mixture of the fuel and the air.
4. The fuel and air must be at the proper temperature for ignition.

Q. Describe the Common-Rail type of solid injection.

A. The fuel pump or pumps directly connected to engine supply fuel to a common rail or header at a constant pressure of 1500 to 10.000 lbs. per sq. in. There is a connecting line leading from the common rail to each cylinder injection valve. The injection valve is mechanically operated and timed to spray fuel into the combustion chamber exactly when needed. Either by varying the lift of the injection valve or by varying the pressure in the common rail, the operator can change the amount of fuel which the engine receives, thus controlling speed and power.

Q. Describe the independent type of solid injection.

A. In the independent type of solid injection an individual pump plunger is used for each individual cylinder of the engine. This plunger must both meter and time the injection of the fuel, therefore pressure is present only at the time when fuel injection takes place. The injection valve is a check valve which is lifted by the pressure of the fuel of the fuel at injection period. Pressures found in this type of injection are 1500 to 50,000 lbs. per sq. in.

Q. What are the advantages of using the solid injection? State the disadvantages.

A. Advantages:

1. No compressor needed.
2. Lower compression pressure.
3. Less weight per horsepower.
4. Simpler construction.
5. Less maintenance and adjustment.
6. Good maneuverability.

Disadvantages:

1. Turbulence not as good.
2. Control of speed and load not so good.
3. Better grades of fuel should be used.

Q. Why must the fuel valve be the most accurately timed of all valves on a Diesel engine?

A. Because the fuel valve regulates the beginning and the duration of combustion. The fuel valve may be called the heart of the Diesel engine.

Q. What are the two major types of fuel-injection systems in common use?

A. 1. Air Injection.
2. Mechanical Injection (Solid Injection).

Q. Describe in detail the operating principles of an air-injection fuel system.

A. An air compressor (usually 3-stage) furnishes air at a constant pressure of 600 to 1000 p. s. i. to all spray valves on the engine.

A metering type fuel pump discharges a measured amount of fuel oil into the spray-valve chamber where it sinks to the bottom. At the proper time in the stroke, the valve gear causes the needle valve in the fuel valve to open and air blows the oil deposited in the chamber violently through the passages into the cylinder, breaking it up into a fine mist.

Q. What air pressures are used in air-injection fuel systems?
A. 600 to 1000 lbs. per sq. in.

Q. What are the major functions that a Diesel fuel-injection system must perform?
A. 1. To supply the proper amount of fuel.
2. To supply fuel at the proper time.
3. To finely atomize the fuel.

Q. What is likely to occur in a modern solid-injection, jerk pump type fuel injection system if the spray nozzle becomes completely clogged?
A. Cylinder will produce no work; therefore, loss of H. P. Fuel line may rupture.

Q. Some plungers of modern constant-stroke solid injection fuel pumps have double helically cut grooves in a portion of their surface. What is the function of these double grooves?
A. The plunger has a constant stroke but the amount of fuel discharged into the cylinder is determined by the degree of rotation of the pump plunger in its barrel. Rotation of the plunger in the barrel causes a higher or lower point of the scroll to uncover the port, thereby cutting off the discharge earlier or later. One helix will control timing injection and the other the oil metering.

AIR COMPRESSORS

Q. Why is it dangerous to operate an air compressor without the proper inlet air filters?
A. Every effort must be made to have only clean, dry air at the compressor intake because dust-laden intake air may cause an explosion within the air compressor, discharge line, or receiver.

Q. Why is it dangerous to operate an air compressor without the proper inlet filters?
A. In the event the compressor was operating in a combustible atmosphere—oil, paint fumes, dust, etc.—there is the possibility of dieselization.

Q. What causes an air compressor to shut off automatically?
A. A pressure switch opens the primary motor circuit when a predetermined pressure is reached.

Q. If you heard groaning noises from air compressor cylinders, what troubles would you look for?
A. This may be caused by excessive pressure against which the compressor is working. Other causes may be a closed discharge valve, an obstruction in a line, lack of lubrication, misalignment, or broken piston rings.

Q. What is the effect of supplying too much oil to the high pressure cylinder of an air compressor?
A. Excess carbon deposits on rings and piston. Possibility of explosion.

Q. What fittings are usually found on the intercoolers and aftercoolers of multistage air compressors?
A. Relief valves; moisture drains.

Q. Describe the construction and operating principle of a multistage air compressor.
A. They are usually of the three-stage type, having a differential or step-cut piston. Usually they are built integral with the engine and are driven by a crank on the engine crankshaft. In the first stage (usually the middle step of the piston), air is compressed to approximately 60 lbs. p. s. i. It then passes through the first-stage discharge valve, through an intercooler and into the second stage of the compressor (usually the lower step of the piston). In the second stage it is compressed to approximately 250 lbs. p. s. i. Then it passes through a second intercooler and into the high pressure or last stage of the compressor (usually the top step of the piston). In the last stage, it is compressed to approximately 1000 lbs. p. s. i. Then it passes through an aftercooler and thence to the storage bottle or engine.

Q. Sketch a 3-stage air compressor, naming all parts.
A. Figure 78.

AIR STARTING

Q. If the pipe leading to a starting air valve becomes hot near the valve, what does it indicate? What should be done?
A. The air starting valve is leaking and may be grooved or warped. It will overheat the starting air line and possibly cause an explosion if any oil fumes are present. Replace the valve.

Q. What is the difference between starting air and injection air compressors and how are they driven?
A. Starting air compressors are usually of the two-stage type with pressures of about 350 lbs. p. s. i., being driven by a separate

drive. Injection air compressors are usually of the three-stage type with pressures of about 1000 lbs. p. s. i., being driven directly from the engine crankshaft.

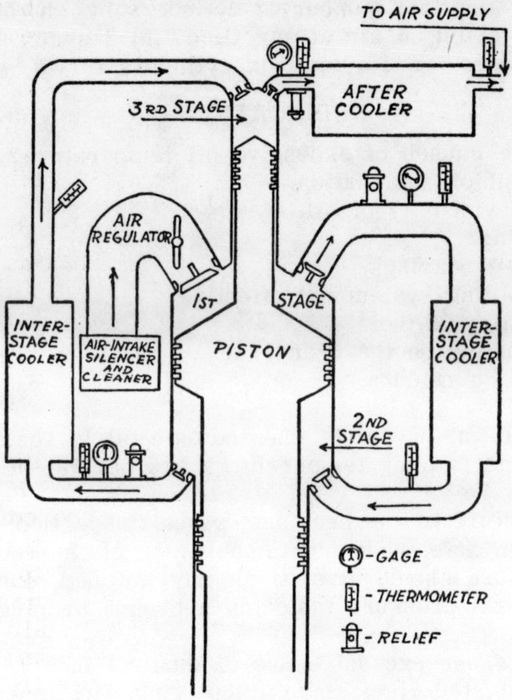

Fig. 78. Three-stage air compressor.

Q. State the air pressure commonly used in air-injection systems and starting-air systems.
A. In air-injection systems, it is about 900 lbs. p. s. i. In starting-air systems, it is about 350 lbs. p. s. i.

Q. How is air prevented from over-heating in an air compressor?
A. By compressing the air in several stages and passing through coolers between stages.

Q. Name 3 methods of reversing a Diesel engine.
A. Sliding camshaft; shifting rollers; rotating camshaft.

Q. What is the starting air-pressure range found in Diesel engine installations? Discuss.
A. It runs from 250 lbs. to 600 lbs. pressure. This depends on the size of the ship and amount of storage space for compressed air. The average is 350 to 400 lbs. pressure.

Q. Discuss the starting of auxiliary Diesels not having air starting valves on all cylinders.

A. The flywheel will be marked for starting position. It is necessary to spot the engine before turning on starting air.

Q. What is the smallest number of cylinders by which a Diesel engine can be started on air at any time? In 2-cycle; in 4-cycle?
A. Three cylinders for 2-cycle; six cylinders for 4-cycle.

LUBRICATION

Q. What are the causes of excessive oil temperature?
A. 1. Insufficient oil circulation.
2. Improper viscosity of oil.
3. Overloading.
4. Overheated bearings.
5. Jacket cooling system not effective.
6. Late burning of fuel.
7. Sludge coating on the crankcase.
8. Oil cooler clogged.

Q. What are the various ways that indicate oil is reaching all principal points and bearing temperatures are normal in a Diesel engine?
A. The temperature of the bearings; visual inspection; feeling of bearings, when safe to do so; temperature of the return oil. The lube-oil pressure should also be closely watched, for any quick change in the oil pressure indicates a broken or plugged line.

Q. What would cause excessive use of lube oil in a Diesel engine?
A. Worn or distorted rings; insufficient ring gap; frozen ring; oil ring worn out or clogged; improper oil viscosity; oil diluted by fuel oil, or excessive temperature; wrist-pin plates not oiltight; excessive bearing play; high oil pressure; excessive suction on crankcase breather. High speed and light load conditions also cause high oil consumption.

Q. Discuss ways of cleaning lube oil in a Diesel ship.
A. In general, it is accomplished by one of three methods, as follows: filtering, centrifuging or settling. Strainers are commonly used and perform a useful function in removing the coarser particles of solid matter, but are not classed as filters. The filters are: the edge type; the cotton bag, waste, or felt type; the Fuller's earth, or activated-clay type. In cleaning by centrifuging, the oil is spun at high speed in a rotating bowl to produce centrifugal force that causes the suspended impurities to separate from the oil.

Settling is the simplest of all methods of purification (cleaning) of oil, but is not readily applicable to Marine service because it involves maintaining the body of oil at rest for a considerable length of time.

Q. Why should the lube pump be run before starting and after stopping a Diesel engine?

A. Before starting, to insure a complete oil film on all bearings, and to give the operator a chance to check to make sure all bearings receive a sufficient supply of oil. In stopping a Diesel engine, the lube-oil pump should run until the difference in temperature between inlet and outlet of the oil is from 2° -5°. This allows the bearings to cool evenly. When the pistons are oil-cooled, a crust of carbon forms in the head, if this practice is not followed.

Q. How may lube oil become contaminated in Diesel engines?

A. Dirt and water in handling and storing aboard ship; metal dust and dust in the atmosphere; water leaks from the cooling system, forming an emulsion; injecting more fuel than will burn; leaky fuel-injection nozzle; oxidation due to excessive temperature; blow-by from stuck piston rings.

Q. What is the effect of over-lubrication in a Diesel cylinder?

A. Excessive cylinder lubrication in a Diesel engine forms carbon, which causes the piston rings to stick, allowing "blow-by." This breaks down the oil film and greatly increases the cylinder wear.

Q. Describe the relationship between the cooling and lubrication systems of a Diesel engine.

A. The lubrication system maintains an oil film, separating the metal in all moving parts, reducing frictional temperature. Insufficient cooling, causing overheated metals, will break down the oil film on the cylinder walls.

Q. Trace the path of lube oil through a Diesel engine.

A. The path is: sump tank, suction strainer, pump, discharge strainers, oil cooler, common or distributing oil header, branch lines to the main bearings, through drilled passage in the crankshaft to crankpin bearings, through drilled connecting rods to the piston (wrist) pin bearing, or crosshead-pin bearing-drilled passage to lubricate the crosshead guide and slipper. With oil-cooled pistons, the oil continues through a drilled passage in piston or crosshead pin, passage in piston, piston head, enclosed return line to engine base, and sump tank.

Q. Describe cylinder lubrication of a Diesel engine.

A. The cylinder of the high-speed Diesel is usually lubricated by bearing oil thrown from the crankpin bearings. In cylinder lubrication of larger low-speed engines (the average Marine Diesel), the oil is fed through from two to eight small openings in the cylinder wall, usually located about the level of the second ring from the top of piston when at bottom center. This oil is supplied by a small pump to each line which measures the amount of oil used.

MATHEMATICS

Q. What is piston displacement?

A. The product of the cylinder bore area and the length of stroke, multiplied by the number of cylinders in the engine.

Piston displacement = $0.7854\ D^2 \times L \times N$ = inches displacement.

N = Number of cylinders

L = Length of piston stroke

D = Diameter of cylinder

Q. What is compression ratio and how is it calculated?

A. The ratio between the total cylinder volume when the piston is at bottom dead center as compared to clearance volume or space remaining when the piston is at top dead center.

$$\frac{\text{Piston Displacement} + \text{Clearance Volume}}{\text{Clearance Volume}} = \begin{matrix}\text{Compression}\\ \text{Ratio}\end{matrix}$$

The result is compression ratio.

$$\frac{PD + CV}{CV} = C.R.$$

(It is always determined in volumes.)

Q. Diameter of cylinder of a 4 cycle engine is 13.5''. Stroke is 1'8'' and MEP is 85 pounds psi. If the IHP is 50, find the RPM.

A.
$$IHP = \frac{PLAN}{33000}$$

$$PLAN = 33000\ IHP$$

$$N = \frac{33000\ IHP}{PLA}$$

$$= \frac{(33000)\ (50)}{(85)\ (1\text{-}2/3)\ (.7854 \times 13.5 \times 13.5)}$$

$$= \frac{1650000}{20268.6}$$

$$N = 81.5$$

Since N equals the number of strokes per minute, then N x 2 = number RPM. Therefore:

81.5 x 2 = 163 RPM

Q. The air valve of a four cycle Diesel engine is open for .635 of a revolution of the engine. For how many degrees is the valve closed in the cycle?

A. .635 of a revolution would be .635 x 360 = 228.6°

The time in degrees the valve would be open.

Then 360° – 228.6° = 131.4° the valve would be closed.

Q. Find the compression ratio of a Diesel engine with a 5-3/4'' bore and a 6'' stroke if the volume of the combustion space at top center is 11.12 cubic inches.

A. CR - Compression ratio
PD - Piston displacement
CV - Clearance volume

$$CR = \frac{PD + CV}{CV}$$

$$= \frac{(.7854)\ (5.75)^2\ (6) + 11.12}{11.12}$$

$$= \frac{155.79 + 11.12}{11.12}$$

$$= \frac{167}{11.12}$$

$$CR = 15$$

Q. Determine the duration of fuel injection in seconds for an engine running at 1400 RPM, if the start of injection occurs at 15.5° BTC and the end of injection at 5.5 ATC.

A. $\frac{21}{360}$ = part of revolution valve is open

$\frac{60}{1400}$ = part of second for one revolution

$\frac{21}{360} \times \frac{60}{1400} = \frac{1}{400}$ of a second

SECOND ASSISTANT ENGINEER

RECIPROCATING ENGINES

CONSTRUCTION

Q. Describe the construction of a piston type valve and explain why it is frequently employed instead of the "D" type slide valve.

A. The piston type valve being round and having steam enter at the center or ends is a better balanced valve than a "D" type valve. Having the steam pressure on its back side forces the "D" type valve against the valve seat, creating a great deal of friction. In triple expansion engines, the H. P. and I. P. are often piston valves and the L. P. a "D"-slide valve. The slide valve at the L. P. end is necessary due to the large volume of steam that must be handled, Figs. 1 and 2.

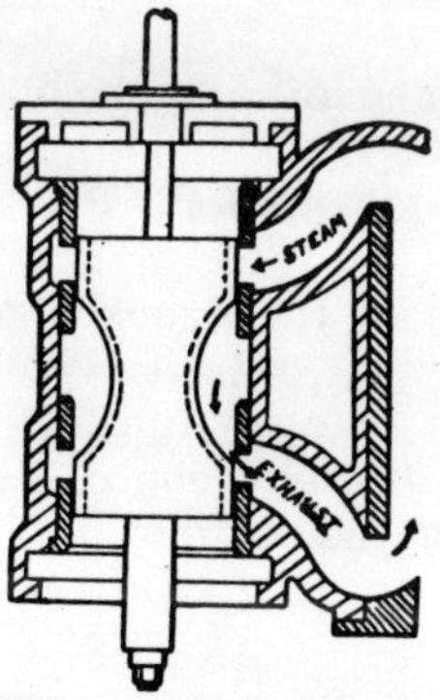

Fig. 1. Piston valve.

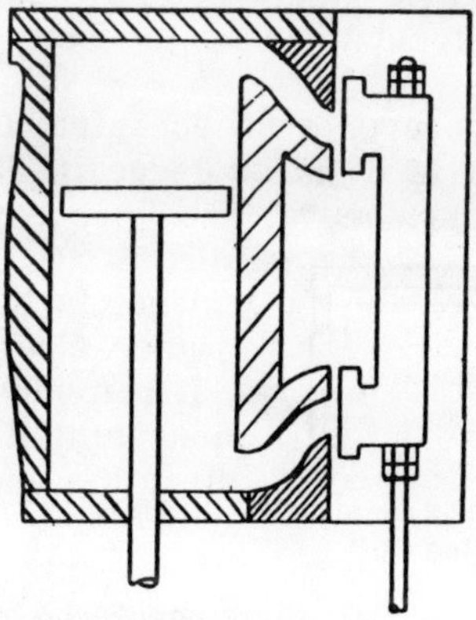

Fig. 2. D-slide valve.

Q. What is a double-ported slide valve?

A. A double-ported slide valve has a system of ports to allow for admission and exhaust of steam from the cylinder through a second port in the valve face. With the same valve travel as a common valve, the double-ported valve gives twice the opening for steam and exhaust. This type of valve is usually adopted for intermediate and low pressure cylinders where the volume of steam is greater. There is less face friction in the double-ported type of valve due to the pressure on the inside port tending to ease the valve off the cylinder face.

Q. Describe the action of a slide valve during one revolution of the engine.

A. The eccentric being 90° ahead of the crank (plus lap and lead) starts to move the valve downward, opening the steam port as the piston reaches the top center. Steam flows in as the valve moves down. As this is happening, the exhaust from the opposite end of the cylinder leaves through the center of the slide valve to the receiver of the next valve chest. The valve coming up again, and having steam lap, cuts off the entering steam at approximately one quarter stroke. The entrapped steam keeps expanding and pushing the piston downward. The exhaust lap on the valve cuts off the outward flow of exhaust steam before the piston reaches the bottom dead center. This entraps a small amount of steam which acts as a cushion for the weight of the downcoming piston. The same events happen on the opposite stroke.

Q. How is an engine bedplate secured to the hull of the ship?
A. The bedplate is bolted to the foundation or sole plate by body-bound bolts which are drive fits. The engineer should check these bolts and nuts periodically.

Q. Could an engine be built with eccentric 90° ahead of the crank?
A. Yes. Valve would not have lap or lead. Valves on winches are of this type in order to obtain full power.

Q. What prevents an eccentric from slipping off its sheave? Sketch.
A. There is a lip or ridge in each side of the strap which overhangs the sheave, thus preventing any side motion, Fig. 3.

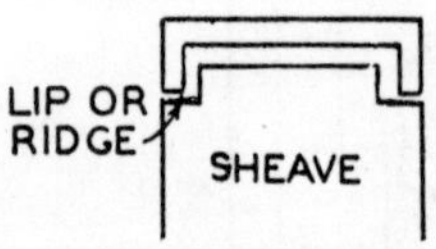

Fig. 3. Eccentric sheave, showing lip.

Q. What is the purpose of the "rock shaft" or "wiper shaft?"
A. It provides a means of operating all the gear links simultaneously, thereby controlling the direction and speed of the engine.

Q. What is the purpose of the crosshead?
A. To connect the piston and connecting rods, thus converting the reciprocating motion of the piston to the rotary motion of the crankshaft.

Q. What provisions are made for adjusting clearances in the open-type crosshead guide?

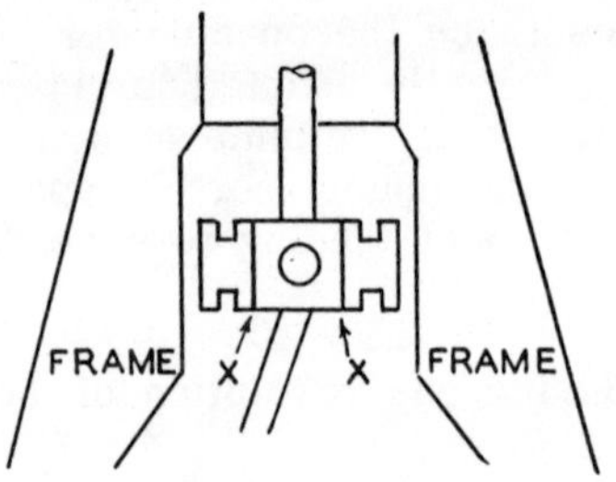

Fig. 4. Open-type crosshead guide.

A. Shims or liners can be inserted between the slipper and the crosshead block, as shown in the diagram at X, Fig. 4.

Q. What provisions are made for adjusting clearances in the closed-type crosshead guide?
A. Shims or liners can be inserted between the backing and ahead guides, as shown in the diagram at X, Fig. 5.

Q. Describe the closed-type crosshead guide. Sketch.

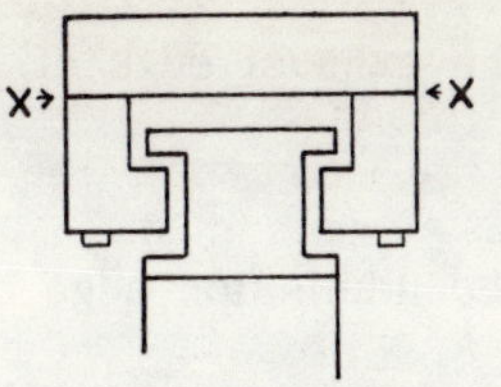

Fig. 5. Closed-type crosshead guide.

A. An angle piece A is bolted to the flat column B and the T-shaped slipper C rides in the groove, as shown in diagram. The flat guide is often cast hollow to allow for cooling-water circulation X, Fig. 6.

Q. Describe an open-type crosshead guide. Sketch.

A. Consists of two flat surfaces X and Y (cast-iron) on which the crosshead (C. H.) slippers ride. They are usually cast hollow to allow for cooling-water circulation, Fig. 7.

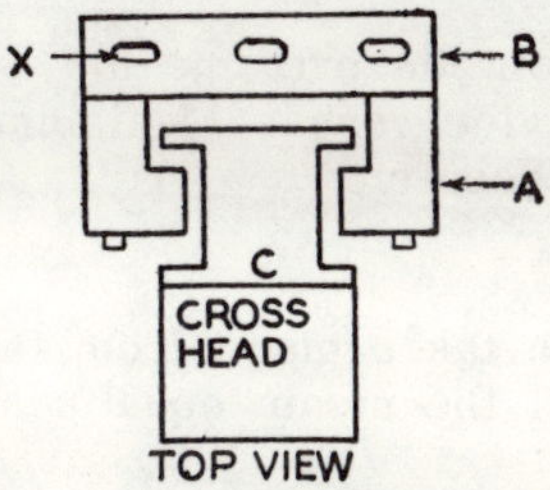

Fig. 6. Closed-type crosshead guide and slipper.

Q. Why is there more piston clearance on the bottom end of a cylinder than on the top end?

A. To allow for bearing wear in the engine.

Q. What is the purpose of jacketing steam cylinders?

A. To reduce condensation in the cylinders by keeping them hot at all times.

Q. What is the purpose of counterbore in a cylinder? Sketch cylinder showing counterbore.

A. To prevent the piston rings from wearing a shoulder in the cylinder. The piston ring partly overrides the counterbore, Fig. 8.

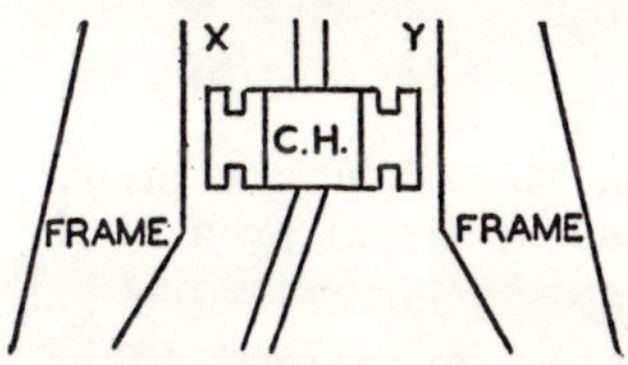

Fig. 7. Open-type crosshead guide.

Q. How are some pistons designed so that the rings may be removed without removing the piston from the cylinder?

A. The ring assembly sets into a space around the body of the piston and is held in place by a follower plate which is secured by studs and nuts.

Q. How is a steam cushion obtained and what is its purpose?

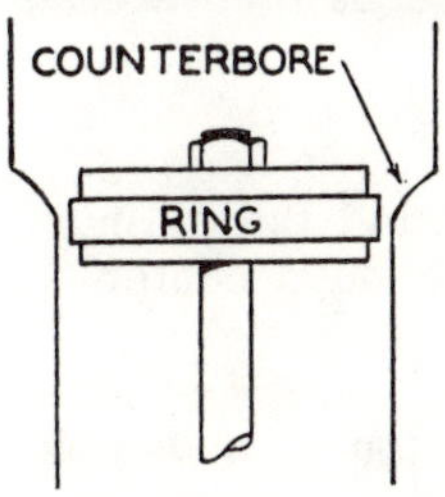

Fig. 8. Piston, ring and counterbore in cylinder.

A. It is obtained by exhaust lap on valve. Exhaust is closed before piston has completed full stroke; thus, some steam is retained to provide a cushioning effect for the heavy piston and its part.

Q. What is the point of admission?

A. It is the point at which the valve starts to open the steam port for the admission of steam to the working cylinder.

Q. What is exhaust lead?

A. It is the amount by which the port is open to exhaust at either end of stroke.

Q. What is exhaust lap?

A. It is the part of the valve which overlaps the exhaust edge of the port when the valve is in mid-position.

Q. What is steam lap?

A. It is the part of the valve which overlaps the admission edge of the port when the valve is in mid-position.

Q. What is steam lead?

A. It is the amount of port opening which permits the admission of steam when the crank is on top or bottom dead center.

Q. What is valve travel?

A. It is the distance traveled by the valve from its extreme top position to its extreme bottom position, or vice versa. Maximum valve travel equals twice the throw of the eccentric.

Q. What is drop in the receiver?

A. It is the drop in the steam pressure which takes place from the instant the exhaust valve starts to open until the steam equalizes with the pressure in the receiver.

Q. What is initial volume?

A. It is the volume displaced by the piston up to the point of cutoff.

Q. What is final volume?

A. It is the volume displaced by the piston up to the instant the exhaust valve opens.

Q. What is piston displacement volume?

A. It is the volume displaced by the piston throughout one stroke and may be found by multiplying the cross-sectional area of the cylinder by the length of the stroke.

Q. What is clearance volume?

A. It is the volume enclosed between the piston and cylinder head plus the volume of steam passage up to the face of the valve, when the piston is at either end of the stroke.

Q. What is cylinder (piston) clearance?

A. It is the shortest distance between the piston and the cylinder head when the engine is on either top or bottom dead center.

Q. What is initial pressure?

A. It is the pressure of the steam on the admission side of the piston at the beginning of the stroke.

Q. What is terminal pressure?

A. It is the pressure of the steam within the cylinder at the end of the working stroke, just as the exhaust port opens.

Q. What is effective pressure?
A. It is the pressure of the steam on the admission side of the piston minus the pressure of the steam on the exhaust side of the piston, or forward pressure minus back pressure.

Q. What is mean effective pressure?
A. It is the average effective pressure throughout the stroke.

Q. What is forward pressure?
A. It is the pressure of the steam on the admission side of the piston at any one point of the stroke.

Q. What is mean forward pressure?
A. It is the average forward pressure throughout the stroke.

Q. What is the position of the eccentric in relation to the engine crank?
A. For an outside valve: 90° ahead, plus lap and lead; for an inside valve: 90° behind, minus lap and lead.

Q. A valve has 1/2" lap and maximum port opening of 1-1/8". What is the travel of the valve?
A. Travel = 2 (lap + port opening)
= 2 (1/2 + 1-1/8)
= 2 (1-5/8)
= 3-1/4"

Q. What is the angle of advance?
A. It is the angle in excess of 90° by which the eccentric leads the crank for an outside admission valve.

Q. What is an inside (indirect) admission valve?
A. It is one which takes steam in the middle and exhausts over the ends.

Q. What is an outside (direct) admission valve?
A. It is one which takes steam over the ends and exhausts through the inside or middle.

Q. What is the point of compression?
A. It is that point at which the exhaust side of the valve closes the port, thereby trapping the remaining steam in the cylinder, which causes a cushioning effect.

Q. What is the point of (exhaust) release?
A. It is the point at which the exhaust side of the valve begins to open the port to permit exhaust of steam from the working cylinder.

Q. What is the point of cutoff?
A. It is the point at which the steam valve closes the port and prevents further admission of steam to the working cylinder.

Q. What is the arrangement which allows removal of the snap piston rings without removal of the entire piston from the main engine of a reciprocating plant?

A. A follower plate is secured to the piston by stud bolts. By removing this plate the rings can be lifted clear of the piston and cylinder.

OPERATION

Q. What is done to cool a hot rod?

A. In the case of a rod that is only slightly hot, application of plenty of cylinder oil may bring the temperature back to normal. Slacking of the stuffing box gland (if possible) will reduce friction on the packing. If the rod is very hot, slow the engine to avoid bending rod. Many engineers have their own pet remedies, such as: application of Albany grease; a mixture of white lead or tallow and oil. If the rod is dirty, swab with kerosene to clean, then immediately apply more cylinder oil.

Q. What is "lap" of a valve and what is the purpose of providing lap? About what part of the valve stroke is given to lap on a main engine?

A. There are two kinds of lap: inside and outside. This extra area, or protrusion of the valve, is for the following purposes: Steam lap closes the steam port when the piston is about one-third or one-fourth of its way on the stroke. This allows the entrapped steam to expand and keep working. In other words, the engine is not using steam for its full stroke: it is using the expansive quality of steam, which makes for a more economical engine. Exhaust lap closes the valve before the piston reaches the end of the stroke. This traps some steam in the cylinder and provides a cushioning effect for the piston at the end of the stroke. On a Liberty ship engine, the H. P. steam lap is 1-3/4" on top and 1-5/8" on bottom. The valve travel is 6". This would mean that the lap would be approximately 27% to 30% of the stroke 1-3/4 ÷ 6).

Q. What is meant by open and crossed eccentric rods?

A. With the link bars thrown to either end of their swing, and crank on bottom center, open rods will appear open, as in Fig. 9A and crossed rods will appear crossed, as in Fig. 9B.

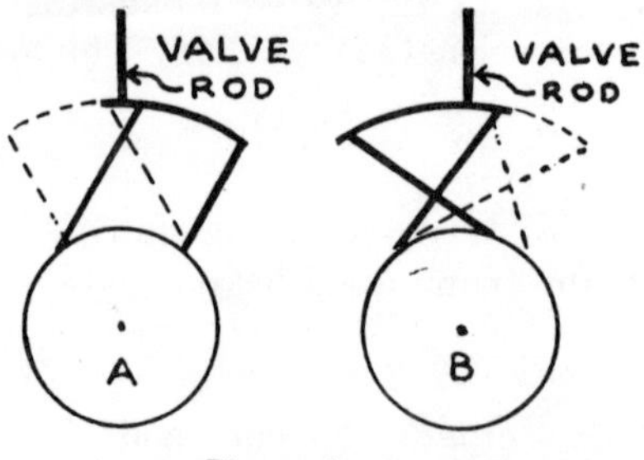

Figure 9.

Q. Explain the effects of linking in and out on open and crossed rods.

A. Linking in refers to the changing of the link bars so that the fulcrum point moves closer to the valve stem. Linking in on either open or crossed rods results in a shorter valve travel, giving an earlier point of cutoff, and earlier release and compression.

With open rods, linking in reduces

valve travel, increases lead, and gives earlier cutoff, release and compression.

With crossed rods, linking in reduces valve travel, decreases lead, and gives earlier cutoff, release and compression.

Q. Explain why throttling of an engine in which saturated steam is used may produce superheated steam in the valve chest. Is throttling economical?

A. The saturated steam, on moving from a high-pressure area to a low-pressure area through a small opening, increases the velocity of the steam through that area, thus drying out the steam and producing a partly superheated steam. This is not economical in a reciprocating engine, as the moisture in the saturated steam helps to lubricate the rings. The steam may also cause "wire-drawing" of the valve disc (eating away of the metal caused by high velocity steam squeezing through).

Q. How are the cylinders of a multiple expansion engine supplied with lubrication and what precaution should be taken by the watch engineer in regard to cylinder lubrication?

A. A dependent oil pump is located on the forward engine column and is driven by a rod connected to the high pressure crosshead. This pump is also fitted with a manual means of pumping the cylinder oil to the engine before it starts to run. Do not pump too much oil into the engine, as it can work its way through the steam and water cycle to the boilers. In some engines, a spray of oil is forced into the steam line just ahead of the H. P. valve.

MAINTENANCE

Q. Explain how you would locate the top and bottom dead centers of an engine.

A. 1. Engage jacking engine.

2. Run piston to approximate top position (1" or 2" from top).

3. Scribe a mark across the slipper and guide.

4. Make a center-punch mark on the column opposite the top of the crank web.

5. Make a tram with 3/8" or 1/2" rod with a 90° turn on one end. Put a taper point on each end of the tram. Tram should reach from center-punch mark on column to approximately the center of the crank web.

6. Place tram in column punch mark and center-punch where other end touches the crank web.

7. Jack the engine over and beyond the top center until the marks on the slipper and guide coincide again.

8. Place tram in column punch mark again and center-punch where the other end touches the crank web.

9. With a pair of dividers, bisect the distance between the 2 center-punch marks on the web and center-punch the center position.

10. Place tram in column punch mark and jack engine until other end of tram coincides with final center position.

11. Engine is now on top dead center.
12. Same procedure is followed for bottom dead center.

Q. How would you locate the striking points of an engine?

A. 1. Remove cylinder head.
2. Remove piston rings.
3. Replace cylinder head.
4. Remove cylinder head relief valve.
5. Disconnect crosshead.
6. Insert eyebolt in top of piston rod.
7. Attach chain fall to eyebolt and lift piston and rod until it touches cylinder head.
8. Scribe mark across slipper and guide.
9. Lower piston and rod until it touches bottom of cylinder.
10. Extend previous mark on slipper over to the guide.
11. Remove cylinder head.
12. Replace piston rings.
13. Connect crosshead.
14. Remove eyebolt.
15. Replace and secure cylinder head.

Another method frequently used is to remove the piston rings, disconnect the connecting rod at the crank, attach chain falls on bottom of cylinder and attach to crosshead, lift the piston and rods until the piston touches the top and lower until it touches the bottom of the cylinder.

Q. How is the lower half of a bearing shell removed for inspection?

A. The top half of the bearing is removed. A special clamp or pin is fitted to the crank web and the engine is jacked over. The clamp or pin forces the bottom half of the shell around to the top of the shaft, from which point it can be removed.

Q. If the H. P. valve stem breaks on a triple-expansion engine and there is no spare on board, what would you do?

A. 1. Remove the valve.
2. Disconnect and remove connecting rod.
3. Remove H. P. piston rings and lower piston to bottom of cylinder.
4. Block steam ports to H. P. cylinder (optional).
5. Reduce boiler pressure and run at reduced speed.
6. If possible, leave valve stem in, as this will save the job of plugging the valve-stem hole in the bottom of the valve chest.

Q. If a knock develops in a reciprocating steam-engine bearing and it is difficult to determine which bearing is at fault, what is done to find the faulty bearing?

A. Flood the suspected bearings with oil, one at a time, and when the faulty bearing is reached, the knock will change in sound.

Q. Name the different points and lines on an indicator diagram.

A. Figure 10.

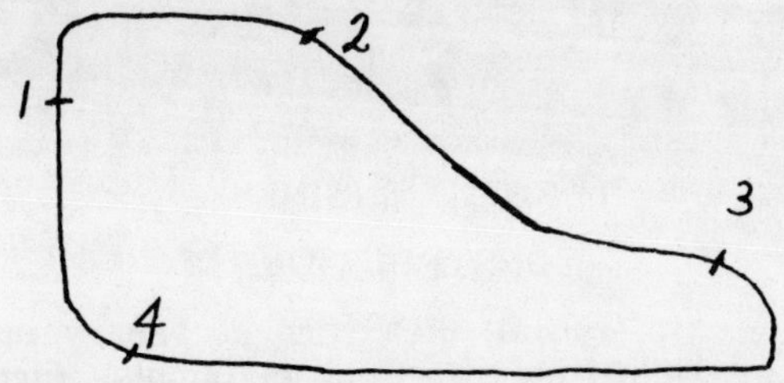

Fig. 10. Indicator diagram. 1) Steam admission; 2) Steam cut-off; 3) Exhaust opens; 4) Exhaust closes; 1-2) Steam line; 2-3) Expansion line; 3-4) Exhaust line; 4-1) Compression line.

Q. How is the wear on the main engine journal bearings determined?
A. By "taking leads".

Q. How would you find top and bottom cylinder clearance if striking points were established?
A. Place engine on top and bottom dead centers, mark slipper guide and slipper or rod at packing box (wherever striking points were established) and measure distance between these dead center marks and striking points.

Q. At which end of the cylinder is the largest clearance allowed and for what reason?
A. On the bottom, to allow for bearing wear.

Q. What is the "lead" of a valve? What is the reason for giving a valve lead?
A. Steam lead is the amount of port opening of the valve which permits the admission of steam when the crank is on top or bottom dead center.

Q. Where does the greatest wear occur on crosshead guides?
A. In reference to the ahead or astern guide, the greatest wear would be on the ahead guide because it gets the greatest use. If referring only to the ahead guide, the greatest wear would be in the middle area of the guide. At this particular part of the stroke, the crank is on its quarter and the direction of force against the guide is, therefore, greater. As the crosshead nears the top and bottom of the stroke, the piston rod and connecting rod come almost into a straight line and the side thrust must, therefore, be less.

Q. About what amount of lead is given to a valve on a main reciprocating engine?
A. On a Liberty ship engine, it would be as follows:

	Top	Bottom
High pressure	1/2"	1/4"
Middle pressure	1/4"	3/8"
Low pressure	1/2"	5/8"

CONSTRUCTION

Q. Why is a small tip (radial) clearance necessary in a reaction turbine? Why is the tip clearance in an impulse turbine irrelevant?

A. Due to the pressure drop across each stage in a reaction turbine, it is necessary to make the tip clearance small to prevent the steam from blowing by the tips without doing useful work.

In the impulse type, the pressure is the same on both sides of the blade and, therefore, the tip clearance may be made larger, allowing a larger safety margin.

Q. What is meant by "gaging" of blades? Why is this necessary?

A. The outlet angle of the blading in the low-pressure end of reaction turbines is increased by twisting or "gaging" the blades. Due to the large steam volume at the L. P. end of the turbine, it is necessary to "gage" the blades, as the only other possible means of handling the excess volume would be to make the blading excessively long and large. "Gaging" allows a swifter flow of steam.

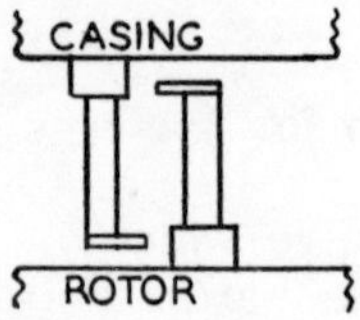

Fig. 11. Seal-stripping.

Q. What is meant by "seal-stripping"?

A. Edges of shrouds are thinned to a fine edge, allowing for closer clearances, yet keeping the safety factor high, Fig. **11**.

A second method is to install a thin-edged strip on the casing, Fig. **12**.

Q. What is the distinguishing difference between impulse blading and reaction blading?

A. Impulse blading is heavier and stronger than reaction blading.

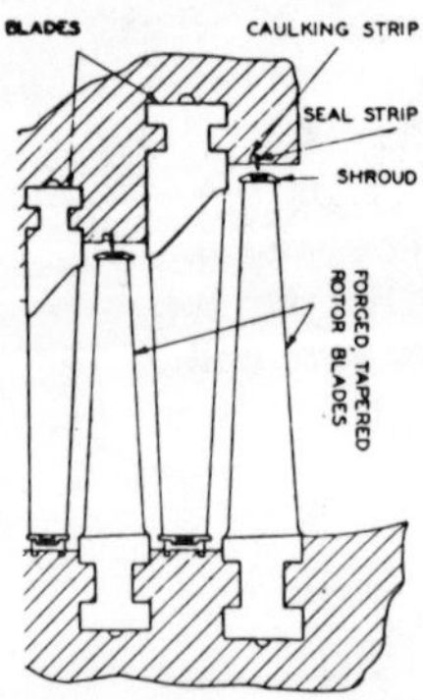

Fig. 12. Special tapered reaction blades for last rows of low-pressure turbine (Westinghouse).

Q. Why is it not necessary to have dummy pistons on impulse turbines?

A. There is very little pressure difference across the blading and, therefore, little axial thrust.

Q. What is the purpose of a dummy or balance piston?

A. To counteract pressure difference across the blading in reaction turbines, which tends to move the rotor axially toward the exhaust end, a balance, or dummy piston is provided on the inlet end. Pressure of steam on the inlet side of the dummy acts to oppose the thrust developed across the blades. Pressure behind the dummy is maintained lower by running a connection to some lower stage through an external pipe or equalizer, or through the hollow rotor. No dummy is necessary in a double-flow turbine.

Q. What are the essential parts of an impulse turbine?

A. 1. Lower and upper casing bolted together at center-line flange.
2. Stuffing boxes and pedestals supporting the bearings.
3. Rotor and rotor shaft.
4. Nozzles and diaphragms.
5. Diaphragm packing.
6. Gland-sealing system.
7. Main bearings.
8. Thrust bearing.
9. Overspeed trip.
10. Relief valve.
11. Speed governor.
12. Lube oil pump and circulating system.
13. Blading.

Q. How are turbine foundations set up?

A. Structural foundations are built up from the hull of the ship to support the turbine. The turbine casings are secured to these foundations by body-bound bolts at one end of the casing, while the other end, usually placed forward in marine installations, is free to move when the turbine casing expands due to heat.

Q. How are turbine casing joints made up?

A. Casing joints are accurately machined to parallel surfaces and made up metal-to-metal without the use of sheet packing. The faces are generally painted with boiled linseed oil, or boiled linseed oil and graphite, before assembling.

Sometimes a simple system of flange-grooving is provided for use in service only, to enable the operator to seal steam leaks while the ship is at sea. The grooves have a radius of about 1/8" and are cut in one face of the joint only. Vents are provided at intervals to insure flow of sealing compound.

Many patented compounds are used, but all have the characteristic of not hardening when in use. They must remain plastic so that it will be possible to break the joints when necessary and to remove the sealing compound with comparative ease.

Q. What are the advantages and disadvantages of turbines and reciprocating engines?

A. <u>Turbine—Advantages</u>

1. Better utilization of high vacuum.
2. Permits use of higher temperatures and pressures.
3. Turbines require less engine-room space and less weight per horsepower.
4. There are fewer wearing parts.

5. Turbine exhaust steam is free of oil.
6. Increased reliability.
7. Less vibration due to the elimination of reciprocating parts.
8. Automatic oiling is used in turbines, saving labor.
9. Can handle extreme overload capacity.
10. Use of superheated steam an asset.

Turbine—Disadvantages

1. Breakdowns are usually serious ones (although infrequent), and require shore experts or special equipment.
2. Reduced power astern.

Reciprocating—Advantages

1. The reciprocating engine has the advantage of simple construction, making it possible to repair any ordinary breakdown at sea, or at least adjust for operation at reduced speed.
2. Heavy-duty construction stands up under terrific treatment.
3. Full power astern.

Reciprocating—Disadvantages

1. Restricted to lower steam pressures and temperatures.
2. Not reliable with superheated steam.
3. Lower efficiency than turbine.

Q. What is a "pressure-compounded" turbine?
A. One consisting of a number of simple stages.

Q. What is a velocity-compounded stage?
A. One wheel on the rotor with 2 or more rows of blades attached. There are stationary blades attached to the casing between the rows for reversing the direction of the steam to the next row.

Q. What is a combination impulse-reaction turbine?
A. One in which there are a number of high-pressure impulse stages followed by a number of reaction stages. This comprises an efficient turbine, as the best features of both types are utilized.

Q. What is the purpose of flexible couplings?
A. To provide for longitudinal flexibility between driving and driven shafts, thereby permitting each shaft to be adjusted axially.

Q. In cross-compound units, do both turbines run at the same speed?
A. No. The H. P. usually runs much faster.

Q. How is this adjusted to give equal ratios at the main gear?
A. By installing correct reductions between pinion gears and the main or bull gear.

Q. What is meant by thin-tipping of blades? What is its purpose? On what type of blading is it found?

A. The tips on ends of blading are reduced in area to approximately .010", Fig. 13. It is necessary to have very small tip clearance in a reaction turbine. This clearance is still obtained but the safety of the turbine is increased due to the fact that if the rotor should drop, only the thin tips would be affected by bending over (in the case of solid blades right to the tip, and in the event that the rotor dropped, the blading very likely would be ripped out).

Q. How are reaction blades usually installed?
A. The serrated blade root fits into corresponding serrations on the rotor or casing and a side locking piece of soft iron is hammered in, acting as an anchor, Fig. 14.

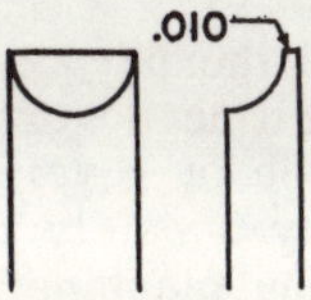

Fig. 13. Thin-tipped turbine blade.

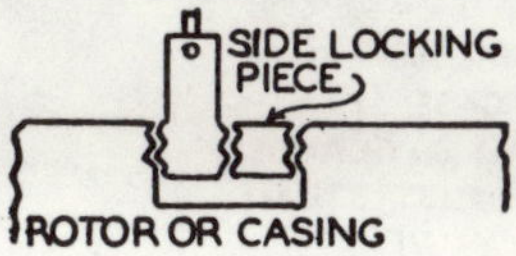

Fig. 14. Reaction blade installation.

Q. Sketch a typical reaction blade.

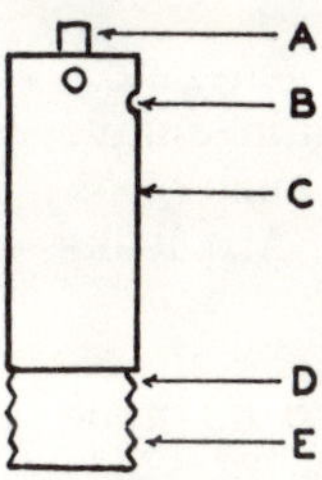

Fig. 15. Reaction blade, showing parts.

A. Figure 15. A) Tenon for shrouding; B) Hole or slot for binding wire; C) Blade (body); D) Root; E) Serrations for securing to casing or rotor.

Q. What is the distinguishing difference between impulse and reaction turbine blading?
A. Impulse blading (A) is heavier in construction and shaped differently, Fig. 16.

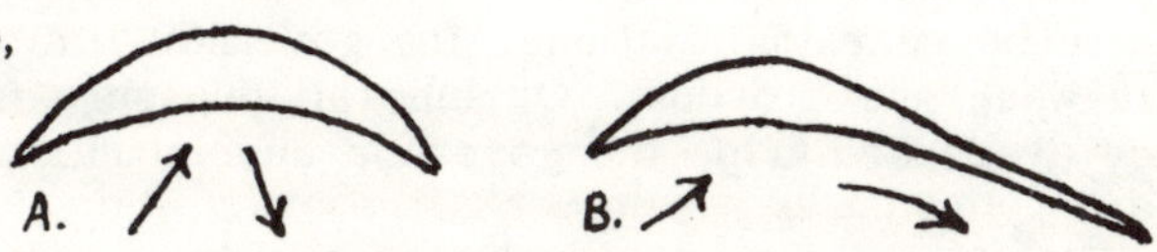

Fig. 16. A) Impulse; B) Reaction.

OPERATION

Q. Sketch a gland-steam seal unit.
A. Figure 17.

Q. What is the purpose of a "guarding valve" in a turbine? How should it be used?
A. To prevent steam leakage to the astern element during ahead operation. Should be opened only while maneuvering.

Q. The pressure in an oil-pump gage rises suddenly. What would you look for?
A. Clogged lines. Inspect all important bearings for heating and

for oil supply. If one is hot, clean out the line leading to it as quickly as possible.

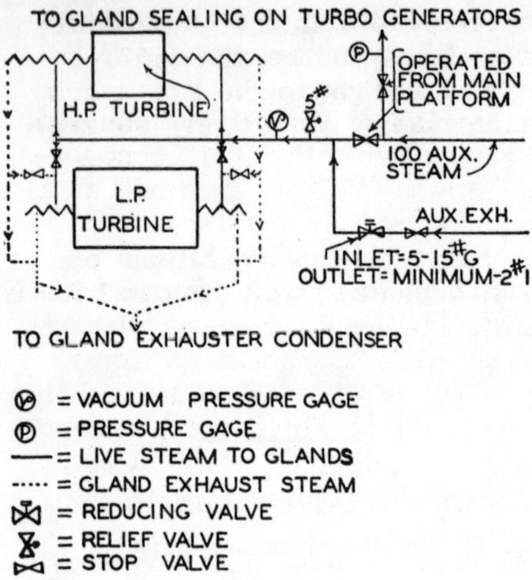

Fig. 17. Gland-steam seals and leakoffs.

Q. What is an overspeed trip? How does it operate?

A. The overspeed trip is an emergency device which shuts off steam to the turbine in the event that the speed-limiting governor fails to function. It may also be operated by hand, and should be tested by hand while warming up a turbine.

Q. What is the purpose of a speed-limiting governor?

A. This type of governor is essentially a safety device to prevent the machine from dangerous overspeeding, due to broken shafts, loss of propeller, etc.

Q. After shutting down a steam-operated turbo-electric generator, you return to the throttle platform and find that the main engine vacuum is dropping rapidly. What is the most likely cause?

A. Failure to close the exhaust valve on the generator. Air leaks through the glands, causing loss of vacuum.

Q. What is the purpose of the governing system installed on most main propulsion turbines?

A. On turbo-gear installations, the governor limits the speed (opens again when speed drops). On lube-oil pressure failure or emergency overspeed trip, the governor closes until reset.

Q. Explain how the throttling losses can be kept to a minimum on the main turbine.

A. By using only the required number of nozzles, fully opened, at any time. Do not use excess nozzles partially opened.

Q. If during your watch a rumbling or sharp metallic sound is heard from a turbine, what would you do and what may be the cause?

A. A rumbling may be caused by water or other foreign material in the turbine. Slow down turbine and check boiler water level and operation. Try bringing up to speed again and, if rumbling continues, shut down.

A metallic sound probably indicates loose blading, loose shrouding, loose sealing strips, etc. If noise continues, shut down immediately.

Q. In general, what is the maximum temperature at which a main bearing should be operated? Name two means which are used to reduce the temperature of turbine bearings.
A. 110°—120° F. By running the oil through a cooler. By circulating oil through the bearing, thereby carrying the heat away.

Q. Why is it desirable on a geared turbine to jack the engine over for several hours after killing the vacuum on the main condenser?
A. To allow the rotor to cool down slowly. Usually run until temperature drops to 125° F.

Q. Other than increased bearing friction, what factor is most likely to change the temperature of the lubricating oil in a circulating oil system, and what should be done when such changes occur?
A. 1. Changing temperature of sea water which is used to cool the oil. Do nothing about the change until temperature and pressure of oil at bearings are changed. If oil at bearings gets too hot, increase the flow of sea water through the cooler, thereby increasing cooling action.

2. If pressure of oil at bearings gets too low, due to cold oil becoming sluggish in distribution pipes, decrease the flow of sea water to the cooler thereby decreasing cooling effects. Open steam coils in oil tanks if necessary.

Q. What alarms or signals are installed on alarm panel of turbine unit?
A. This will vary in many ways. Some of the alarms would be: low pressure oil, sump tank oil level indicator, salinity indicator, steering engine motor failure, reverse telegraph direction alarm signal, turning gear engage or disengage.

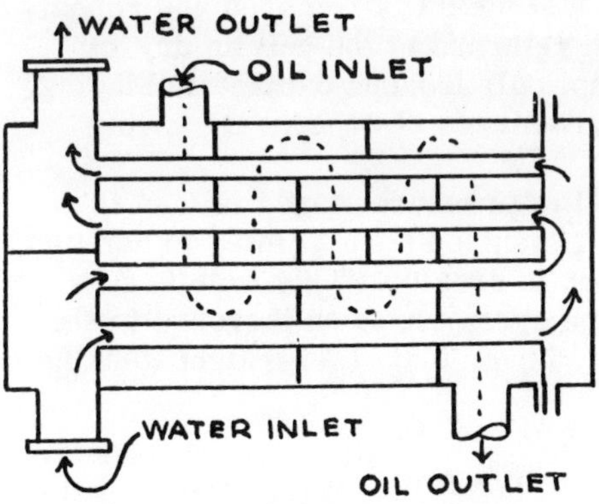

Figure 18.

Q. Sketch a lube oil cooler. Explain its operation and its location in relation to the turbine.
A. The lube oil cooler is placed between the lube oil pumps and the turbine. Cooling water passes through the tubes and the oil passes around the tubes. The oil is caused to take a circuitous path by a series of baffles. The pressure of the cooling water is always less than that of the oil. Thus, if a tube ruptures, no water can leak into the oil, Fig. 18.

Q. Describe the method and frequency of cleaning the metal edge-type filter frequently used in a turbine lube system.
A. It consists of either a stack of thin metal disks separated by

spacers mounted on a central shaft or a metallic ribbon wound on a corrugated frame, the ribbon having spacer lugs. Oil enters the body around the strainer element, passes through the spaces between the disks and into the center section and then to the discharge. Impurities are lodged on the outside surface of the strainer element. The size of the particles removed depends on the spacer thickness, which ranges from .0005" to .0035".

Cleaning blades are provided to remove accumulation. These blades are mounted on a post and the stack is rotated by means of a handle extending through the top of the casing. Impurities scraped off by the blades fall to the bottom and are removed through drain plugs.

Q. How would you continue to operate a turbine at sea, maintaining a vacuum, if an air leak developed that could not be corrected?

A. Reduce the speed of the unit so that the operation will maintain safe turbine limits which will be relative to the amount of vacuum that can be maintained. Operate main circulator pump at increased speed. If possible, make temporary repairs to carry the best vacuum possible. It may be necessary also to reduce steam pressure and temperature to minimize chances of overheating the unit. As long as some vacuum can be maintained with the air ejector, the uncondensable gases and vapors will be removed.

Q. Explain the use and advantages of reheaters in connection with turbines.

A. In expanding through a turbine, steam becomes wet. This moisture causes both corrosion and erosion, especially in the later stages. Therefore, some turbines are fitted with reheaters between the high-pressure and low-pressure units. Wet steam is removed and reheated by an outside source. It is then returned to the unit as dry or superheated steam. This feature not only lessens damage to blading, but combines the thermodynamic advantages of superheat, giving increased economy.

The moisture in the steam in the low pressure stages is less than it would have been had the original expansion been allowed to complete itself, and the added equivalent work is greater. This results in a net gain in thermal efficiency for the cycle due to an increase in the average temperature of heat received over that of a straight Rankine Cycle.

Q. What are the rotation losses in a turbine?

A. Leaving loss: The residual velocity of the steam passing through exhaust to the condenser.

Windage loss: Caused by fluid friction as the turbine blades rotate in the surrounding steam.

Blade loss: Caused by friction as the steam passes over the blades.

Tip Leakage loss: The loss of steam over the tips of both fixed and moving blades.

Mechanical Friction loss: As a result of friction through the bearings, and reduction gears.

Q. What procedure would you follow when using a bearing for the first time?

A. When a new bearing has been installed it should be run in at a low speed and the temperature checked frequently. Should the temperature rise more than 10° to 15° F in the first 10 to 15 minutes of operation, the unit must be stopped and the bearing taken down and all high spots carefully scraped and smoothed down so that a good surface is made. The above process is repeated until the bearing can run steadily with a steady temperature. A new bearing under normal operation should be able to run within a 40° F temperature spread.

Q. What is the purpose of vents on gear casings?

A. Vents are placed on gear casings and certain bearings and couplings to prevent static air pressure from building up. It has been found that the fanning of the gears builds up the pressure which causes leakage through bearings and couplings. Vents also reduce the amount of air entrained in the oils and reduces the tendency to foam. Fire hazard is kept to a minimum in that the free admission of atmospheric air keeps the air-oil vapor mixture clean.

Q. What is the purpose of a thermometer placed in the steam chest of a turbine? What will a variation in temperature indicate?

A. A variation in temperature over or under the design for a given power requirement will give indication of the condition or quality of the steam entering the turbine. (Total heat, superheat priming etc.) When the inlet temperature is compared to the exhaust, the heat drop and the resulting pressure drop can be computed. From this computation the amount of work and efficiency of the unit may be determined.

Q. Explain how to find the total expansion of steam in a turbine.

A. The total expansion of steam in a turbine is the ratio of the total volume of the final stage to that of the first stage. The ratio of expansion in a turbine whether single or multi-stage is equal to the final volume ÷ initial volume and this is approximately equal to the absolute initial pressure ÷ absolute terminal pressure.

Q. What is the purpose of an emergency governor and state how it operates?

A. The function of the emergency governor is to shut down the turbine automatically if the speed increases to approximately 10% above normal full load speed. It consists of a plunger, which is set in the governor hub perpendicular to the rotor axis. This weight is placed with its center of gravity offset so that the centrifugal force exerts an unbalanced force, tending to throw it outward. Normally, it is held in its inner position by the compression of a spring and retainer. If the speed increases to the tripping point, the centrifugal force of the weight overcomes the compression of the spring and the weight flies outward and strikes the trip lever. Through a bolt and lever the trip lever rotates the trigger, releasing the overspeed trip piston which pushes the steam chest valve shut. The steam chest valve trip can be reset by pulling out the reset lever until the trigger re-engages the piston. Resetting cannot be done until the turbine speed has decreased sufficiently to allow the overspeed trip weight to return to its normal position, which will occur at slightly below normal running speed. The turbine should be overspeeded occasionally to check the speed at which the weight flies out and disengages the trip rod.

Note: When the driving apparatus to which the turbine is connected is such that the load cannot be removed, it may be found difficult (or even impossible) to increase the speed to 10% above normal. In such cases it is advisable to disconnect the driving apparatus when running the over-speed test. In order to increase the speed gradually push inward on the governor lever. The speed should be watched carefully so that it does not go much over 10% overspeed.

Q. Briefly describe the operation of a hydraulic speed-limiting governor on a main propulsion turbine.

A. The overspeed control valve is usually held open against spring pressure by oil at constant pressure which is supplied by the lubricating oil service pumps. This oil is supplied to the operating cylinder through a pilot valve which, when actuated by the governor, will shut off the oil flow and drain the operating cylinder, thereby allowing the spring pressure to close the overspeed control valve. The pilot valve may be actuated hydraulically by a small centrifugal oil pump directly connected to the turbine shaft or by mechanically connected speed governor weights or flyballs.

Q. What is a dummy micrometer used for?

A. To measure the axial clearance of a turbine rotor.

MAINTENANCE

Q. What are the purposes of the jacking or turning engine?

A. 1. To warm up the rotor evenly for equal expansion and prevent sag of rotor and shaft.

2. To cool rotor evenly when securing turbine.

3. To turn over the rotor for inspection while secured.

4. To turn over frequently while secured to prevent rotor and shaft sag.

Q. Are oil grooves cut in turbine rotor bearings? Why?

A. No. In high speed engines it is essential that the journal run on an unbroken surface. The rotor actually floats in the oil. The oil must not be allowed to churn up by running in grooves. The bottom half of the bearings has no grooves. The upper half is usually relieved, leaving babbitt only at the edges to seal in the oil.

Q. What is meant by the lapping-in of gears?

A. During manufacture or after shop repairs, the mating gears and pinions are set up and run together with a fluid containing a compound running over the teeth. This method is gradually being replaced with the process whereby the teeth are cut with a hob and and then machined with a machine that shaves each tooth individually to the correct shape.

Q. Give reasons for the use of reduction gears with turbines rather than direct drive.

A. In order to preserve the steam speed-blade speed ratio, turbines should operate at high speeds. Propellers are efficient only at low speeds due to churning, eddies, and cavitation. With direct drive, the turbine shaft is joined directly to the propeller shaft through a flexible coupling. The rotor must turn very slowly, making the same revolutions per minute as the propeller. In order that the blade peripheral speed ratio, the corresponding turbine diameter must be enormous. To accommodate the high speed of the turbine to the low speed of the propeller, mechanical reducing gears are employed. Since direct drive has no outstanding advantage over the geared turbine but does have many disadvantages such as weight, space, and low economy, it has become obsolete.

Q. Describe and explain the use of the axial clearance indicator on a turbine.

A. The indicator is located at the forward end of the turbine and provides an easy means of obtaining a rough check on the axial position of the rotor. It consists of a pointer connected at one end of a shaft that passes through the turbine casing and a lever connected to the other end of the shaft, which bears upon a shoulder of the rotor. When not in use, the lever is held out of contact with the shoulder by a spring. The indicator is used by manually turning the shaft into contact with the shoulder. The pointer registers on the indicator plate which carries markings indicating the central and danger positions of the rotor in the forward and aft positions. The same information is given on some vessels by the thrust bearing indicator which operates in much the same manner.

Q. What are the two major adjustments of the main propelling turbine?

A. In all main propelling turbines installed in vessels, the two major adjustments are the fixing of the rotor in its proper radial and axial position. The radial position of the rotor is maintained by the main bearings and the axial position by the thrust bearing.

Q. What are the usual principal parts of a complete governing mechanism (oil) for a large turbine?

A. An oil-actuated piston connected to the turbine steam trip valve. The lube oil flowing to the turbine holds the piston in such a position as to hold this steam valve open. If the oil pressure drops below the set pressure, the piston drops and releases the steam valve which closes and shuts off the steam supply to the turbine. The pump is located on the end of the shaft.

Q. Name three methods of determining the quantity of steam consumed by a turbine.

A. The quantity of steam consumed by the turbine is generally determined by one of the following methods:

1. By measuring the condensate.

2. By measuring the feed water.

3. By a steamflow meter.

The first method—that of weighing the condensate—will, generally, result in greater accuracy than will any of the other methods. Consequently, where practicable, it should be used.

Q. Name some other kinds of steam turbine governors.

A. Spring-loaded emergency tripping device. Centrifugal-hydraulic governors.

Q. What is meant by scoring or galling of reduction gears?

A. Scoring pertains to the scratches or gouges caused by foreign materials in the oil. Galling is the chafing of metal from one gear tooth and depositing it onto the mating tooth of another gear. This leaves a pit in one tooth and builds up metal on the other mating tooth.

Q. When in port, what attention should be given the main turbine?

A. Turn rotor with jacking gear periodically. Remove carbon packing rings if unit is to be shut down for any length of time. Check oil in sump tank to be sure that it is in good condition.

Q. Describe the turning gear arrangement as used in connection with steam turbines. What are the two main purposes for which this gear is used?

A. In most installations, the turning gear arrangement is bolted to the end of the pinion shaft (HP turbine side of gear). This facilitates and reduces the expense of replacing it in case of damage, Fig. 19.

To keep the rotor turning over slowly while warming up to prevent rotor sag. To keep the rotor turning over for a cooling-off period after shut down to prevent rotor sag. To turn over rotor after upper casing has been removed to check blading, shrouding, etc.

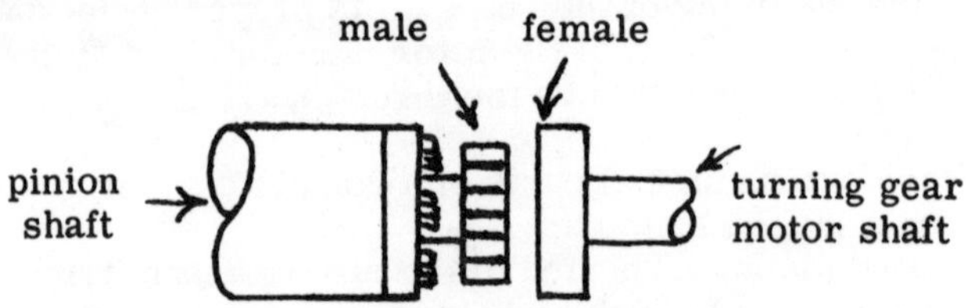

Fig. 19. Turning gear arrangement.

BOILERS

FIRE-TUBE—CONSTRUCTION

Q. Sketch a cut-away view of a Scotch boiler, showing construction features. Name parts. Show furnace connections clearly.
A. Figures 20 and 21.

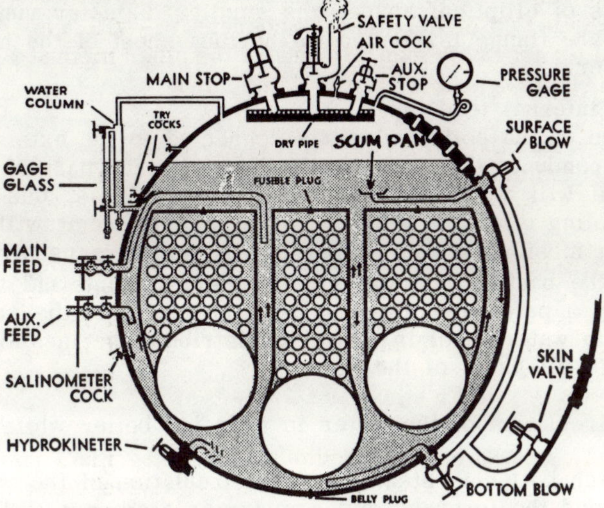

Fig. 20. Front view of Scotch boiler with fittings attached.

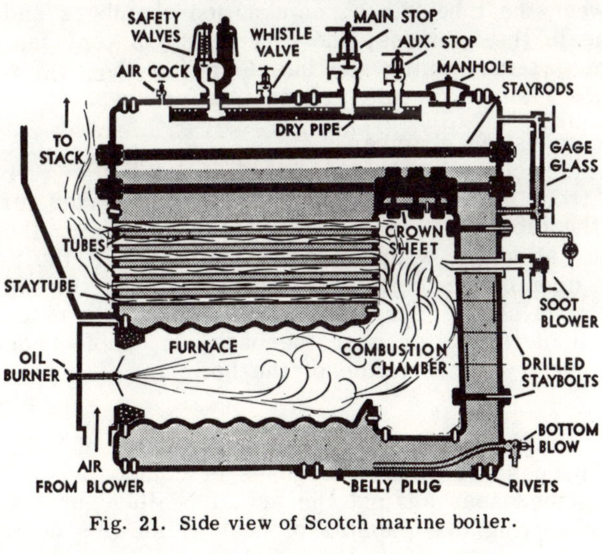

Fig. 21. Side view of Scotch marine boiler.

Q. What is a course in a boiler shell?
A. When the length of a boiler shell exceeds the width of a single plate, that section which comes between the circumferential joints is termed a course.

Q. What is a horsecollar on a furnace?

A. The combustion-chamber end of the furnace is flanged to a horsecollar shape to facilitate removal and replacement of the furnace. There is a deep throat at the bottom of the flange between the flange and the furnace and an extending flange on the top. The sides of the flange are made small enough to fit through the opening in the front head.

The flange is of elliptical shape, the smaller diameter running horizontally. The flange is riveted to the tube sheet of the combustion chamber.

Q. What is the internal feed pipe?

A. A pipe inside of the boiler which is connected to the external feed line and conducts the incoming feedwater to the part of the boiler where it will do no harm to the heated surfaces, due to the sudden cooling effect it would produce, and where it will aid the circulation most. In a Scotch boiler, the water is usually led between the tube banks. In a watertube boiler, the internal feed pipe is usually a perforated pipe running the length of the steam drum below the water level in order to distribute the incoming water evenly to all parts of the boiler.

Q. Trace the circulation of the water in a Scotch boiler while it is steaming.

A. While a Scotch boiler is steaming, the circulation of the water is upward around the furnaces and combustion chambers and between and around the tubes in the tube banks. The circulation is downward between the tube banks, combustion chambers and furnaces. Underneath the furnaces, the circulation is confused and sluggish because of the mixing and interfering of the downward currents and the rising currents around the furnaces.

Q. Describe a combustion chamber.

A. It is roughly a rectangular box made up of steel plates riveted together. The front sheet has openings for the tubes and furnaces and is called the tube sheet of the combustion chamber, the sides and bottom are called the wrapper sheet, the plate at the rear of the combustion chamber is called the back sheet and the top plate is called the crown sheet. The combustion chamber is placed inside of the shell of the boiler near the back. In double-ended boilers, it is placed at the middle of the boiler.

Q. What are tube retarders?

A. Long twisted strips of metal inserted in the tubes of a fire-tube boiler for the purpose of giving the gases of combustion a rotary motion, forcing the gases against the actual heating surface of the tube and improving the transfer of heat by slowing down the gases of combustion.

Q. What is a diagonal stay?

A. Another form of stay used to support the heads of a boiler. It consists of a bar with a flat palm on each end. One palm is riveted to the shell and the other palm to the head. The bar runs diagonally across the angle formed by the head and shell. A diagonal stay is often called a palm stay, Fig. 22.

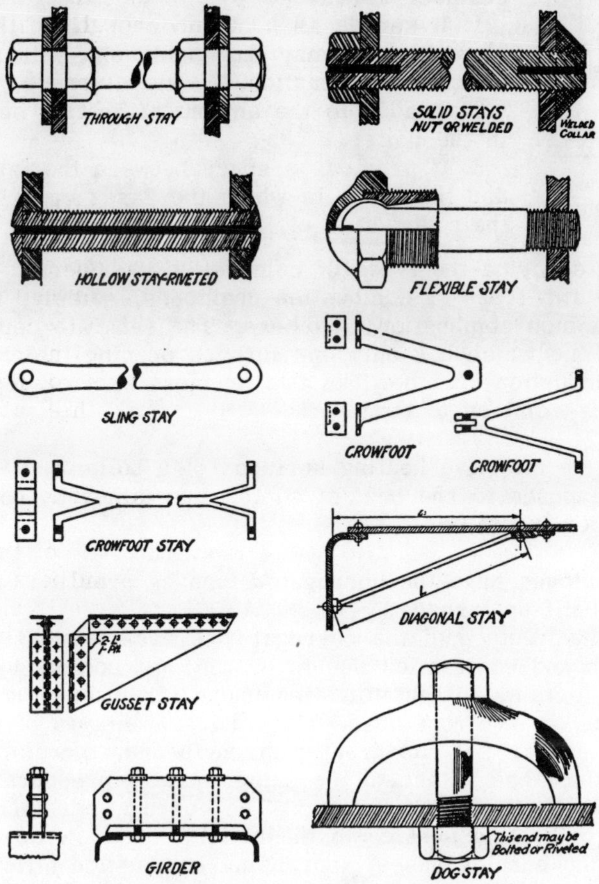

Fig. 22. Types of boiler stays.

Q. How is a combustion chamber stayed?
A. The tube sheet of the combustion chamber is stayed by means of stay tubes to the front head. The wrapper sheets are stayed by stay bolts connecting them either to the wrapper sheets of adjacent combustion chambers or to the shell of the boiler. The back sheet is stayed to the back head of the boiler by stay bolts. The crown sheet is stayed by stay bolts which are connected to the crown bars spanning the top of the combustion chamber.

Q. Why are the backs of combustion chambers sloped? Show, by sketch, how this slope would look.
A. This is done to promote better circulation of the warm water

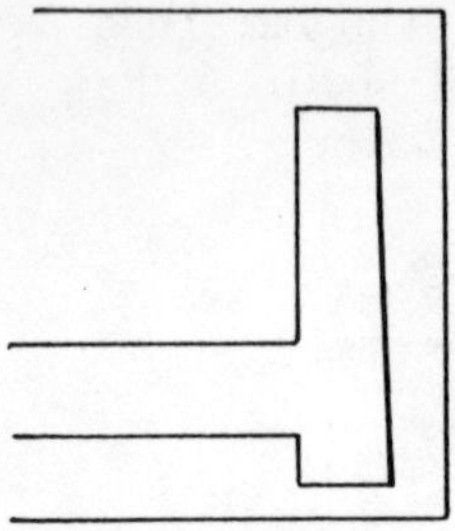

Fig. 23. Combustion chamber with sloping back.

which naturally tends to rise. The opening is larger at the top, as shown in Fig. 23.

Q. What three purposes does the combustion chamber serve?

A. 1. It serves as a chamber in which the unburned gases may thoroughly mix with air and complete combustion.

2. It adds to the amount of heating surface in the boiler.

3. It provides a space between the furnace and the tubes in which the gases can reverse their direction of flow.

Q. Name and describe the types of combustion chambers.

A. There are two types of combustion chambers, namely, the separate and common combustion chambers. The separate combustion chamber is one which has only one furnace opening into it. The common combustion chamber has two or more furnaces opening into it and is common to them.

Q. What are the effective heating surfaces of a boiler?

A. Surfaces exposed to the heat of combustion either by conduction, convection, or radiation.

Q. What advantages have the corrugated furnace over the riveted, sectionally built-up furnace?

A. The initial advantage of the corrugated furnace is that it takes up expansion and contraction due to heating and cooling by its accordion-like action, without throwing undue strains on the combustion chamber or the front head. The other advantages of the corrugated furnace are: It has greater strength and, therefore, can be made lighter and withstand the same working pressure as a sectionally built-up riveted furnace. The corrugations increase the amount of heating surface in the boiler, and slow down the passage of the burning gases, increasing combustion efficiency.

Q. Where is the drain cock or belly plug located and what is its purpose?

A. The drain cock or belly plug is located at the very bottom of the shell of a Scotch boiler. Its purpose is to allow the water to drain from the bottom of the boiler and to allow the scale, mud, etc., to run out of the boiler when it is being washed down and cleaned.

Q. Name three types of corrugated furnaces.

A. Leeds, Morison, Fox, Adamson, Brown, Purves, Fig. 24.

Q. How is the rear head of a Scotch boiler supported?

A. By means of through stays connecting it to the front head and by stay bolts connecting the back sheet of the combustion chamber to the rear head.

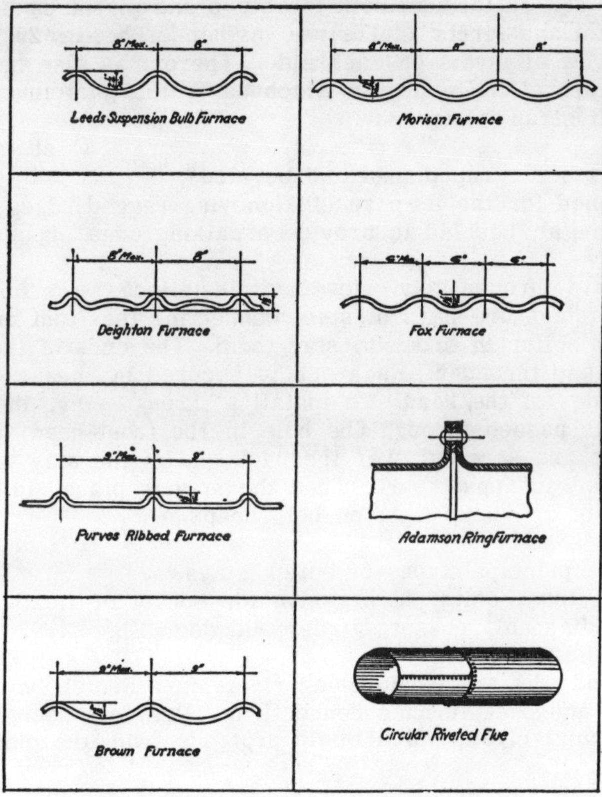

Fig. 24. Types of furnace construction.

Q. How is the front head of a Scotch boiler supported?

A. By means of through stays connecting the front head and back head, stay tubes which connect the front tube sheet with the tube sheet of the combustion chamber, and to a certain extent by the furnaces. Sometimes portions of the head are stayed by gusset or diagonal stays to the shell.

Q. How is longitudinal expansion of a Scotch boiler taken care of?

A. The foundation of a Scotch boiler is constructed so that there are sliding feet at one end. This allows the boiler to expand longitudinally when heated.

Q. Name and describe the types of riveted joints used on a Scotch boiler.

A. Lap Joint: In this type of joint, the plates to be riveted together are lapped over each other and holes are drilled through both of the plates where they are lapped. Rivets are driven in the holes. There may be one or more rows of rivets.

Butt-Strap Joint: In this type of joint, the plates to be riveted together are placed so that their edges butt against each other. Short pieces of plate called straps are placed over the seam.

Holes are drilled through both the strap and plates on either side of the seam and rivets are driven in them. There may be more than one row of rivets on each side. There may also be a strap on both sides of the plate, in which case, the joint is called a double butt-strap joint.

Q. Why is a butt strap planed and beveled?
A. It is planed for the purpose of removing ragged edges and stressed metal; beveled to provide a calking edge.

Q. Describe a through stay. How is it installed?
A. Through stays are bars of steel connecting the front and back heads of a boiler in order to stay them. The ends of the stays are upset and threaded. Each end is secured by means of nuts on both sides of the head. To install a through stay, the back end of it is passed through the hole in the front head, and both inside nuts are screwed over the back end of the stay to beyond the threaded and upset part. Then the stay is placed in position and the nuts drawn up tight on both heads.

Q. Name the principal types of boiler stays.
A. Through, tube, solid, hollow-riveted, socket or flexible, sling, crowfoot, diagonal, gusset, girder and dog, Fig. 22.

Q. Where and why are countersunk rivets in a Scotch boiler?
A. At front and back furnace connections. Prevents heat from contacting normal rivets which would protrude from the plates.

Q. Explain how you would blow down and empty a Scotch boiler?
A. Cut out the fires and close the steam and feed stops, then allow the boiler to cool off. When a pressure of about 25 lbs. or less per square inch shows on the steam gage, open the bottom blow and allow the pressure to blow the water out. When there is no longer any pressure in the boiler, open the air cock and remove the belly plug, and let rest of water run out of boiler into the bilges.

Q. In which of the shell seams of a cylindrical boiler is the strongest riveting used? Explain why this is necessary.
A. The strongest riveting is used in the longitudinal seam.
Let P = internal pressure per sq. in. T = thickness of plate
D = diameter of boiler L = length of boiler

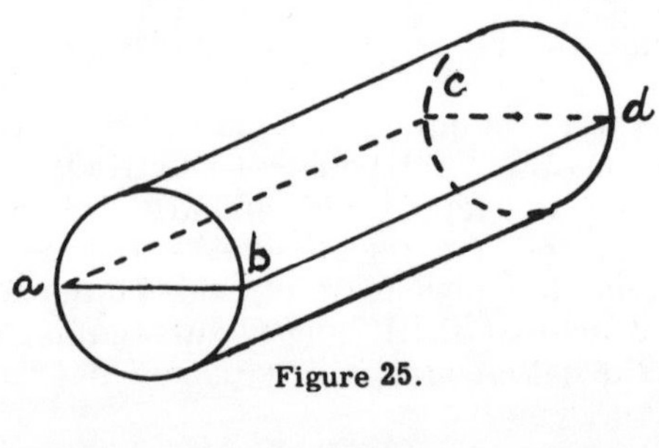

Figure 25.

Assuming a section abcd through the center of the boiler (Fig. 25), the force trying to separate these two parts will be equal to P X D X L.

The area of the material trying to resist the tendency to burst along the sections ab and cd will be equal to (ac + bd) T, which equals 2LT.

Therefore, the stress per square inch on the plate can be expressed as:

$$S = \frac{\text{load}}{\text{area}} = \frac{PDL}{2LT} = \frac{PD}{2T}$$

Now, to find the pressure tending to tear the boiler in two lengthwise or, in other words, to blow the heads off:

Area of head = $.7854\ D^2$

Total pressure on head = $.7854\ D^2\ P$

The area of the material to resist this tendency equals the circumference of the shell X thickness of plate, or πDT or 3.1416 DT.

This can now be expressed as:

$$S = \frac{\text{load}}{\text{area}} = \frac{.7854\ D^2\ P}{3.1416\ D\ T} = \frac{PD}{4T}$$

Comparing the results of these two answers, we see that the stress is only half as great in the latter case. In other words, the plate is twice as likely to give away in the direction of the length of the boiler as circumferentially. For this reason, longitudinal seams are made stronger than girth or circumferential seams.

Q. What are the usual causes of flame impingement upon one side of the furnace of an oil burning Scotch boiler? What are the dangers associated with such continued flame impingement?

A. A partially plugged atomizer tip causing the spray to shoot out on an angle or a scratched hole in the tip causing the same results. (Never use metal wire to clean the holes in an atomizer tip.)

Q. State in detail how you would prepare a Scotch boiler for the regular annual inspection. What temperature should the water be?

A. For internal inspection, the manholes, and handholes (if there are any) and washout plugs should be removed. The furnaces and combustion chambers should be thoroughly cleaned and cool. In preparing the boiler for the hydrostatic test, it should be filled with water not more than 100° F.

Q. In what relation in regard to thickness must a butt strap compare with the boiler plate?

A. Thickness must be sufficient to permit efficient calking of edges. Where the pitch of rivets is so great that effective calking is doubtful (as in quadruple riveted), the edge should be scalloped around the outer rivets. The outer butt strap is figured by formula but is never less than 1/4". The inner butt strap must equal thickness of outer strap plus .125", but is not required to exceed thickness of the shell plate.

Q. Why are blowdown cocks sometimes fitted with spanner guards?
A. To prevent unshipping spanner before completely closing cock.

Q. What is the proper procedure when a fusible plug blows out on a Scotch boiler?
A. 1. Secure fires and feedwater.
2. Secure main and auxiliary stops.
3. Open safety valves slowly with the hand-lifting gear.
4. Secure air to the boiler, and allow it to cool slowly.
5. After the boiler has cooled, inspect the crown sheet and upper rows of tubes for sag. If these surfaces are in good condition, renew the fusible plug, fill the boiler with water, and light off. Submit a report to the inspectors in the next port where there is an inspector. If there has been damage done to the tubes or crown sheet, call the inspector before attempting repairs. Make complete report in logbook.

A common practice used by many engineers is to secure the main steam stop and open the auxiliary steam stop, letting the auxiliaries draw off the steam and thus conserving water.

Q. Sketch a triple-riveted zigzag, double butt-strap joint.
A. Figure 26.

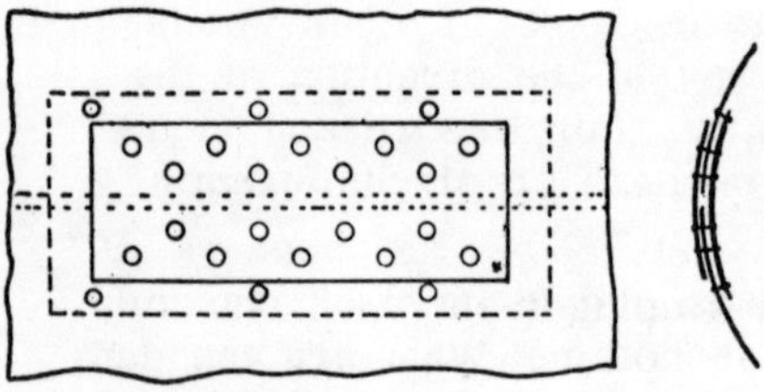

Fig. 26. Triple-riveted zigzag, double butt-strap joint.

Q. What are the advantages of superheated steam?
A. The superheating removes the moisture from the steam, making it dry and giving the steam more expansion. It increases plant efficiency and reduces size of boiler and engine.

It is better suited for turbines, as wet steam affects blading.

Q. How would you tram a furnace to see if it deviates from a true circle?
A. A tram is used which consists of a sliding scale which can be extended across the inside diameter of the furnace to measure that diameter. The tram is used at several angles at each point of measurement and the difference between the largest and the smallest measurements is the amount the furnace deviates from a true circle at that point. When tramming a furnace, this is done every two or three corrugations along the length of the furnace.

Q. What is the purpose of a stay tube?
A. To provide the main staying for the tube sheets and also to add more heating surface to the boiler.

Q. What may rusty streaks on the metal covering of a boiler indicate?

A. They may indicate leaking joints, or that steam or water is coming out of the telltale hole of a stay bolt, and the bolt is broken.

Q. What is a hard patch? How and where should it be applied to a boiler?

A. A hard patch is a permanent repair to a boiler. It is made by cutting away the defective part and riveting a patch of certified boiler plate over the hole. The patch must be applied to the pressure side of the boiler and be of equal thickness with the original plate. It is also calked around the edges in order to make it steam- and watertight.

Q. What is a soft patch? How and where should it be applied to a boiler?

A. A soft patch is a temporary repair made by cutting a piece of boiler plate the same thickness as the damaged plate and large enough to cover the defective area. It is shaped to conform with the part of the boiler to which it is to be applied. Holes are drilled around the edge of the patch at a distance apart of about 4 to 6 times the thickness of the patch. Holes are then drilled and tapped in the damaged plate, using the patch as a template. Before the patch is set in place, a coat of heavy red-lead putty about 1/16" thick is placed on the patch so that it will be between the patch and the damaged plate. The patch is secured, using tap bolts with washers and grommets under the washers. If it is at all possible, the patch should be placed on the pressure side. This type of repair should be used only in an emergency.

Q. What is meant by "sewing a crack?"

A. This is a temporary means of repairing a crack in a boiler. The process of sewing a crack is as follows: First, a hole about 3/8" in diameter is drilled so that one end of the crack is drilled out, the resulting hole is tapped out and a screw plug screwed into the hole. The stud is cut off near the surface of the plate and peened over. Then another hole, lapping over into the plug which was put in first, is drilled along the line of the crack. This second plug is then cut off and peened over. The process is repeated until the entire line of the crack has been filled with plugs. This should be done only when absolutely necessary, and the pressure on the boiler should be reduced in proportion with the seriousness of the crack.

Q. What usually causes furnace collapse? What usually causes fractured boiler plates?

A. Usually oil and scale deposits on the water side of the furnace act as an insulator, the metal overheats and the furnace sags. If this sag becomes too great, it may lead to the collapse of the furnace. Fractured plates may be caused by one or a combination of any of the following: failure of rivets; failure of weld; results of undue expansion and contraction; pitting or honeycombing carried to a point that has weakened the plates; caustic embrittlement.

Q. What would you look for in examining a boiler?
A. In examining the water side of a boiler, look for signs of scale, grease and corrosion, and check all stays to see that none are broken or cracked. (Stays, if broken or cracked, give a dull sound when struck by a hammer.) Check the furnaces, tubes and combustion chambers for deflection, sag or blisters, and carefully examine all joints for signs of cracking between the rivet holes, pulling apart, etc. All internal fittings, such as dry pipes, internal feed pipes, hydrokineters, scum pans, surface blow lines, etc., should also be examined.

In examining the fire sides of a boiler, tram the furnaces, examine the ends of all stay bolts for any signs of steam or water leakage either around threads or through the telltale holes. The combustion chambers should be checked for sag, blister or bulge and the accumulation of soot should be noted. Also examine the riveted joints.

In examining the shell of a boiler, if practical, remove the lagging and give the shell a good once-over. If it is impractical to remove the lagging, examine the outside of the lagging for rusty streaks, indicating leaks from the telltale holes or riveted joints. Examine the foundation of the boiler for corrosion and rust.

Q. How would you renew a stay bolt?
A. To renew a stay bolt it is first necessary to remove the defective one. This is best done by drilling out the end of the stay bolt with as large a bit as will not touch the thread in the stayed sheet. The hole should be drilled well beyond the inside of the stayed sheet. Then, with a blunt-nosed tool, the ends are collapsed inward and broken off. It will then be a fairly easy matter to drop the stay bolt to the bottom of the boiler, to be removed later. The threads in the stayed sheet are then cleaned thoroughly and carefully examined. If they are found to be undamaged, a new stay bolt may be screwed into place. If the threads are damaged, it will be necessary to ream out the hole, tap new threads, and install a larger stay bolt.

Q. What part of a Scotch boiler is first affected by low water?
A. The crown sheet of the combustion chamber.

Q. Describe a screw stay bolt. How is it secured?
A. Screw stay bolts are threaded steel bolts, screwed into two flat surfaces that are near each other. They have telltale holes drilled into either end. These holes are 3/16" in diameter and must be drilled into the bolt far enough so that they extend at least 1/2" beyond the stayed sheet into the water side. Stay bolts may be secured by riveting the ends over, by welded collars or by nuts.

BOILERS

WATERTUBE — CONSTRUCTION

Q. Sketch a B & W (Babcock and Wilcox) three-pass boiler. Name all parts.

A. Figures 27 and 28.

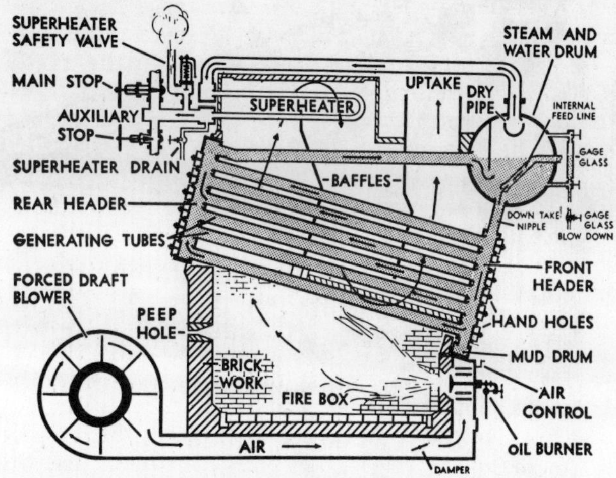

Fig. 27. Side view of B and W straight tube, cross drum watertube boiler—sectional header.

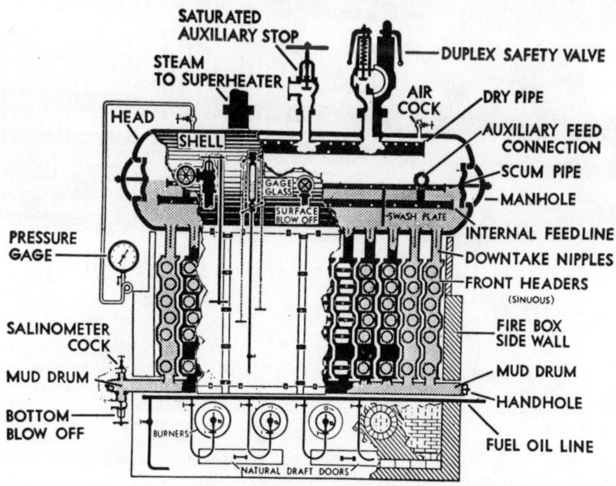

Fig. 28. Front view of B and W straight tube, cross drum watertube boiler.

Q. Name all the parts of the B & W (Babcock and Wilcox) two-drum boiler shown in Fig. 29. Explain the steam flow.

A. Figure 29. Figure 30 shows a schematic of the same boiler.

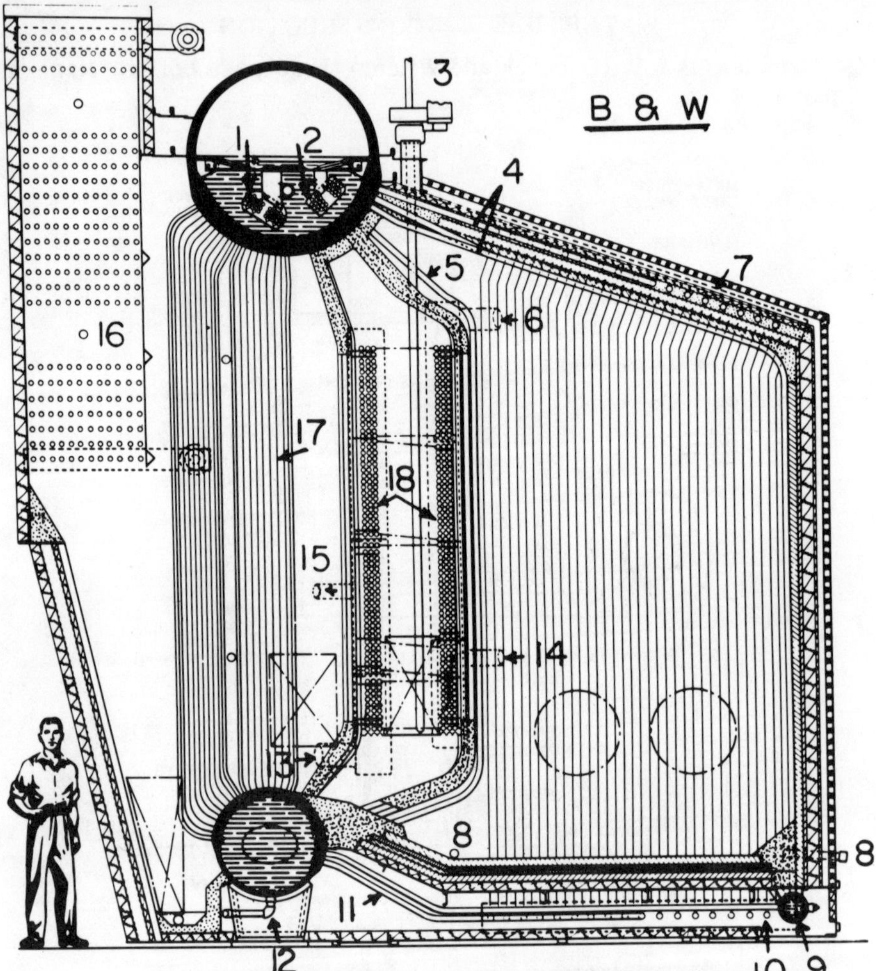

TWO-DRUM INTEGRAL FURNACE MARINE BOILER WITH HORIZONTAL SUPERHEATER

Figure 29.

1. Auxiliary desuperheater
2. Controlled desuperheater
3. Soot blowers (retractable)
4. Side wall tubes to steam drum
5. Screen tubes
6. Superheater inlet
7. Upper rear waterwall header
8. Furnace drains
9. Side waterwall header
10. Lower rear waterwall header
11. Side and rear waterwall supply tubes
12. Bottom blow line
13. Superheater outlet
14. From controlled desuperheater
15. To controlled desuperheater
16. Economizer
17. Generating tubes
18. Superheater tubes

STEAM FLOW

From steam drum to superheater inlet 6; through three passes of superheater to outlet nozzle 15; to the controlled desuperheater 2; to inlet nozzle 14 and fourth pass of superheater; through fifth pass of superheater to outlet 13.

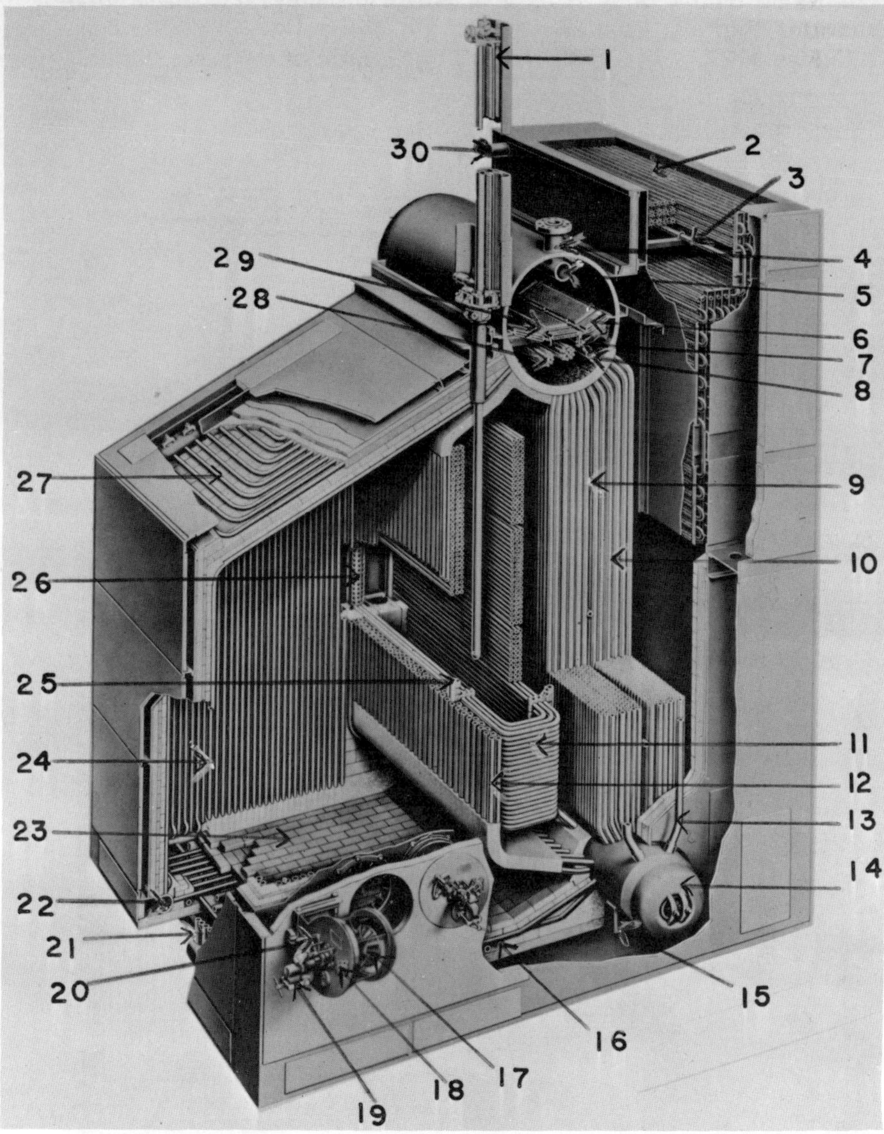

Fig. 30. A schematic diagram of a B & W two-drum boiler as shown in Fig. 29.

1. Retractable soot blower
2. Economizer tubes
3. Economizer soot blowers
4. Main saturated steam nozzle
5. Dry pipe
6. Baffle
7. Chemical feed pipe
8. Internal feed pipe
9. Generating tubes soot blower
10. Generating tubes
11. Superheater tubes
12. Screen tubes
13. Downcomer tubes
14. Water drum
15. Water drum saddle
16. Rear wall floor tubes
17. Impeller plate
18. Peep hole
19. Atomizer
20. Air door handle
21. Side waterwall tubes
22. Side waterwall header
23. Furnace floor firebrick
24. Rear waterwall tubes
25. Spacer tubes, spacer and support fingers for superheater tubes
26. Superheater header
27. Riser tubes
28. Desuperheater
29. Surface blow line
30. Economizer inlet

Q. Name all the parts of the B & W two-drum integral furnace boiler shown in Fig. 31.

A. Figure 31.

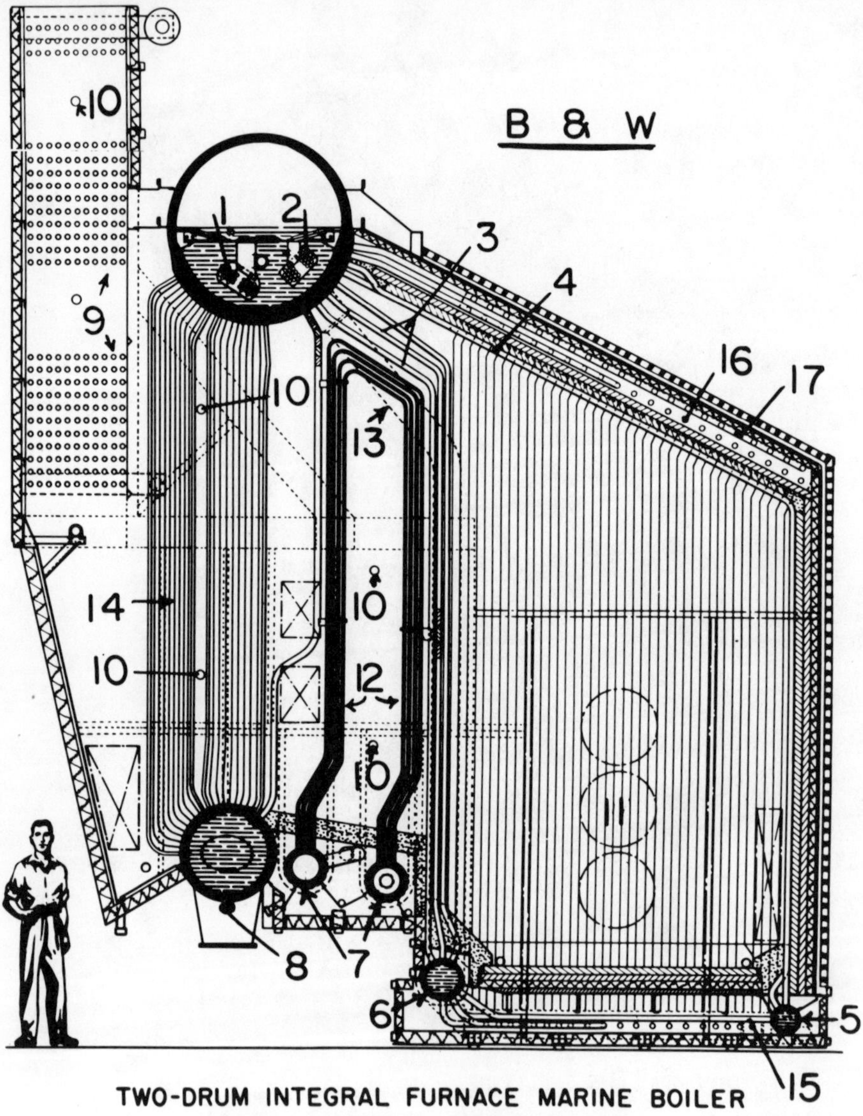

TWO-DRUM INTEGRAL FURNACE MARINE BOILER WITH VERTICAL SUPERHEATER

Figure 31.

1. Auxiliary desuperheater
2. Controlled desuperheater
3. Screen tubes
4. Side wall tubes in roof area
5. Side wall header
6. Screen header
7. Superheater headers
8. Bottom blow valve
9. Economizer
10. Soot blowers
11. Burners
12. Superheater tubes
13. Downcomer tubes
14. Generating tubes
15. Side and rear waterwall supply tubes
16. Rear wall risers
17. Front wall riser tubes

Q. Name all the parts of the B & W top-fired two-drum boiler shown in Fig. 32. Explain the steam flow.

A.

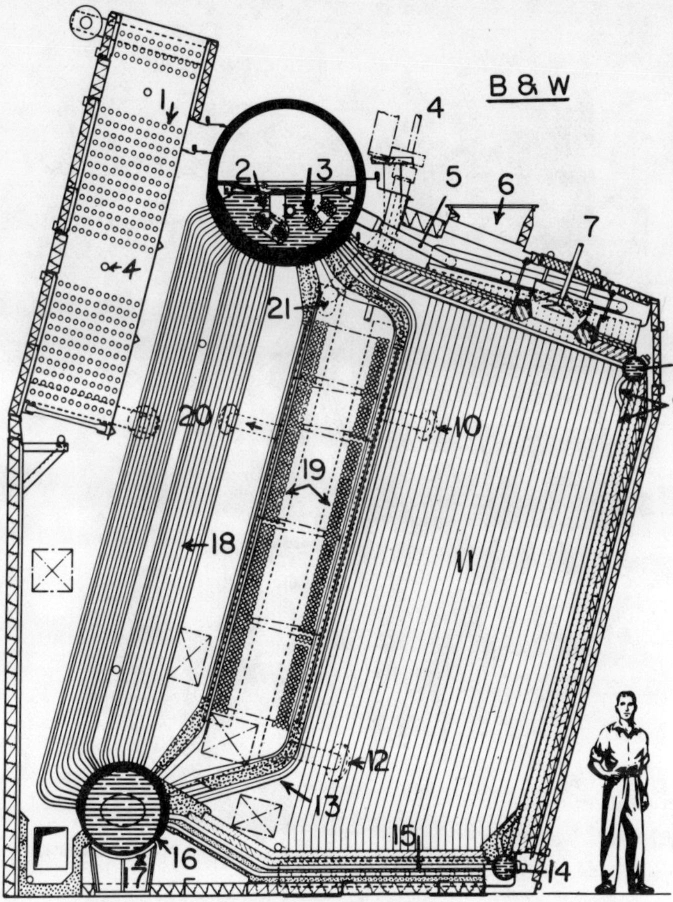

TOP FIRED TWO-DRUM INTEGRAL FURNACE MARINE BOILER

Figure 32.

1. Economizer
2. Auxiliary desuperheater
3. Controlled desuperheater
4. Soot blower
5. Waterwall riser tubes
6. Air inlet
7. Oil burner
8. Upper side waterwall header
9. Side wall tubes
10. From controlled desuperheater
11. Waterwall tubes
12. Superheater inlet
13. Screen tubes
14. Lower side waterwall header
15. Side and rear waterwall supply tubes
16. Water drum
17. Bottom blow valve
18. Generating tubes
19. Superheater tubes
20. To controlled desuperheater
21. Superheater outlet

STEAM FLOW

From steam drum to superheater inlet 12; through three passes of superheater to outlet nozzle 20; to controlled desuperheater 3; to inlet 10 through the fourth pass of the superheater; through the fifth pass to superheater outlet nozzle 21.

Q. Name the parts on the Combustion Engineering two-drum boiler shown in Fig. 33.
A. Figure 33.

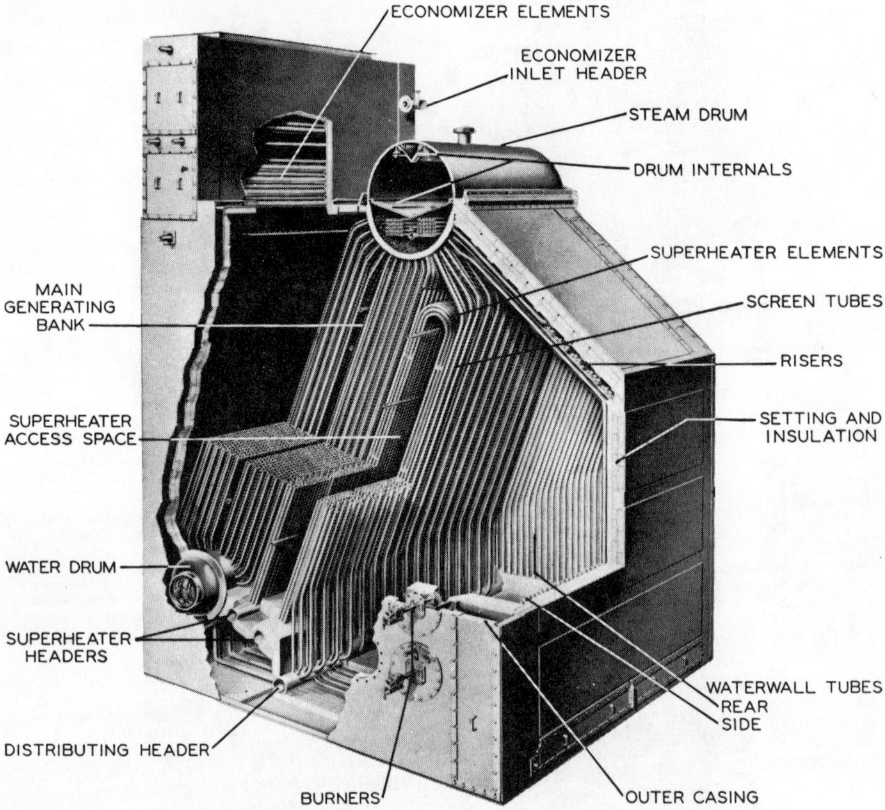

Fig. 33. Combustion engineering-V2M-vertical superheater boiler.

Q. Name the parts on the Prototype Marine Steam Generator.
A. Figure 34.

Q. Name the parts on the ESRD Marine Reheat Steam Generator.
A. Figure 35.

PROTOTYPE MARINE STEAM GENERATOR
Foster Wheeler Corporation

Figure 34.

1. Economizer
2. Dry pipe
3. Soot blower
4. Baffle plates
5. Desuperheater
6. Surface blow line
7. Controlled desuperheater
8. Internal feed line
9. Riser tubes
10. Rear waterwall tubes
11. Side and roof waterwall tubes
12. Side waterwall headers
13. Floor tubes
14. Furnace superheater screen tubes
15. Bottom blow line
16. Water drum
17. Superheater
18. Generating tubes
19. Economizer outlet
20. Economizer inlet

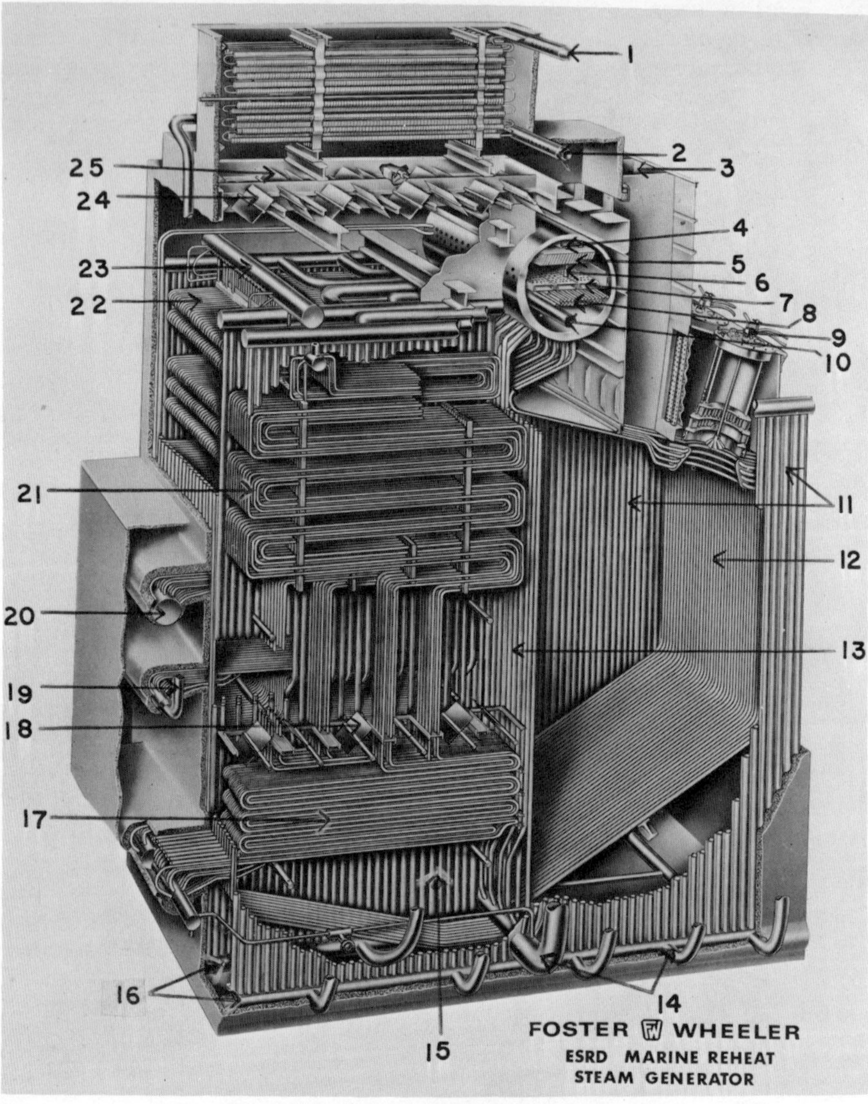

Figure 35.

1. Economizer outlet
2. Economizer inlet
3. Air inlet
4. Dry box
5. Chevron steam drier
6. Surface blow line
7. Perforated plates
8. Desuperheater
9. Oil burner
10. Internal feed line
11. Side waterwalls
12. Front, floor and roof waterwalls
13. Division waterwalls
14. Waterwall downcomers
15. By-pass waterwalls
16. Lower waterwall headers
17. Superheater by-pass economizer
18. Superheater control dampers
19. Superheater inlet header
20. Reheater outlet header
21. Reheater by-pass economizer
22. Reheater
23. Reheater inlet header
24. Reheater by-pass dampers
25. Reheater shut-off dampers

Q. What is the purpose of a controlled-circulation boiler?

A. With the advent of higher pressures and temperatures it is necessary to have a more positive water circulation than is found in the ordinary boiler. Figure 36 shows a controlled-circulation Combustion Engineering (CE) boiler. Superheater outlet conditions of 2000 psig and 1050°F. temperature are obtained.

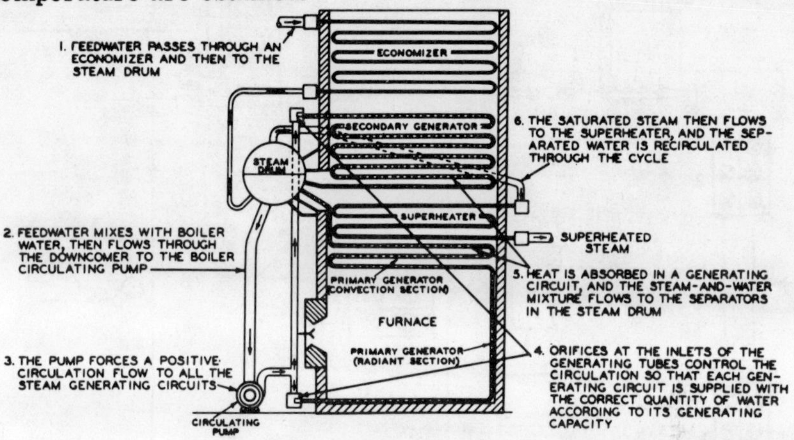

Figure 36.

Q. Explain the operation of a steam temperature control arrangement on a boiler. What is the purpose of this device?

A. Figure 37 represents one type of such control. The purpose of this device is to protect the last passes of the superheater. An air-operated control valve is actuated by the steam temperature at the superheater outlet. Below the set temperature all steam generated passes through the orificed line, the control valve being closed. When the temperature tends to rise above the set temperature, the control valve opens to permit a portion of the steam to flow to the desuperheater and have its temperature reduced. The control valve is provided with a hand jack for manual operation.

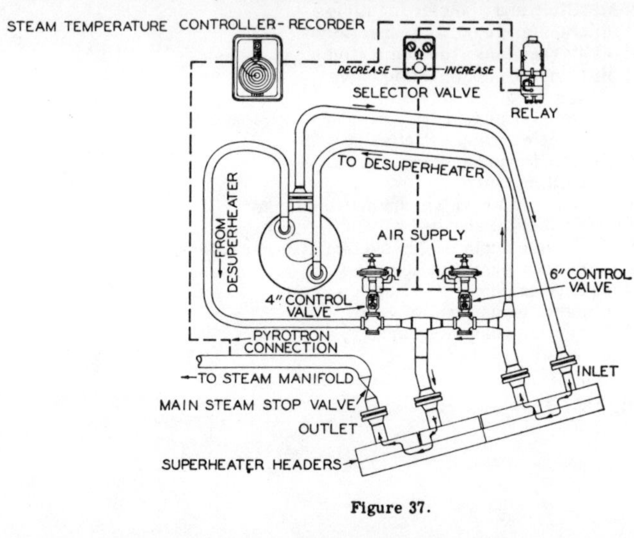

Figure 37.

Q. Sketch a detailed fuel oil cycle naming all parts.

A. Figure 38.

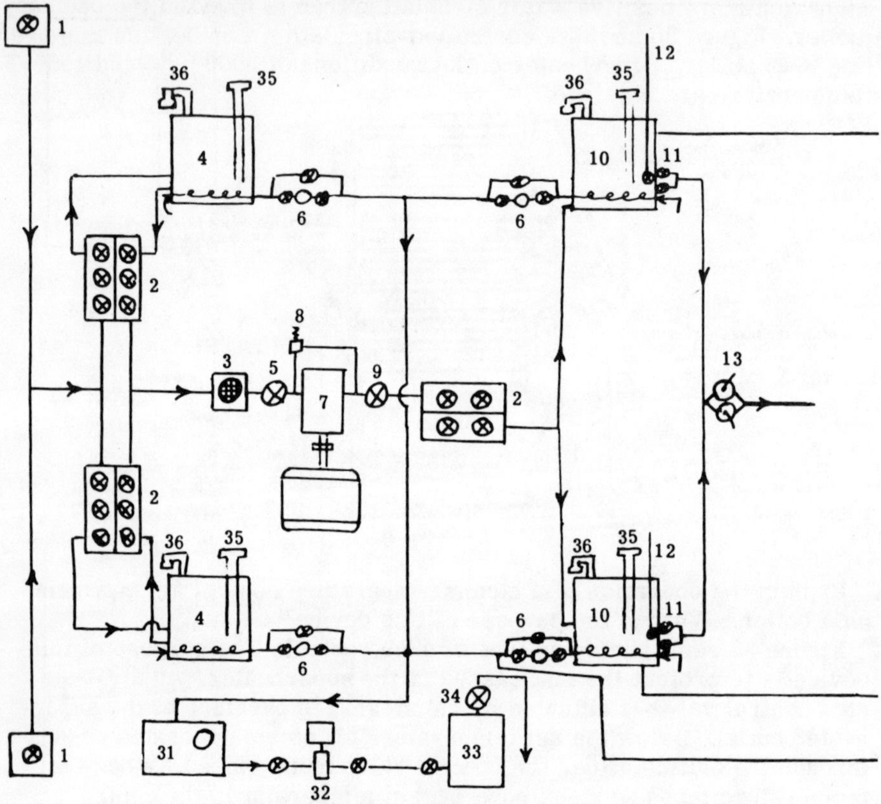

List of Parts

1. Fuel oil connections on deck (port & starboard)
2. Fuel oil filling and transfer manifolds
3. Fuel oil transfer pump suction strainer
4. Fuel oil storage tanks (double bottoms and deeps)
5. Fuel oil transfer pump suction valve
6. Steam traps
7. Fuel oil transfer pump
8. Fuel oil transfer pump relief valve
9. Fuel oil transfer pump discharge valve
10. Fuel oil settling tanks
11. High and low suction valves on settling tanks
12. Emergency fuel oil shut-off valves (internal gate valve)
13. Fuel oil service pump duplex suction strainer (coarse-cold)
14. Fuel oil service pump suction valve
15. Fuel oil service pump
16. Fuel oil service pump relief valve
17. Fuel oil service pump discharge valve
18. Fuel oil meter

Fig. 38. Fuel oil service system.

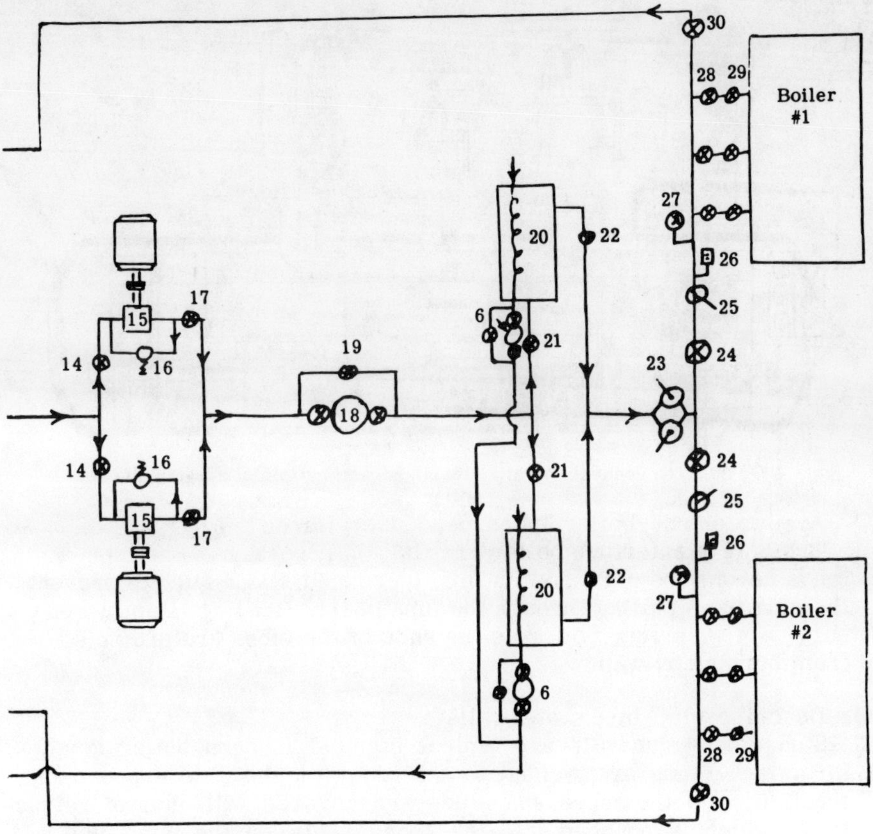

List of Parts (cont.)

19. Fuel oil meter by-pass valve
20. Fuel oil heaters
21. Fuel oil heater inlet valves
22. Fuel oil heater outlet valves
23. Fuel oil service pump discharge strainer (hot-fine)
24. Fuel oil master shut-off valves
25. Fuel oil quick shut-off valves
26. Thermometers
27. Pressure gages
28. Burner root valves
29. Burner valves
30. Recirculating valves
31. Contaminated system inspection tank
32. Contaminated evaporator feed pump
33. Contaminated evaporator
34. Contaminated evaporator vapor discharge valve to auxiliary steam heating system
35. Sounding tubes
36. Vents

Fig. 38. Fuel oil service system (cont.).

Q. Sketch the drum of a watertube boiler and locate all fittings.
A. Figure 39.

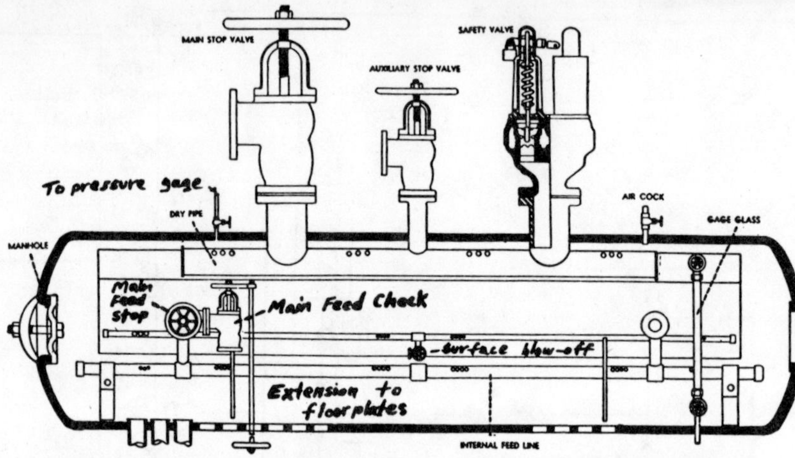

Fig. 39. The connections in a steam and water drum of a watertube boiler.

Q. Are watertube boiler tubes beaded or flared? Why?
A. Tubes in a watertube boiler are flared. A portion of the tube extends beyond the plate and its size increased about 1/8" to prevent the tube from pulling through the tube sheet. Beading is not necessary on a watertube boiler as the ends of the tubes are protected from burning by water.

Q. Describe stud-tube waterwalls.
A. Stud-tube waterwalls are typical in most modern boiler construction. Studs are welded to the tubes on the side exposed to the fire, and the tubes and studs are covered with plastic refractory, which is pounded into the spaces between the tubes and studs. The studs hold the refractory in place and the water in the tubes carries away a large part of the heat absorbed by the refractory, but there is still sufficient heat reflected and reradiated to preserve a high furnace temperature which is necessary for good combustion.

Q. What is a downcomer?
A. A large tube or pipe connecting the water space of the steam drum to the water drums or to the waterwall headers. It is placed outside the boiler casing so that there will be no possibility of any heat transfer through the downcomer walls which might result in the reversal of the downward circulation in them. The purpose of a downcomer is to insure a large and plentiful supply of water to the generating tubes and waterwalls under all conditions of operation.

Q. What is a corbel in the furnace of a watertube boiler? How is it installed?

A. The corbel is built around the base walls of the furnace to prevent slag erosion and undercutting of the lower rows of fire brick in vertical furnace walls. (When the boiler is in operation, there is an amount of liquid slag on the furnace floor. As the ship rolls, this hot slag, in washing from side to side of the furnace floor, eats away the lower rows of fire brick.) The corbel is usually built of chrome ore similar to that used on stud tubes. A heavy sheet of brown paper is placed on the floor bricks before building the corbel. The corbel is then built up along the floor and wall. When the boiler is lit off the paper burns away leaving a space for the floor bricks to expand beneath the corbel, Fig. 40.

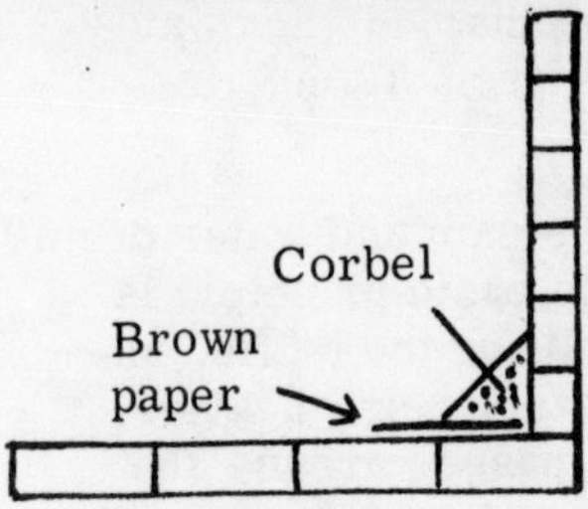

Figure 40.

Q. Describe the brickwork and insulation of the furnace of a typical watertube boiler.

A. The insulation of a typical watertube boiler furnace consists of a layer of uncalcined diatomaceous earth blocks 1" thick next to the steel casing. Over that is a layer of calcined diatomaceous earth blocks 2-1/2" thick, and next to the fire is a wall of refractory brick 4-1/2" thick. Sometimes this brick is as much as 9" thick. The brick walls are laid up in the conventional staggered form, and have a thin layer of mortar between them. A small space is left at one end of each wall to allow for expansion. Anchor bolts are used to secure the brick walls against the rolling of the ship. One end of an anchor bolt fits into a slot between the bricks, and the other end extends through the insulating material and the steel casing and are secured by nuts, metal washers and asbestos grommets. The floor of the furnace has insulation similar to the walls except that there are no anchor bolts and usually no mortar between the bricks. A space is left all around the floor to allow for expansion of the bricks.

Q. What is plastic refractory and what is it used for?

A. It is similar in composition to refractory brick except that it is furnished in a workable or plastic form, containing enough moisture to make it soft.

It is used wherever it is necessary to get refractory into corners or irregular shapes, such as on the tops of walls, around burner openings, for patching eroded portions of walls, and around studs and tubes of stud-tube waterwalls.

Q. Why are furnaces lined with refractory?

A. To prevent metal furnace walls and floor from overheating. Help maintain high furnace temperature which aids in rate of combustion and, therefore, efficient combustion. Protects drums and headers from radiant heat of fire and flame impingement.

Q. What are the purposes of the refractory lining in the furnace of a boiler?

A. Used to line the furnace with a substance that will withstand heat and thereby confine heat to the furnace. This, in turn, aids combustion. It also shields metal boiler parts from flame impingement.

Q. What is the purpose of the steam baffle in the steam and water drum?

A. It helps to cause steam-water separation. The basic principle is that of inertia. The steam-water mixture, on hitting the baffle, undergoes a sudden reversal of direction. The water, being heavier, continues in a straight line, whereas the steam passes around the baffle quickly.

Q. Trace the path of steam from the generating tubes to the main and auxiliary steam lines on a sectional header boiler fitted with an internal superheater and desuperheater.

A. From the generating tubes to the rear headers, to the horizontal return tubes to the steam and water drum, dry pipe, through the superheater tubes to the main steam valve. Usually the desuperheated steam is used for some auxiliaries and this is removed through a separately controlled valve and run through the desuperheater which consists of one or two pipe coils placed in the bottom of the steam and water drum.

Q. Explain how baffles and flame plates are installed in a 3-pass sectional header, low superheat, watertube boiler to give the path of the gases of combustion their necessary course.

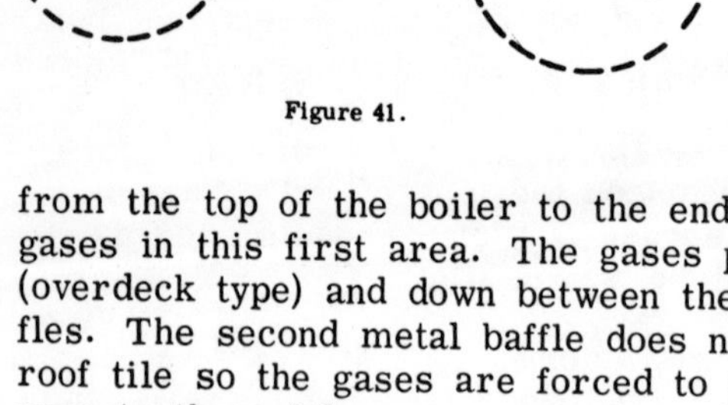

Figure 41.

A. The first baffle would be the roof tile, (See Fig. 27). These tiles are placed over the bottom row of tubes about two thirds of the way back in the furnace. (Roof tile shown in Fig. **41.**) The gases of combustion are forced to the back of the furnace where they make their first upward pass. A steel plate coming down from the top of the boiler to the end of the roof tile keeps the gases in this first area. The gases pass through the superheater (overdeck type) and down between the first and second metal baffles. The second metal baffle does not go all the way down to the roof tile so the gases are forced to pass upward in their third pass to the uptake.

Q. Describe a smoke indicator.

A. This apparatus usually consists of a series of mirrors so located that the fireman can easily see the condition of the stack, a strong lamp being placed on one side of the uptake and the reflecting unit located exactly opposite.

Q. What are the requirements of a steam-stop valve? Why?

A. 1. It must be closed with a right-hand motion. If some valves are closed with a right-hand motion and some with a left-hand motion, there would be confusion and a possibility that someone would twist off the valve stem trying to turn the valve the wrong way.

2. It must seat against the pressure. If the opposite were true, it might be impossible to open the valve against the pressure on the top of the disc.

3. It must indicate at a glance whether open or closed. This is important because it is necessary to tell by inspection if the valve is open or closed, otherwise damage may result due to someone's trying to close an already closed valve or to open one which is already open.

4. Steam-stop valves over 6" in diameter must be fitted with by-pass lines.

Q. What is a steam non-return valve? Where is it placed?

A. In some cases, non-return valves are placed between the boiler and the steam-stop valve. In one type of non-return valve, the disc of the valve is connected to a floating piston. The valve will remain open so long as the boiler pressure under the disc exceeds the sum of the discharge pressure and the weight of the floating piston. The function of the floating piston is to eliminate the chattering characteristic of an ordinary check valve when there is intermittent flow. In addition to this, the valve can be closed or opened by hand by means of the usual handwheel and screw. These valves are used to prevent steam from other boilers from entering a boiler in case of tube or other failure and to cut the boiler off the line when the fires are cut out.

Q. What is a water column?

A. A short pipe or casting of comparatively large diameter, attached to the boiler shell or head with brackets. A smaller pipe is led from the top of the water column to the upper part of the steam space of the boiler, where it is connected to a small stop valve. A similar pipe is led from the bottom of the water column to a stop valve well down in the water space of the boiler.

Q. Describe a seatless-bottom blow valve.

A. It consists of a hollow piston, which has ports in it, attached to a valve stem. This piston is inside of a cylinder which also has ports in it. When the ports in the piston coincide with the ports in the cylinder, there is a connection between the boiler, through the valve and the blow line. When the ports do not coincide, there is no connection between the boiler and the blow line and the valve is closed. There are several packing rings which are compressed by a gland which prevent leakage while the valve is in a closed position.

Q. Describe "extended surface" on economizer tubes.
A. The "extended surface" is made up of cast iron or aluminum gilled rings which are shrunk onto the tube throughout its length.

Q. What are the benefits of "extended surface?"
A. Extended surface tubes have from 6 to 9 times the heating surface of bare tubes. Extended surface also protects the tube from external corrosion.

Q. Describe a waterwall as installed in a boiler. What purposes are served by the waterwall?
A. A waterwall consists of a series of vertical or inclined water tubes installed along one or more walls of the combustion chamber and exposed to direct radiant heat of the fire. These tubes are connected directly, or through headers and connecting nipples, to the circulatory system of the boiler.

Two purposes are served by the waterwall: (1) Added heating surface and boiler capacity are secured. Revamping and adding waterwalls to an existing boiler may even double its capacity. (2) Maintenance of the refractory wall is reduced greatly. This protection of refractory walls is extremely important when high combustion rates are desired. The practical combustion rate is about 18,000 to 25,000 B.t.u. per cu. ft. furnace volume per hr. when bare refractory walls are exposed to the furnace. But combustion rates of 40,000 to 200,000 B.t.u. per cu. hr. have been attained without difficulty with all refractory surfaces protected by waterwalls.

The circulation of water is upward through the wall tubes exposed to radiant heat. These tubes are supplied by water through headers connected by piping from one or more of the boiler drums and located outside of the boiler setting. The supply system is known as the "downcomer."

Blowdown valves are required for each header at the bottom of a series of waterwall tubes, for the same reasons that the boiler itself needs them. Sediment accumulation in a header supplying wall tubes might cause interruption of circulation with consequent overheating and failure of the tubes.

Q. What determines the location of the superheater in a watertube boiler, and how is it protected against excessive heat?
A. The exact location of the superheater depends upon the degree of superheat desired. The higher the degree of superheat for which it is designed, the closer to the furnace will be its location. The superheater is usually protected against excessive heat by screening rows of generating tubes placed between it and the furnace.

Q. Is the volumetric capacity of each gas-pass the same in a watertube boiler?
A. No; it decreases in each succeeding pass. The gases contract as they cool, and in order to maintain the high gas velocity necessary to

sweep off stagnant gas films to effect good gas transfer, the cross-sectional area of the passes must decrease as the gases require less space.

Q. What is a dry pipe and where is it located?

A. A dry pipe is a perforated or slotted pipe placed in the highest part of the steam space of a boiler to prevent priming. The steam outlet valves are connected to the dry pipe. The steam must pass down through the holes (or slots) into the dry pipe on its way to the steam valves. On making these turns, the moisture in the steam is thrown off by inertia.

Q. Show diagramatically the arrangement of an attemperator used to control steam temperature.

A. Figure 42 shows two methods.

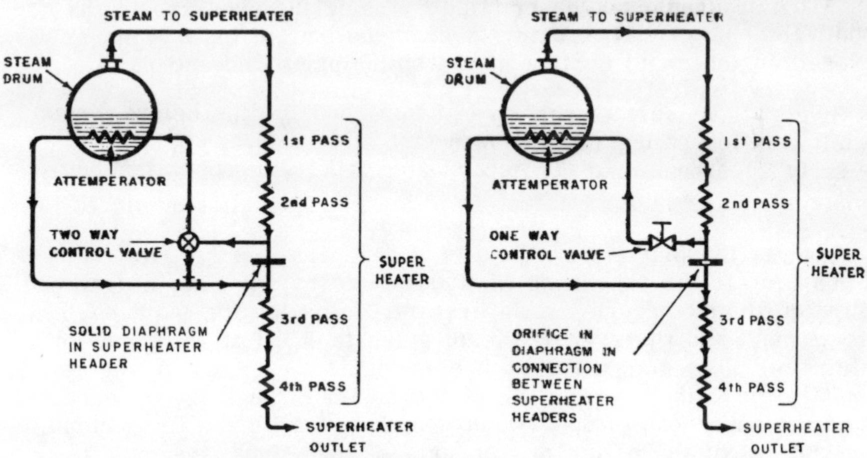

Fig. 42. Diagrammatic arrangements of attemperators and valves.

WATERTUBE —OPERATION

Q. What precautions must be taken in blowing down a modern boiler?

A. A modern watertube boiler should never be blown down while the boiler is steaming at high rates. In fact, it is recommended that the boiler not be blown down to remove sediment unless the fires are cut out. The reason for this is that, because of the small tubes and parts of a modern boiler, it is quite possible that blowing while under pressure will result in the emptying of some part of the boiler for a few seconds and, if the boiler is steaming, overheating of the exposed surfaces might result.

Waterwall headers should never be blown down until the boiler is cut out and the furnace thoroughly cooled. If the waterwalls are blown while the boiler is steaming, damage is almost certain to result.

A boiler should not be emptied by using the pressure in the boiler to blow out the water through the bottom blow. This may result in too rapid cooling of the boiler parts and setting.

Q. Give reasons for temperature of superheat falling below normal while load on boiler remains constant.

A. Dirty superheater; leaky superheater tubes; improper combustion.

Q. What is the primary purpose of a desuperheater?

A. To protect the superheater. It does this by assuring a continual flow of steam through the superheater, although the main engine is stopped and there is no steam going to it. If it were not for this flow of steam, the sudden stopping of the engine might result in the complete stopping of the steam through the superheater, and the superheater tubes would overheat before the fires could be cut out of the boiler or, in spite of this, from the heat radiated from the glowing brickwork in the furnace.

Q. Explain the effect of an increase in the excess air on the superheater outlet temperature, the steam pressure, and the stack temperature of an integral superheater boiler while operating at a constant rate of combustion.

A. At a constant rate of combustion an increase in the amount of excess air will result in a decrease in the furnace temperature. This decrease in the furnace temperature will reduce the quantity of steam formed in the generating tubes. The reduction of the steam so formed will cause a pressure drop in the boiler. The reduced flow of steam through the superheater will result in a higher temperature of the outlet steam. The reduction in the rate of heat transfer, due to the decreased furnace temperature, results in a higher stack temperature.

Q. To what temperature is the air heated in an air heater?
A. In marine boilers, the temperature of the air leaving the air heater is about 300°F.

Q. What danger is there in allowing soot deposits on air heaters?
A. Soot deposits on air heaters are a serious fire hazard. Such fires have entirely destroyed air heaters and all the boiler tubes directly below the unit.

Q. What is secondary combustion; how do you detect it; what harm can it do; how is it dealt with?
A. Secondary combustion is ignition of gases after they have passed the combustion chamber. It can cause excess heat in the stack, thus increasing natural draft and drawing out still more gases before they are ignited. It can do serious damage to air heaters and other devices intended to make use of stack gas heat, since these devices are not intended to stand the temperature of actual flames. It is detected by rapid increase in stack temperature, and by shortage of air due to improper operation of air preheaters. It is dealt with by operating the soot blowers, thus extinguishing the flames with steam, and by correcting the adjustments of the oil burners so that combustion in the furnace is more complete.

Q. How often should tubes be blown?
A. Tubes should be blown at least once in every 24 hours. When the boilers are carrying a heavy load or combustion is poor, the tubes should be blown more often. It is good practice to blow tubes whenever the stack temperature increases 25°F. above normal.

Q. What is done when a fire occurs in an air heater?
A. 1. Secure burners and air.
2. Operate soot blowers to smother fire.
3. Use steam lances or CO_2 extinguishers (never air) through uptake doors, if possible.
4. Do not operate boiler until soot is cleaned from air heaters.

Q. (a) Explain why boilers require extra feed. (b) What is indicated by a constantly rising water level in a manually fed boiler?
A. (a) Extra feed is required to make up the losses of steam or condensate throughout the vessel. These losses may come from leaking glands, joints, steam relief or safety valves, and boiler tubes; also by overflowing of open heaters and the use of the steam whistle.
(b) A constantly rising water level would indicate a leaking condenser tube or plate, a leaking or stuck open extra feed valve, a jammed or excessively open check valve, or excessive pressure or speed of the feed pumps.

Q. Describe or sketch a strainer assembly used in the suction and discharge lines of a fuel oil system.

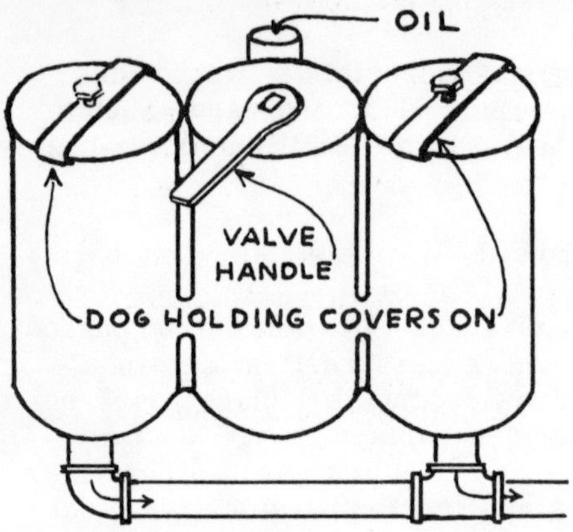

Figure 43.

A. These strainers are of the duplex type. The whole apparatus is one casting, as shown in Fig. **43.** The valve in the center is a three-way cock arranged so that oil can be directed to either one of the strainers by shifting the handle on the valve. This allows the oil to flow through one strainer while the other is available to open for cleaning.

Q. Give the steps taken in blowing tubes.

A. 1. Increase the air pressure at the burners above normal.

2. Open soot blower manifold drain.

3. Open steam valve to blower manifold.

4. When manifold is hot and dry, close drain valve.

5. Operate blowers using the one nearest the furnace first and proceed following the path of gases through the boiler.

6. Rotate the blower slowly, making at least three revolutions with each.

7. When finished blowing tubes, close the manifold stop valve and open the drain valve.

Q. What precautions should be taken when using soot blowers?

A. 1. Never use wet steam.

2. Never use soot blowers on cold boiler.

3. Never stop rotating blower while steam is flowing through the unit.

Q. What are the most important points in connection with the operation of soot blowers?

A. See that the steam line to the soot blower is thoroughly drained of moisture. If wet steam is used in the soot blower, it will cause soot to cake on the tubes. Water admitted to hot soot-blower elements may fracture them.

Blow steam line down well before operating.

Have a regular schedule for operating soot blowers. Regular, frequent cleaning is necessary, as soot acts as an insulator and cuts down the transfer of heat from the gases to the steam or water and, therefore, reduces efficiency.

Regular inspections should be made to see that there is no impingement of the steam from the soot-blower nozzles on any tube or header.

Q. What would be the probable cause of a constantly decreasing alkalinity and chlorinity of the boiler water in a modern watertube boiler fitted with a superheater and a desuperheater, and operating under normal conditions of make-up feed?

A. A constantly decreasing alkalinity and chlorinity of the boiler water indicates a leak in the waterside of the boiler. If no excess extra feed was being used, the leak would probably be caused by a poorly fitted joint in the desuperheater line allowing the water from the drum to enter the desuperheater.

Q. What are the causes of slag accumulations in the furnaces of oil burning watertube boilers?

A. The mineral residues of most fuels are ordinarily incombustible, but at sufficiently high temperatures they combine with firebrick materials, causing fusion below the normal melting point of the brickwork. Slag formed by this combination is sometimes sufficiently fluid to flow back and forth on the furnace floor and even to be drawn into the pores of the brick.

Q. In preparing a watertube boiler for annual inspection, what should the temperature of the water be for the hydrostatic test?

A. Not less than 70°F. and not more than 160°F.

Q. Explain why it is poor practice to use dissimilar sizes of sprayer plates in a boiler furnace.

A. Poor combustion effects. Some of tips will collect carbon deposits due to lack of oil pressure.

Q. Why would the temperature of the superheated steam fall below normal while the load on the boiler remains constant?

A. Dirty or leaky superheater.

Q. What operating precautions should be taken in connection with economizers?

A. Watch for corrosion (due to sulphur in fuel). Watch out for fires caused by soot accumulations.

Q. How does superheater temperature vary with load conditions?

A. With radiant type superheater, the steam temperature decreases with an increase in boiler load since furnace temperature varies only a slight amount under all conditions and the flow of steam through the superheater varies with the load. With the convection type superheater, the opposite occurs, i.e., the steam temperature increases with an increase in boiler load.

Q. After establishing proper alkalinity, why is it necessary to continue adding compound?

A. To maintain the alkalinity due to extra feed water being added continually.

Q. Explain in detail how to set the safety valves on a boiler so as to operate at the allowable pressure.

A. 1. Check boiler steam gage with test gage or gage-testing machine.

2. Gag all valves on the boiler except the one to be set.

3. Disconnect hand-lifting gear, remove key and cap from top of valve spindle.

4. Turn compression screw right-handed (down) to increase spring pressure and consequently to increase the lifting pressure of the valve.

5. Handle fires to increase steam pressure in boiler to lifting pressure. Note pressure on steam gage at which safety valve lifts and also pressure at which it reseats.

6. Readjust compression screw up or down and test again with steam until correct lifting pressure is obtained.

7. Set up compression screw nut.

8. The difference between lifting and reseating pressures (blowdown) should be about 3 percent of set pressure. If the difference in pressures, noted on the steam gage, was greater than this, back out the adjusting ring lockpin, insert a small screwdriver or other instrument through the hole and turn the adjusting ring until it has moved down a small amount. Replace lockpin.

9. Test again with steam and readjust the adjusting ring until, by trial and error, the proper blowdown has been obtained.

10. Gag this valve and go ahead with the same procedure on the other valves of the same boiler until all have been adjusted.

11. Replace caps and keys. Connect up hand-lifting gear and test hand-lifting gear to see that all valves lift by hand. When shaft is rotated the system of levers lifts up on the cap, which is keyed to the valve spindle. As the valve spindle is also keyed to the valve, the valve is lifted from its seat. The hole through the end of the cap key is for the U.S. seal to prevent unauthorized persons from tampering with the safety valve setting.

Q. How is air eliminated from feed water?

A. By running the water through a properly operating deaerator. One of the largest sources of oxygen comes during the plant starting-up procedure. It is during this time that the deaerating tank is not properly warmed up and, therefore, cannot remove the excess air that has been accumulated in the tank itself as well as from the incoming feed. It is essential that the deaerating tank be up to operating temperature as soon as possible.

Q. Why is it desirable to exclude air from boiler feed water? Why is this more important in high pressure boilers than in low pressure boilers?

A. The oxygen in the air attacks the boiler steel. At higher pressure and temperatures it becomes harder to eliminate the oxygen from air. Sodium sulfite is used to absorb oxygen in plants from 1000 to 1800 psi. Over 1800 psi., hydrazine is used.

Q. What is the most positive method of judging furnace efficiency?
A. By proper use of the stack gas analysis apparatus. The two most popular types aboard ship are the Ranarex, which measures CO_2 only, and the Orsat, which measures the amounts of CO_2, O_2, and CO.

Q. What may a low CO_2 reading indicate?
A. A low CO_2 reading usually indicates excess air, since the gases are diluted with air. It will be accompanied by a high O_2 reading.

A low CO_2 reading may also indicate insufficient air. This will be accompanied by smoke and a high CO reading.

Q. Does the highest possible CO_2 content indicate ideal combustion conditions? Explain your answer.
A. No. Too high CO_2 usually means high CO or waste, plus the cooling effect of too much excess air. Best efficiency usually from 12-1/2 to 14% CO_2.

Q. What amounts of the different gases must be present in the stack for excellent combustion?
A. For excellent combustion, the stack gases should contain: 14% CO_2, 0% CO, 4% to 6% O_2—the remainder of the gases consisting mostly of nitrogen.

Q. Why should air registers not be throttled?
A. They are designed to supply sufficient air to the burners when operating, therefore throttling them will effect combustion.

Q. (a) Explain how the water level in a watertube boiler is affected when the load is taken off the boiler, the fires cut out and relighted as when maneuvering, the feedwater being under the control of a single element automatic feedwater regulator.

(b) How is this condition overcome in actual operation aboard ship?

A. (a) Single element feedwater regulators are designed to maintain a constant water level in the boilers at all times. When the load is taken off the boiler and the fires extinguished, the steam bubbles in the water collapse and cause a lowering of the water level. The regulator immediately functions and raises the water to its former level. When the fires are lit again, the level will rise rapidly, possibly causing priming and damage to the machinery.

(b) When maneuvering or operating under varying loads, the hand-controlled feed-check valve should be used and the level maintained slightly below the normal steaming level unless a surge chamber or other dampening device is used in conjunction with the single element feedwater regulator to reduce the effect of sudden changes in the water level.

Q. What is the usual operating oil pressure from the fuel oil service pump to the wide-range variable capacity oil burner? How are small variations in the firing rate adjusted?

A. 300 psi. The amount of oil admitted to the furnace is controlled by the amount of back-pressure in the oil return line (line connecting with the annular passage in the burner barrel. The greater the pressure in this line, the greater will be the amount of oil admitted to the furnace and less will go to the return line, Fig. **44**.

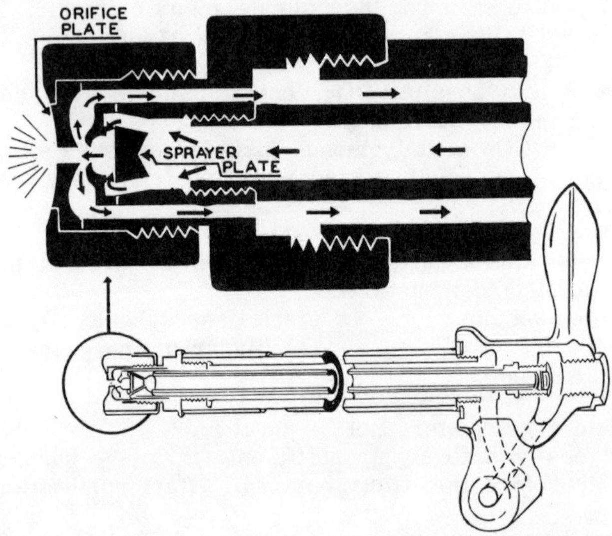

Fig. 44. Variable capacity atomizer and detail sketch of sprayer plate.

Q. Explain the precautions that should be observed in connection with blowing down the waterwall header of a boiler?

A. The fires should be secured and the boiler taken off the line. Waterwall headers should never be given a bottom blow while the boiler is steaming because the normal circulation of water in the tubes might be stopped, endangering the tubes. These valves should be labeled, warning against opening, and some method used to lock the valves closed and prevent their being inadvertently or accidentally opened.

Q. What is the proper care for economizers while raising steam in a boiler?

A. While raising steam in a boiler or operating at very low loads it is desirable to maintain some flow of water through the economizers. This was more critical with the older ships than today in order to minimize the temperature swings that lack of circulation would cause with the resultant possibility of leaks. It is still a good policy to avoid temperature swings and for this reason the recommendation to maintain some circulation through the economizer during any firing

period is sound. This means, perhaps, some blow-down during raising steam pressure if the water level raises too much and making certain that any chemical feed line that by-passes the economizer is secured. The economizer is an integral part of the boiler and requires just as much consideration as the rest of the boiler.

Q. Describe and give the purpose of the diffuser on the mechanical atomizing oil burner.

A. The diffuser (impeller) is a slotted cast metal cone mounted at the end of the jacket tube. It prevents the oil spray and flame from being blown away from the atomizer tip and to cause the first mixing of the fuel oil particles with the incoming air, Fig. **45**.

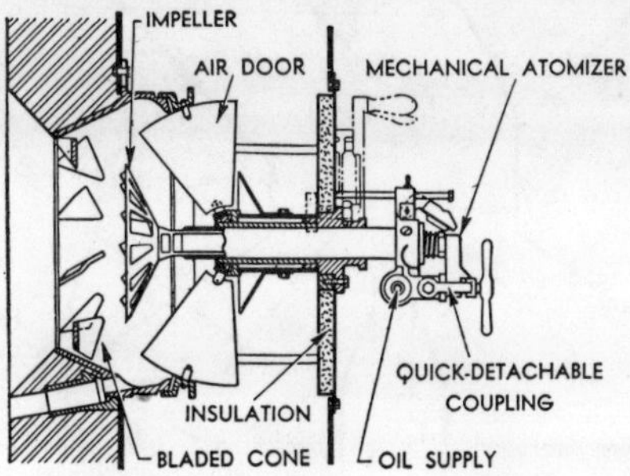

Fig. 45. Diffuser on the B & W mechanical atomizing oil burner.

Q. What is an air casing alarm?

A. It is a fire alarm unit located in the windbox or uptake to warn of a fire. It consists basically of a sensing element whose impedance decreases with rising temperature thus permitting more current to flow in the alarm system thus setting off the alarm.

Q. List the common causes of high final steam temperature.

A. Assuming a two-drum boiler with automatic combustion control, convection superheater, and superheat control valve which is either open or shut, the following represent possible causes of high final steam temperature:

1. **Excess load on the boiler.**
2. **Superheat control valve closed.**
3. **Low feedwater temperature.**
4. **Too much excess air.**
5. **Low combustion-air temperature.**
6. **Superheater tubes clean, generating tubes dirty.**

The basic rule is that anything which causes the firing rate to be greater than normally necessary for the existing boiler load will cause the final steam temperature to be higher than normal.

Q. Explain briefly the general operating principle of a Tjungstrom regenerative air heater.

A. Figure 46. Lightweight synchronized air hoods are rotated above and below stationary heating surfaces. Because the heavy heat transfer elements are kept stationary, warpage, vibration and wear are substantially reduced and added protection is offered from the effects of hull motion.

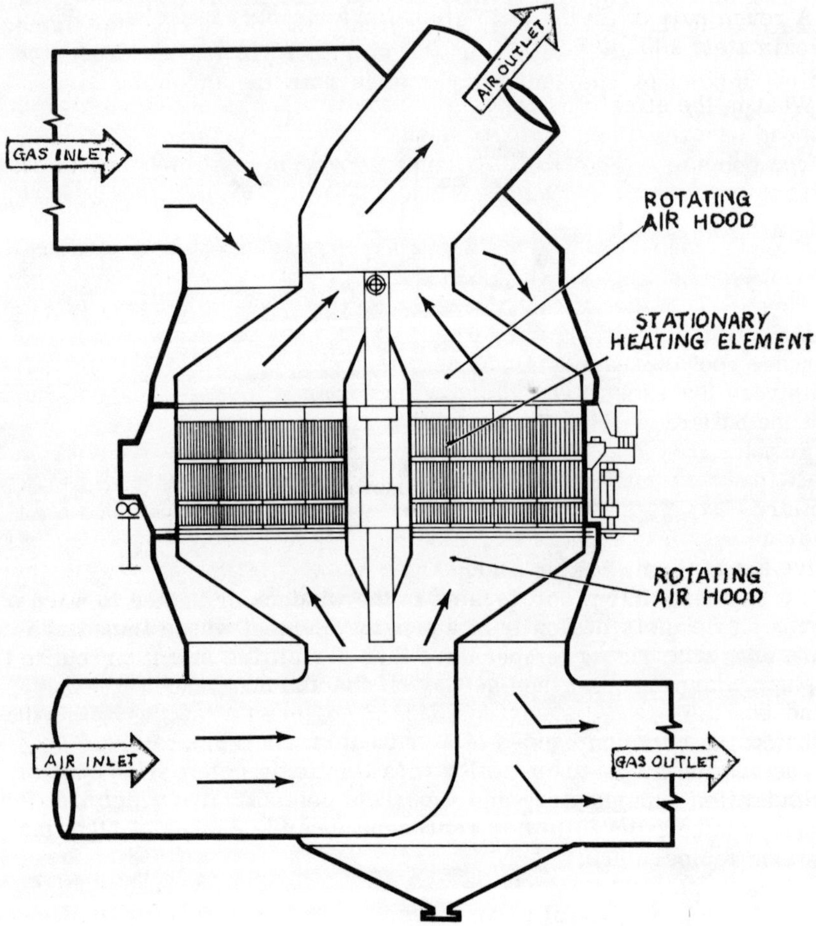

Figure 46.

Q. What is a boiler cooling curve? What is an approximate time relation when starting up or cooling down a boiler in reference to temperature?

A. Modern boilers with their higher pressures must necessarily be made of thicker materials. The thicker drums and headers heat and cool more slowly than the thinner tubes. Expanded joints become overstressed and can lead to leaking and failure if heated or cooled too rapidly. Every boiler maintenance manual will include a "steaming-up" and "cooling-off" chart for that particular boiler.

A rough rule of thumb is to allow for a temperature change of approximately 100° per hour.

Q. What is the effect on "popping pressure" of a safety valve that is popped several times in succession?

A. The popping pressure will become progressively lower due to the expansion of valve parts due to heat.

Q. Explain briefly the general operating principle of a Foster Wheeler reheat type boiler.

A. Figure 47. A single row of widely spaced burners assures proper distribution of air for combustion. Location of the burners on the furnace roof increases residence time for more complete combustion with very low excess air and assures more uniform gas flow throughout the boiler.

Furnace gases pass through the boiler bank and down the division wall, over the lower drum, into the cavity below the superheater and upward through the superheater and reheater. The gases then pass over an economizer which limits the gas temperature to the regenerative air heater. During maneuvering or in-port operation when there is no steam in the reheater, the gases are rerouted by automatic damper arrangement through a by-pass zone which is cooled by a bare tube economizer. Simultaneously, cooling dampers operating on the same shaft as the shut-off dampers direct air from the burner wind-box through the reheater. This precludes thermal shock to the reheater when steam flow is later resumed.

The air control dampers are automatically operated by pneumatic cylinders.

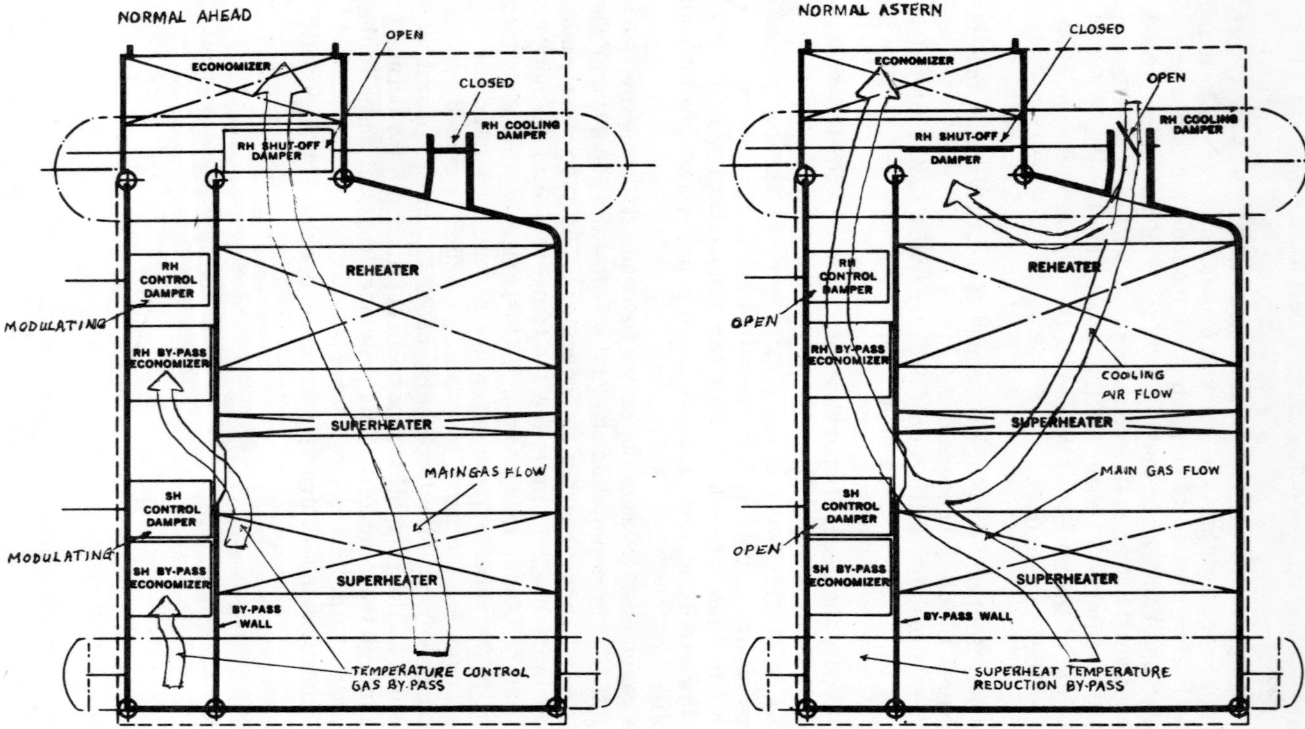

WITH STEAM flow in reheater, the shut-off damper is open and the air cooling damper is closed. The superheat and reheat control dampers modulate to maintain the desired temperature.

WITH NO STEAM flow in the reheater, the reheat and superheat control dampers open and the shut-off and air cooling dampers rotate 90 degrees to stop gas flow and introduce cooling air.

Figure 47.

Q. What are the causes and means of preventing corrosion on the water side of boilers?

A. Corrosion on the water side of boilers may occur in the following ways: by acid attack, by galvanic action, and by oxidation. Combinations of these processes also are possible. The best protection against acid attack and galvanic action is to insure that the boiler water is always maintained between its prescribed limits of alkalinity, usually between pH 10 to 12, taking care not to exceed the upper limit. Oxidation is best controlled by preventing the introduction of air to the feedwater and removing the dissolved gases by heating and the use of deaerators. Every precaution should be taken to keep the feedwater as pure as possible. The boiler water should be regularly tested and treated chemically to neutralize acid-forming salts and remove traces of oxygen that may escape the deaerator.

Q. Which is the most serious of the common scale-forming salts normally present in feedwater? What chemicals are used to prevent the formation of this scale?

A. Calcium sulphate is usually the most serious of the scale-forming salts. Scale is prevented from forming by adding to the water one or more highly soluble chemicals, such as sodium carbonate or the sodium phosphates, which will react with the scale-forming calcium sulphate and form a harmless sludge which may be blown out of the boiler.

Q. What care should be taken when cleaning or renewing a gage glass?

A. Be careful not to scratch glass. Always install new gaskets. Use the thick piece of mica near the water side and the thin piece on the glass. Tighten nuts from center outwards. Tighten up on nuts after glass and casing have warmed up.

Q. What precautions are taken in water-washing firesides of boilers?

A. No water-washing should ever be started unless it is possible to properly dry out the unit following the washing. Start at the top and work down. Wash thoroughly so as to minimize external corrosion. Use canvas or a suitable covering to protect the furnace brickwork as much as possible. Dry out thoroughly by intermittent light firing for 8 to 12 hours (never water-wash a hot boiler).

Q. Is "Hydrazine" ever used in marine boilers to eliminate oxygen from the feedwater?

A. Yes. It reduces the iron oxide. It does not add to the total solids like sulphite does.

Q. After working on the water side of a boiler, what would you inspect before closing up the boiler?

A. Examine it very carefully, making sure that there are no rags, tools, lights, spare gaskets, etc., inside the boiler, and that all inner surfaces are clean. See that all workers are out. Make sure that all the manhole and handhole seats are clean and in good condition. Provide manhole and handhole plates with new gaskets to insure tight joints.

Q. How would you fit a manhole plate gasket?

A. 1. Clean the manhole plate and boiler seat thoroughly.

2. Smear a new gasket on one side with white lead and place this side on the manhole plate. (This burns the gasket to the plate, allowing easier removal.)

3. Smear the exposed side of the gasket with a graphite and oil mixture. (This side will break away easily from the boiler when removing the plate.)

4. Place the plate in the boiler and lift it into position.

5. Secure dogs to tighten nuts, being sure to allow an even space between the edge of the manhole and the shoulder on the manhole plate (in other words, center the plate).

6. When steam has been raised on the boiler, set up again on nuts.

7. Always use the manhole wrench provided for this job, and do not use a pipe or any other form of persuader.

Q. How would you make a hydrostatic test on a boiler?

A. 1. Clean the boiler thoroughly both on the fire sides and the water sides.

2. See that all workers are clear.

3. See that all foreign matter, tools, rags, etc., are removed from the boiler, and close up the manholes and handholes.

4. All valves on the boiler must be in good condition. All valves except those through which the water for the test is to be let into the boiler should be closed.

5. The main steam line may be tested along with the boiler. In this case, it may be necessary to put a blank at the far end of the line immediately before the throttle valve to the main engine or auxiliary engines. If there are two stop valves in the line, it may be unnecessary to blank off the line.

6. Hydrostatic pressure should not be put on one side of a valve which has steam pressure on the other side. This is important when steam is up on one boiler or on a donkey boiler while another boiler is being given a hydrostatic test. Blanks have to be used if there are not two valves on the line. Two valves may be considered sufficient.

7. Provision must be made for a test gage.

8. The safety valves must be gagged.

9. The air vent (air cock) at the top of the boiler must be left open until water runs out of it. Then it may be closed.

10. The pressure, applied by means of a pump, must be applied and released slowly.

11. A hydrostatic test must never be applied to a hot superheater or boiler.

Q. What are the general steps in rebricking a boiler?

A. 1. The fire side of the boiler should be thoroughly cleaned of soot and dried of all moisture, in order to prevent corrosion of the metal while the boiler is cut out.

2. Old brickwork and insulation must be removed.

3. The inside of the steel furnace casing must be cleaned and painted.

4. Set the insulating material blocks in place.

5. Mix fire-clay with water to make a thin mortar into which the bricks are dipped, making only a thin layer of the mortar between the bricks when they are set in place.

6. In laying up the walls, a space is left at one end of each wall. This space is covered by the end of the adjoining wall.

7. Burner openings should be carefully located according to the specifications of the manufacturer of that particular boiler and burner, so that the oil spray will not hit upon the brickwork or the lower rows of tubes.

8. Burner openings may be made with specially-shaped bricks or with plastic refractory.

9. If plastic refractory is used, it is customary to punch it with nail holes about 1" deep and about 1" apart, to facilitate the drying of the refractory.

10. Brickwork should be dried out with a moderate fire such as a wood fire or an electric heater. Regular brickwork dries out in about 24 hours, plastic refractory takes longer, possibly from 72 to 100 hours.

Q. How should an atomizer be set in a burner? What bad effects result from improper setting?

A. The atomizer should be set so that the flame will barely clear the furnace opening.

If the atomizer is set back too far, the oil will strike the furnace opening and carbon will be formed.

It the atomizer is set in too far, air will enter the furnace without properly mixing with the oil, and excess air will be used. In extreme cases, carbon will be formed on the atomizer tip due to eddy currents throwing oil back on the tip.

Q. How are sprayer plates stowed?

A. They should be stowed in a wooden box, each tip in a separate compartment, in a bath of non-acid lubricating oil.

Q. What precautions should be taken when mechanically cleaning (turbining) the tube interiors of a watertube boiler?

A. If the rotor heads will not keep moving ahead in the tube, stop it or pull it out immediately. If allowed to remain revolving at the obstructed point, it will start to chew away the tube.

Q. How should a boiler be laid up for a short time?

A. Fill it with water until the water runs out of the air cock; cut off the water and secure all checks and stops. While the boiler is still hot, all tubes should be blown, and after the boiler has cooled, all the fire sides should be cleaned with wire brushes and scrapers. The air registers should be closed and the uptake dampers, if any, should be closed.

Q. Why must the fire sides of a boiler be cleaned?

A. They must be cleaned frequently to prevent corrosion from sulphurous acid and to improve the efficiency of the boiler.

Q. Give the necessary steps in assembling the various parts of a high-pressure boiler gage glass.

A. The gasket goes between the mica and the water chamber. The mica goes between the glass and the gasket.

Q. What is the danger if a burner flexible connection bursts? How would you handle such a casualty? How would you prevent such an accident?

A. The escaping hot oil will burn any crew members in the vicinity of the broken line. If the break is large, the excess loss of oil in the system will affect the automatic appliances. There is danger of an open fire if a naked flame is close to the break. Shut off the master oil valve immediately.

To prevent such accidents, inspect and check all oil lines and fittings in the oil service system at frequent intervals.

Q. How would you wash and clean a boiler? What examinations should be made while cleaning?

A. Clean water side of tubes and furnaces with scaling tools and wire brushes. Wash down with hose and clean out. Check all stays, hangers, braces, etc. Look for scale or grease deposits. Look for pitting and grooving. Check zinc plates. Punch tubes. Clean combustion chambers and furnaces (wire brush). Clean air-heater tubes. Check for leaky stay bolts and blisters or bags in the furnaces.

BOILERS—GENERAL

Q. What percentage of excess air is considered excellent, good, and poor?

A. Excellent: 20% excess air; good: 50% excess air; poor: 200% excess air.

Q. How are sprayer plates cleaned?

A. When cleaning sprayer plates, steel knives, drills or any metal which might nick or scratch the oil passages, should not be used. Use copper knives or wire, or wooden sticks.

Q. Name three ways to express alkalinity.
A. PH value; PPM (parts per million); "A" and "B" alkalinity.

Q. Flat surfaces of a boiler must be braced or stayed. Why doesn't shell need bracing?
A. In a cylindrical shell, the pressure is equal in all directions.

Q. What are the benefits of an economizer?
A. 1. It increases boiler efficiency, since the temperature of the stack gases is reduced and the temperature of the feedwater entering the boiler is increased.

2. It increases boiler output, since the higher-temperature feedwater requires less heat to generate steam.

3. It will result in less stress on boiler due to a hotter feedwater entering boiler.

Q. What are the average temperatures of boiler furnaces?
A. 2500°—3000° F.

Q. Define the following: Tensile strength; yield point; elastic limit; elongation; reduction in area.
A. Tensile strength is the ability of material to resist stresses of tension, such as stretching or pulling it apart.

Yield point is a term used to denote the sudden yielding of the material while under tensile stress, without any increase in load.

Elastic limit is the maximum stress to which the material can be subjected without causing its permanent deformation.

Elongation is the increase in length of a specimen under tensile stress and is expressed as a percentage of the original length.

Reduction in area is the amount the original cross-sectional area of a test specimen is reduced at the point of rupture. Reduction in area is expressed as a percentage of the original area.

Q. What are the benefits of an air heater?
A. 1. It increases boiler efficiency, since the stack temperatures are reduced.

2. It increases furnace efficiency, since hot air increases furnace temperature, which promotes better combustion.

Q. What is likely to happen in any of the coil type of heat transfer units if the coil drain trap jams in an open position?
A. Steam will leave the coils before the latent heat is removed and will result in an increase in the amount of steam used.

Q. In preparing a watertube boiler for annual inspection, what should the temperature of the water be for the hydrostatic test?
A. Not less than 70° F. and not more than 160° F.

Q. What is an attemporator?
A. Another name for a desuperheater.

Q. What is meant by "quality of steam?"

A. It is a measure of the moisture content of the steam. It varies with the temperature and the pressure, and is determined by a calorimeter. It is a measure of its weight as compared with that of dry steam.

It is expressed as a percentage, that is, if saturated steam has a moisture content of 5%, as determined by a calorimeter, its quality or dryness factor would be 100 minus 5, or 95%.

Q. What is saturated steam?

A. Steam produced by boiling water in a closed vessel. It has the same temperature as the water. In other words, saturated steam is steam at the temperature of the boiling point of the water from which it was generated, and may be either wet or dry saturated steam, depending upon whether or not it contains moisture.

The term saturated steam refers merely to the fact that the steam is under the same conditions, temperature and pressure, as the water, and does not indicate whether it is wet or dry.

Q. What is wet steam?

A. Wet steam is saturated steam, but is usually considered as containing more entrained moisture by volume than is natural to its "saturation" pressure, volume and temperature. In this case, it may be called wet steam or wet saturated steam.

Q. What is dry steam?

A. Dry steam is a term frequently used to distinguish between wet saturated steam and so-called dry saturated steam, that is, steam which contains little or no moisture. Some boilers supply reasonably moisture-free or dry steam, due to the arrangement of their dry pipes or other component parts.

Q. What is superheated steam? How is it obtained?

A. Steam heated to a temperature above that due to its pressure. This is accomplished by passing saturated steam as it is generated in a boiler through a series of so-called superheated tubes or coils, which are, in most instances, heated by the boiler furnace.

The reason for the use of superheater tubes or coils is that saturated steam cannot be superheated as long as it remains in contact with the water from which it was generated, since to add more heat to the water merely increases its temperature and pressure and will in no way superheat the steam generated by its evaporation.

Q. What is the cause and effect of corrosion between headers of sectional header boilers?

A. If the asbestos rope used for caulking between the headers gets wet, corrosion sets in. (Moisture can be given off from leaky handholes.) The corroding asbestos will hold dampness and corrode the headers.

Q. Describe an oil burner system with which you are familiar and trace the flow of fuel oil through each valve and fitting, from the time it leaves the settling tank until it discharges at the burner.

A. In a typical system, the oil leaves the settling tank through a valve located on the bulkhead, inside or outside the tank, passes through the suction strainer to the suction valve on the fuel-oil pump, through the pump, out the discharge valves on the pump, through the oil meter, through the intake valve into the heater, out the heater discharge valve to the discharge strainers, to the master valve on the oil line, through the valve on each burner line, into the burner and atomizers on each burner, and to the furnace to be ignited and burned.

Q. How would you subdue an oil fire in the fireroom of an oil burning vessel?

A. Use sand and foam extinguishers immediately. Never use water. If there is any water in the bilges, pump it out immediately. Notify the engineer on watch.

Q. What is galvanic action and how is it prevented?

A. A form of corrosion in which the metal is eaten away by the galvanic (voltaic) action of non-homogeneous parts of the iron and steel. Zinc plates are fitted into boilers, the zinc being acted on more readily than the boiler shell, thus leaving the latter untouched. Iron plates are installed in condenser water boxes for the same reason.

Q. What is caustic embrittlement and where does it occur?

A. Metal is made up of molecules which are held together by a substance known as cementite. High alkaline solutions attack the cementite and intergranular cracks occur.

Caustic embrittlement occurs at points where a high concentration of alkalinity and mechanical stresses exist, such as at riveted joints.

Q. Name the principal parts of a steam plant where impurities may enter the feedwater and the impurities generally found in them.

A. Reciprocating engine: cylinder oil used for swabbing valve stems and piston rods enters the cylinder and is carried by the exhaust steam to the feed and filter tank.

Steam pumps: same as above.

Main and auxiliary condensers: if a leak occurs, salt water will enter the system.

Piping: Internal corrosion of piping will produce rust and scale which will be carried along by the water.

Q. What is meant by "boiler capacity?"

A. Refers to the normal quantity of steam in pounds per hour that the boiler is designed to generate. It is a convenient means of expressing the amount of work a boiler can do.

Q. What amount of oxygen in boiler water is considered dangerous?
A. Water containing more than .05 cc. of oxygen per liter is considered dangerous in high-pressure boilers.

Q. What are the three bad effects of impurities in feedwater?
A. 1. Scale formation.
2. Corrosion.
3. Carry-over.

Q. Give the steps in making the salinity or chloride test.
A. The following test is made, using standard Navy chemicals. Before making this test, the boiler water should first be neutralized by the alkalinity test.
1. Pour 50 cc. of boiler water into a dish.
2. Add 3 or 4 drops of potassium chromate, which will give the water a yellow color.
3. Fill silver nitrate burette to zero.
4. Add silver nitrate from burette, stirring continuously, until water turns a faint pink.
5. Read burette.
6. Burette reading X 2 = salinity in grains per gallon.

Example: Burette reading 5. 5 cc.
5. 5 X 2 = 11 grains per gallon.

Q. Explain corrosion by electrolysis.
A. A solution capable of conducting electrical currents is called an electrolyte. When an externally-applied electrical current flows between two poles in an electrolyte, ions leave the positive pole and move toward the negative pole. Corrosion then occurs at the positive, and the negative is protected. This decomposition is known as electrolysis, and the corrosion resulting directly from externally-applied electric currents is generally called electrolytic corrosion or stray-current corrosion.

Q. Explain corrosion by galvanic action.
A. Although a boiler material may be composed of the same amounts of identical metals, it may not be uniform, due to different heat treatments or mechanical stresses. Under these conditions, the part of higher potential (+ part) will tend to go into solution and plate out on the part of lower potential (— part) in the presence of an electrolyte.

Q. What are temporary and permanent hardness?
A. Temporary hardness is hardness which will disappear upon boiling the water. Permanent hardness is hardness which will not disappear upon boiling the water.

Q. How is boiler foaming prevented?
A. Dissolved solids are a prime factor in causing boiler foaming. Limiting the concentration of chloride and alkalinity reduces the probability of sufficient dissolves solids to cause foaming. Bot-

tom and surface blows, or sometimes a complete change of boiler water, are often necessary.

Q. What is carryover?
A. Carryover consists of particles of water leaving the boiler with the steam.

Q. What causes carryover?
A. Oil, high alkalinity, and high concentration of dissolved or suspended solids.

Q. What causes oil to adhere to the bottom tubes in a watertube boiler inasmuch as oil is lighter than water and will float?
A. Oil will float on water that is in a dormant condition. However, the water in the boiler follows a definite path of flow. The hot water, being less dense, rises and the cooler entering water drops, forming a definite moving condition at all times. If there is any oil in the water, it will adhere to the lower rows of tubes because they are the first tubes that the water will flow through. Also, these tubes are the hottest and oil has an affinity for hot surfaces.

Q. Give the steps taken in the alkalinity test.
A. The following test is made, using standard Navy chemicals:

1. Draw 50 cc. of sample water.
2. Pour sample into casserole or dish.
3. Fill acid burette to zero.
4. Add 2 or 3 drops of phenolphthalein indicator to sample, which will give it a deep pink color.
5. Add acid solution until pink color disappears.
6. Read burette.
7. Burette reading multiplied by .2 will give alkalinity in per cent of normal solution.

Q. How would you make a hardness test of feedwater?
A.
1. Place 50 cc. of cooled sample water in shaker bottle.
2. Add soap solution, shake vigorously and let stand.
3. Lather should last at least 5 minutes.
4. Amount of standard soap solution required (less 0.5 cc. for lather factor) to produce this permanent lather, is the hardness of the water in grains per gallon.

Q. What is the test for concentration of dissolved solids in boiler water?
A. Dissolved solids are measured by testing for sodium chloride (common salt), which makes up the greater part of the total dissolved solids.

Q. What are the limits of salinity in all types of boilers?
A.
1. Firetube: 100 grains per gallon.
2. Watertube (above 2" tubes): 50 grains per gallon.
3. Express (2" tubes and smaller): 30 grains per gallon.

Q. Give 5 causes for boiler explosions.
A. Low water; furnace rupture (Scotch boiler); tube rupture; plate rupture; seam rupture; furnace backfire large enough to rupture boiler; safety valve stuck, allowing pressure to rise above safe load.

Q. What are the two types of corrosion?
A 1. General: Corrosion which occurs over the entire surface of the metal.
2. Local: Corrosion (pitting) which occurs in a few small areas.

Q. What is the most common scale-forming salt?
A. Calcium sulphate.

Q. Sixteen lbs. of sea water with density 1/32 contain how many ounces of salt?
A. One-half oz. per pound, or 8 ozs. (32 lbs. of sea water with density of 1/32 would contain 1 lb. of salt).

Q. How would you make a salinometer?
A. 1. Take a long narrow tube (gage glass will do), close one end and weight it so that it floats upright.
2. Heat a bucket of fresh water to 200°F. Place the tube in the water and mark "O" at the water level.
3. Heat 1 gallon of sea water to 200°F., place tube in water and mark 1/32 at the water level.
4. Take 2 gallons of sea water, boil down to 1 gallon, heat to 200°F., place tube in water, mark 2/32 at water level.
5. Take 3 gallons of sea water, boil down to 1 gallon, heat to 200°F., place tube in water and mark 3/32 at the water level.

Q. What instrument is always used in connection with the salinometer?
A. Thermometer.

Q. What is a safety valve?
A. A device that will open automatically to relieve excess pressure in the boiler.

Q. How should a safety valve perform?
A. Open without simmering to full lift at definite set pressure; remain open until pressure has dropped definite amount (blowdown); close tight without chattering; remain tight when closed.

Q. Why is a safety valve installed on the outlet header of a superheater?
A. To protect the superheater from overheating, in case excess pressure in the boiler causes the drum safety valves to blow and the surge of steam from the drum out the escape pipe causes a momentary stop in the flow of steam from the drum through the superheater. The superheater safety valve is set at a lower pressure than the drum safety valves and, if the steam pressure in

the boiler becomes excessive, the superheater safety valve will blow first, insuring that there will be a flow of steam through the superheater.

Q. Why should a superheater be drainable?
A. To prevent water which may be trapped in it during idle periods from being carried over into the engine when it is started up. The drains are also used to circulate steam through the superheater when raising steam, to carry away the heat and prevent overheating of the superheater tubes.

Q. Explain in detail how and when a safety valve is tested by the engineer.
A. Before lighting off, test hand gear by lifting as far as possible without lifting valve, and when steam pressure gets within 100 lbs. of popping pressure, the valve is actually lifted. Test once a day while steaming.

Q. What is the advantage of outside spring-loaded valves over that of the enclosed type?
A. The outside-spring type is kept free from the intense heat of the enclosed space.

Q. A safety valve is just large enough for a boiler with maximum working pressure of 200 lbs. The working pressure is cut to 100 lbs. Can the same safety valve be used? Why?
A. No. The valve will not be able to handle the excess volume of steam at the lower pressure.

Q. What is a "gag" for a safety valve?
A. A specially-made clamp for holding the safety valve closed during a hydrostatic test.

Q. Upon what does the size of a safety valve depend?
A. The size is determined by the capacity and the steam pressure of the boiler.

Q. What effect does reducing the boiler pressure have on the size of the safety valve?
A. If the boiler pressure is decreased, the size of the safety valve must increase, since steam at lower pressure has greater volume.

Example: In solving the lever safety-valve formula for A,

$$A = \frac{W_1L_1 \ W_2L_2 \ W_3L_3}{PL}$$

It is obvious that, as P is decreased, the value of A will increase.

Q. What is "blowdown" of a safety valve?
A. The "blowdown" of a safety valve is the difference between the

lifting pressure and the seating pressure. It may also be termed the "pressure drop" when the safety valve opens.

Q. What is meant by "extended area" of a safety valve?
A. The valve disc is constructed with a blowdown ring of greater diameter just above the seat, which provides a greater area for the steam pressure to act upon. When the valve is closed, the upward force is the steam pressure acting on the area of the valve. After the valve starts from its seat, the steam pressure also acts on the extended area, causing the valve to pop open.

Q. What type of safety valve is used in connection with superheated steam?
A. An exposed-spring type of safety valve.

Q. What is meant by "gagging" a safety valve?
A. Gags or clamps must be carried on all ships to be used for holding the safety valve on its seat while the hydrostatic-pressure test is being applied.

Q. What is the purpose of a "blow-back" or "adjusting" ring on a safety valve?
A. This is a movable ring which regulates the opening for the escape of steam from under the "extended area." When this opening is decreased, the amount of "blow down" is increased. When the opening is increased the amount of "blow down is decreased.

Q. Describe an accumulation test.
A. This is an actual test of the ability of the boiler safety valves to relieve the boilers of all the steam they are capable of generating when fired to full capacity. In this test, all steam outlets from the boiler, except such as may be necessary to operate the boiler, are shut off. The fires are then forced to the maximum capacity for a period of 15 minutes for fire-tube boilers and 7 minutes for watertube boilers. During the test, the steam pressure must not exceed 6% above the maximum allowable working pressure.

Q. How is the superheater protected while testing the safety valves?
A. The superheater drains are used to protect the superheater while testing the safety valves under steam. This is done by opening the drains wide enough so that sufficient steam passes through the superheater to carry away the heat and prevent overheating. Since safety valves are customarily tested with all other steam outlets from the boiler closed, it is not necessary to force the fires in order to get the steam pressure up to the blowing-off pressure. Forcing the fires may possibly save a little time, but it is better to take a little more time and not endanger the superheater.

Q. What is a hydrostatic test on a boiler?

A. A test in which the boiler is filled with water, and pressure is put on the boiler. Its purpose is to determine whether the boiler is tight and capable of safely holding its working pressure.

Q. Describe a Bailey feedwater regulator.

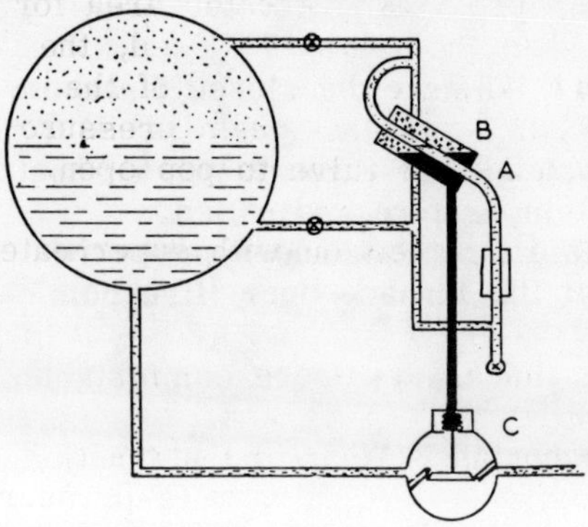

Fig. 48. Bailey feedwater regulator.

A. This regulator consists of an inner tube A enclosed by the generator B, and the bellows-operated regulating valve C, Fig. **48**. The inner tube is connected as a gauge glass and keeps the boiler water level. As the water level drops, the upper end of the generator is internally exposed to steam which heats the water in the generator, causing it to expand, and expand the bellows, which open the regulating valve. If the water level rises, the reverse occurs.

Q. Describe a B & W feedwater regulator.
A. This apparatus consists of a hollow cylindrical valve which is rotated on its axis inside of a cylinder liner. The valve has ports cut in its walls. The liner has corresponding ports. The valve is connected by a lever with a float which is in the steam drum. As the float rises with the water level, it rotates the valves and closes the ports. As the float lowers, it rotates the valve and opens the ports, thus maintaining a constant water level.

Q. Why is a strong definite circulation required in a modern water-tube boiler?
A. A strong, definite circulation is required in a modern watertube boiler because of the large quantities of heat liberated in the furnace and the large amounts of steam generated. This necessitates a high rate of heat transfer through the heating surfaces, and a strong flow of water or steam is necessary to carry away the heat and prevent overheating of the heating surfaces.

Q. What is meant by "rate of evaporation?"
A. The amount of water evaporated into steam per square foot of heating surface per hour. It is a means of indicating how hard a boiler is being forced.

Q. What is meant by "rate of combustion?"
A. The amount of fuel burned per square foot of heating surface per hour.

Q. What are the requirements of efficient oil burning?

A. 1. Proper amount of oil (sufficient pressure).
2. Proper amount of air (minimum excess air).
3. Proper atomization (good mixture of air and oil).
4. Proper oil temperature.
5. High furnace temperature.

Q. What is the Howden draft system? What is the closed fireroom system? Which is best? Why?

A. Howden system: Air is forced by a blower through the air heater and into the furnace, volume and temperature controlled.

Closed fireroom: Air is forced into fireroom under pressure although the air is required only at the furnace door (fireroom must be kept airtight).

Howden system is more efficient and always more comfortable.

Q. Of what two parts does a burner consist? What is the function of each part?

A. A burner consists of an atomizer and a register.

The atomizer breaks the oil into fine foglike particles which completely fill the furnace opening in the form of a hollow cone.

The register allows entry of a strong blast of whirling air, which catches the oil fog, mixes with it, and enters the furnace where combustion takes place.

Q. Define combustion.

A. Combustion is the rapid combination of oxygen with a combustible (fuel oil) accompanied by the evolution of heat.

Q. Define complete and perfect combustion.

A. Complete combustion occurs if the combustible is completely consumed.

Perfect combustion is the result if the combustion is complete without use of excess air.

Q. Describe a simple form of manometer for measuring natural draft.

A. A glass U-tube, both ends open, partially filled with water, one end connected to boiler stack, the other end open to atmosphere. Back draft creates suction, causing the water to rise in the stack end of the tube and to be lowered accordingly in the opposite end, due to atmospheric pressure. This change is interpreted to measure the intensity of the draft.

Q. How is carbon formed in a furnace?

A. Carbon is formed when unburned oil strikes a heating surface. The relatively cool heating surface brings the temperature of the oil below the ignition point and the more combustible components of the oil burn, leaving carbon.

Q. What are the advantages of small tubes vs. large tubes?
A. The evaporation rate is greater. See next question.

Q. Why is the evaporation rate greater in small diameter tubes than in large diameter tubes. Show this ratio, using 4", 2" and 1" tubes.
A. The evaporation rate is the ratio of generating surface area to the volume of contained water. Therefore, the smaller the tube, the greater the ratio of generating surface to the volume of contained water. (Inasmuch as π appears in all the computations, we will leave it in that form to make the math easier.) Assume all tubes 10 feet long (120").

4" tube

Surface area $= \pi D\,(120)$
$= \pi(4)\,(120)$
$= 480\pi$

Volume $= \pi r^2(120)$
$= \pi(2)^2\,(120)$
$= \pi(4)\,(120)$
$= 480\pi$

RATIO 1:1

2" tube

Surface area $= \pi D\,(120)$
$= \pi(2)\,(120)$
$= 240\pi$

Volume $= \pi r^2\,(120)$
$= \pi(1)^2\,(120)$
$= 120\pi$

RATIO 2:1

1" tube

Surface area $= \pi D\,(120)$
$= \pi(1)\,(120)$
$= 120\pi$

Volume $= \pi r^2\,(120)$
$= \pi(.5)^2(120)$
$= \pi(.25)\,(120)$
$= 30\pi$

RATIO 4:1

Q. What are the basic factors that determine the rate of heat transfer between heat exchangers?
A. 1. The BTU content of both the cooling or heating medium and the medium being heated or cooled.
2. The rate of flow of both of the above mediums.
3. Whether the unit is uniflow or counterflow.

Q. Why does a safety valve continue to blow off after the boiler pressure has dropped below the pressure for which the valve is set?

A. Blowdown ring needs adjusting.

Q. What would cause a safety valve to chatter or rumble when blowing off?

A. Worn blowdown ring; broken or weak spring; distorted disc.

Q. A draft gauge registers 2-1/2" of water. What is the psig?

A. Atmospheric pressure = 14.7 psi

Atmospheric pressure will support a column of water 34 feet high.

$$34\overline{)14.700} = .432$$

.432 lbs. per ft.

or .036 lbs. per in.

2-1/2 X .036 = .09 psig

Q. Explain the change in temperature, pressure and condition of steam as it passes from the drum, through the superheater, and then through the drum type desuperheater.

A.

	Pressure psig	Temperature °F
Drum outlet	480	465
Superheater inlet	475	465
Superheater outlet	450	750
Desuperheater inlet	445	750
Desuperheater outlet	400	500

Q. Name the different salts which are found in sea water, and state what percentage by weight of sea water is composed of these salts.

A.

Sodium chloride	NaC1	77.4
Magnesium chloride	MgC1	10.5
Magnesium sulphate	$MgSO_4$	4.5
Calcium sulphate	$CaSO_5$	3.6
Potassium sulphate	KSO_4	2.2
Calcium carbonate	$CaCO_3$	.4
Silicous matter		.7
Iron and aluminum oxides		.7

Q. Make a simple sketch showing the top section of a safety valve with a gag properly installed for a hydrostatic test.
A. Figure 49.

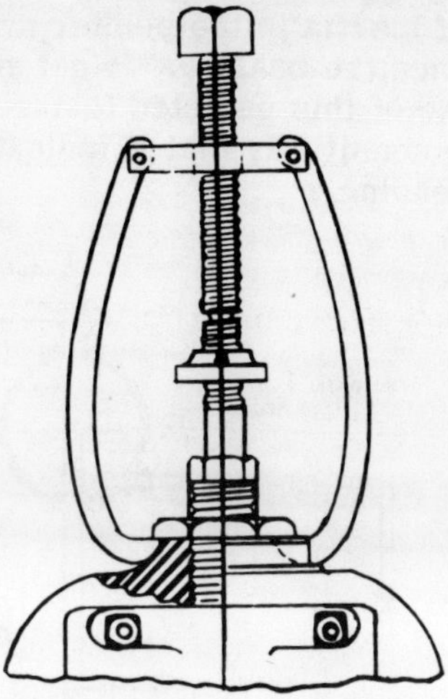

Fig. 49. Safety valve.

Q. Explain the operation of a Jerguson Truscale boiler water gage.
A. Figure 49A. The Jerguson Truscale is essentially a mercury-filled manometer with a stainless steel float in one leg which is actuated by the movement of mercury resulting from changes in the pressure differential across the manometer. The complete installation consists of three main parts; a datum column, a dirt trap and valve manifold, and the gage itself. The datum column is connected to the boiler drum or vessel and is the means of impressing a fixed head on the gage. Tubing, run from the datum column to the manifold, and thus through to the Truscale gage, transmits these heads to the mercury-filled manometer. The manifold, which is mounted above the gage, incorporates dirt traps, shutoff valves, blowdown valves, and an equalizing valve in a single, convenient and compact unit.

The movement of the float in the manometer is transmitted to the pointer on the scale by means of a magnetic coupling drive. This coupling consists of an internal magnetic armature on the end of a float shaft which is enclosed in a stainless steel capped tube, and an external yoke with magnetically energized arms.

The yoke forms a part of the pointer system, maintaining alignment with the armature which is attached to the float shaft, and pivoting on a fulcrum with precision bearings as the armature moves up and down.

This specially designed magnetic coupling is unique in that it has a fixed mounted Alnico magnet so arranged that the arms of the pivoted yoke, to which the pointer is attached, are highly magnetized by induction. This arrangement provides a maximum of available coupling force with a minimum of inertia in the pointer system, as well as a minimum of friction, since the bearings do not support the weight of the magnet. It is because of this patented feature that the gage is capable of such extreme sensitivity that it indicates level changes to 1/2 of 1% of full scale reading.

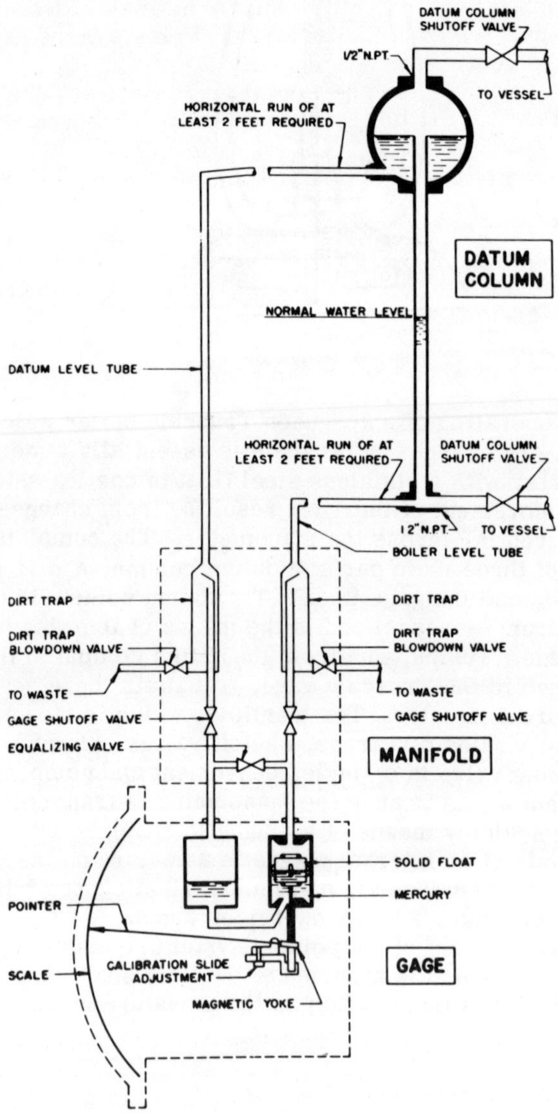

Fig. 49A. Jerguson Truscale boiler water gage.

Q. What is required by regulations with respect to examination of boiler mountings?

A. All valves and other boiler mountings and attachments shall be opened up and examined by the inspector after each 4 years of service. Such valves, boiler mountings and attachments shall be removed from the boilers after each 8 years of service for examination of the studs or bolts connecting them to the boiler and to permit examination of those parts of the valves, mountings and attachments which are not susceptible to examination by simply opening up the valves. The Officer in Charge, Marine Inspection, may require the examinations prescribed to be made at more frequent intervals if, in his opinion, such action is necessary to be assured of the safety of the boiler.

Water columns, gage glasses, and gage connections shall be examined to determine that they are in satisfactory working order.

Q. What are the requirements for feed lines and feed valves on boilers?

A. Feed piping must be designed for not less than 125% of maximum allowable pressure of the boiler or the feed pump relief valve setting, or shut-off head when a pump relief valve is not fitted, whichever is greater. There shall be at least two separate means of supplying feedwater to the boilers. Feed stop and check valves shall be fitted in the feed lines as close as possible to the drum feed inlet nozzles or to the economizer feed inlet nozzles on boilers fitted with integral economizers.

Q. What are the factors that tend to reduce the overall size and weight of modern marine boilers?

A. Minimum overall size and weight are obtained by: (a) Accelerated water circulation; (b) Increased velocity of gases of combustion, hence increased heat transfer per unit area of heating surface; (c) Increase in furnace loads, i. e., rate of fuel consumption per cubic foot of furnace volume.

Q. What are the functions of the water drum on the two-drum watertube boiler?

A. The functions of the water drum are to equalize the distribution of water to the generating tubes and to provide a receptacle for the accumulation of scale and other solid material which may be present in, or may be precipitated from, the boiler water. Removal of this solid matter is provided for by the bottom blow valve.

Q. Explain how the temperature throughout the furnace is partially equalized through the use of studded waterwalls.

A. At high steaming rates the rear of the furnace tends to be the hottest portion. To equalize this condition the waterwalls in the front section are fully studded and covered with chrome ore refractory while the waterwalls in the rear section are only partially studded and covered. This results in more heat being transferred to the rear waterwalls and a lowering of the furnace temperature in that section.

Q. What are the data markings on approved safety valves?

A. Approved safety valves shall be marked by the manufacturer either by means of a plate attached to the body of the valve or by stamping or casting on the body of the valve itself the following data:

1. Name or registered trademark of manufacturer.
2. Serial number of safety valve.
3. Inlet diameter of safety valve.
4. Operating pressure and guaranteed discharge capacity—pounds of steam per hour at that pressure.
5. Safe working pressure of the body of the valve.
6. Blowdown, in pounds per square inch.

Q. (a) What is the principle of operation of a combustible gas indicator? (b) What precautions would you observe in order to assure proper readings from a combustible gas indicator?

A. (a) A combustible gas indicator operates on the principle that when a mixture of air and combustible gas or vapor is brought into contact with hot platinum wire, rapid combination of the combustible gas with the oxygen of the air takes place at the surface of the wire, raising its temperature and therefore its electrical resistance. This change in resistance, the amount of which depends on the concentration of combustible gas, actuates a sensitive meter.

(b) All instructions issued by manufacturers of such instruments must be carefully followed and all necessary calibrations carefully made. Unless all necessary precautions are taken, incorrect readings might lead to dangerous situations. In ports of the United States and wherever available, qualified chemists should determine gas hazard in accordance with Coast Guard regulations.

The switches for a combustible gas indicator should not be operated in a gaseous atmosphere as the case is not explosion-proof.

Q. What determines the location of the superheater in a watertube boiler, and how is it protected against excessive heat?

A. The exact location of the superheater depends upon the degree of superheat desired. The higher the degree of superheat for which it is designed, the closer to the furnace will be its location. The superheater is usually protected against excessive heat by screening rows of generating tubes placed between it and the furnace.

Q. Is the volumetric capacity of each gas-pass the same in a watertube boiler?

A. No; it decreases in each succeeding pass. The gases contract as they cool, and in order to maintain the high gas velocity necessary to sweep off stagnant gas films to effect good gas transfer, the cross-sectional area of the passes must decrease as the gases require less space.

Q. Explain in detail how to set the safety valves on a boiler so as to operate at the allowable pressure.

A. 1. Check boiler steam gage with test gage or gage-testing machine.

2. Gag all valves on the boiler except the one to be set.

3. Disconnect hand-lifting gear, remove key and cap from top of valve spindle.

4. Turn compression screw right-handed (down) to increase spring pressure and consequently to increase the lifting pressure of the valve.

5. Handle fires to increase steam pressure in boiler to lifting pressure. Note pressure on steam gage at which safety valve lifts and also pressure at which it reseats.

6. Readjust compression screw up or down and test again with steam until correct lifting pressure is obtained.

7. Set up compression screw nut.

8. The difference between lifting and reseating pressures (blowdown) should be about 3 percent of set pressure. If the difference in pressures, noted on the steam gage, was greater than this, back out the adjusting ring lockpin, insert a small screwdriver or other instrument through the hole and turn the adjusting ring until it has moved down a small amount. Replace lockpin.

9. Test again with steam and readjust the adjusting ring until, by trial and error, the proper blowdown has been obtained.

10. Gag this valve and go ahead with the same procedure on the other valves of the same boiler until all have been adjusted.

11. Replace caps and keys. Connect up hand-lifting gear and test hand-lifting gear to see that all valves lift by hand. When shaft is rotated the system of levers lifts up on the cap, which is keyed to the valve spindle. As the valve spindle is also keyed to the valve, the valve is lifted from its seat. The hole through the end of the cap key is for the U.S. seal to prevent unauthorized persons from tampering with the safety valve setting.

Q. Describe a waterwall as installed in a boiler. What purposes are served by the waterwall?

A. A waterwall consists of a series of vertical or inclined water tubes installed along one or more walls of the combustion chamber and exposed to direct radiant heat of the fire. These tubes are connected directly, or through headers and connecting nipples, to the circulatory system of the boiler.

Two purposes are served by the waterwall: (1) Added heating surface and boiler capacity are secured. Revamping and adding waterwalls to an existing boiler may even double its capacity. (2) Maintenance of the refractory wall is reduced greatly. This protection of refractory walls is extremely important when high combustion rates are desired. The practical combustion rate is about 18,000 to 25,000 B.t.u. per cu.ft. furnace volume per hr. when bare refractory walls are exposed to the furnace. But combustion rates of 40,000 to 200,000 B.t.u. per cu.hr. have been attained without difficulty with all refractory surfaces protected by waterwalls.

The circulation of water is upward through the wall tubes exposed to radiant heat. These tubes are supplied by water through headers connected by piping from one or more of the boiler drums and located outside of the boiler setting. The supply system is known as the "downcomer."

Blowdown valves are required for each header at the bottom of a series of waterwall tubes, for the same reasons that the boiler itself needs them. Sediment accumulation in a header supplying wall tubes might cause interruption of circulation with consequent overheating and failure of the tubes.

Q. Explain the effect of an increase in the excess air on the superheater outlet temperature, the steam pressure, and the stack temperature of an integral superheater boiler while operating at a constant rate of combustion.

A. At a constant rate of combustion an increase in the amount of excess air will result in a decrease in the furnace temperature. This decrease in the furnace temperature will reduce the quantity of steam formed in the generating tubes. The reduction of the steam so formed will cause a pressure drop in the boiler. The reduced flow of steam through the superheater will result in a higher temperature of the outlet steam. The reduction in the rate of heat transfer, due to the decreased furnace temperature, results in a higher stack temperature.

Q. Why does a safety valve continue to blow off after the boiler pressure has dropped below the pressure for which the valve is set? What would cause a safety valve to chatter or rumble when blowing off?

A. Blowing off continues after the boiler pressure is reduced below the "pop" pressure because the discharge passages guide the escaping steam in such a manner that it helps the boiler pressure to hold the valve away from its seat. The greater this assistance while the valve is open the lower the boiler pressure will be reduced below the "pop" pressure before the valve will be closed. The reduction in boiler pressure below the "pop" pressure is called the "blowdown" and most direct spring-loaded safety valves are provided with an adjustable "blowdown ring" or other means for varying the amount of blowdown. When the valve is raised from its seat the pressure exerted by the spring is greater than when the valve is closed. Hence, when the valve opens, unless the escaping steam exerts enough force to overcome the greater spring resistance, for continuance of a little more than pop pressure in the boiler there are rapid alternations of partial openings and closings, due to the ricocheting of the spring and small sudden changes of boiler pressure that give rise to chattering of the valve with intermittances of discharge, causing vibrations of the atmosphere that produce a chattering or rumbling sound.

Q. Explain why the safety valve escape piping is fitted with an open drain.

A. The open drain is to remove any condensate formed from the escaping steam. After a safety valve has blown many times, it is not uncommon for slight leakage to develop. Condensation of this leakage may gradually fill an undrained escape pipe with water. This condition would prevent the safety valve from blowing at its set pressure. Also, when blown, the water hammer may damage the escape piping.

ELECTRICITY

DIRECT CURRENT

Q. What is meant by residual magnetism? Why is it important in the case of self-excited DC generators?
A. It is magnetism remaining in the fields after all exciting current is cut off. It is important because the DC generator could not build up a voltage without it.

Q. What is meant by "staggering the brushes" of a generator?
A. It means placing them on the commutator so that they are not in line, so as to prevent them from grooving the commutator.

Q. What is meant by the neutral point of the brushes of a generator?
A. It is the point of best commutation. It is the point where the voltage is the highest and the sparking at the brushes is the least.

Q. How would you determine the neutral point for setting the brushes of a DC generator?
A. With the machine running, loosen the locking gear on the brush rigging and, by means of the handle on the rigging, shift the brushes back and forth a bit, watching the voltmeter at the same time. The voltmeter will read its highest when the brushes pass over the neutral point. At the same time, the sparking at the brushes will be the least, or none at all. Lock the brush rigging at the point where the voltage is the highest.

Q. What is the effect of brushes ahead or behind neutral in a DC generator?
A. The voltage will not be its highest; more field current must be used to get the required voltage, thus heating up the field poles. Sparking at the brushes will cause the commutator to run hot.

Q. What are interpoles? Why are they used on DC generators? How are they connected?
A. They are small auxiliary magnetic poles placed between the regular fields of the generator. Usually there are the same number as there are main poles, but in small generators there are sometimes half as many. They are used to assist commutation and prevent sparking at the brushes. They are always connected in series with the armature. The interpole would be connected so that its polarity would allow the armature conductors to pass under it first and then under a main pole of the same magnetic polarity.

Q. What is the maximum temperature to which an armature should be exposed when drying out by heating with heaters or air blast? When field coils are dried out by passing a current through them, what maximum current should be used?

A. No more heat than the name-plate reading (usually 40°C. rise). The circulating current should be limited by its voltage to a value not greater than full-load current. A very low voltage is all that is needed to circulate this current.

Q. Should the field strength of a shunt motor be increased or decreased to reduce the speed? Why?

A. Increased. Increased field generates a greater counter EMF, thus reducing current input (and speed) to the motor.

Q. In a shunt generator, an increase in load causes a voltage drop. How is this compensated?

A. A rheostat is placed in the field circuit. If the load increases and voltage drops, some of the field resistance is cut out. This raises the field voltage and increases excitation. This raises voltage back to normal.

Q. Why do most DC motors designed to operate at various speeds have compensating coils and/or interpoles?

A. These poles are placed midway between the main poles to induce a voltage in the coil being commutated. This helps in a quick reversal of current in that coil which reduces sparking.

Q. How does a DC generator differ from an AC generator?

A. The DC generator has a commutator. The AC generator has slip rings.

Q. How does a DC motor differ from a DC generator?

A. No difference.

Q. Explain why compound wound DC generators are most generally used on ships and how the voltage is controlled.

A. This type of generator develops a constant voltage at all loads. Voltage is controlled by means of a rheostat.

Q. Why are conductors of large area and brushes of large contact area used in DC generators which have a large current rating?

A. To reduce the amperes per square inch of contact area and the consequent heating.

ALTERNATING CURRENT

Q. What is meant by "slip" of an induction motor?

A. The difference in speed of the rotor and the synchronous speed of the changes in the stator. It is usually expressed as a per cent of synchronous speed. Slip varies from 3% to 8%, according to the load.

Q. How is a 3-phase induction motor reversed?
A. By interchanging any two external leads to the stator.

Q. List five conditions which would cause an AC generator to overheat.
A. Overload, short circuit in coils, damp coils, dirty windings, fan or air cooler defective, low power factor load.

Q. If a loaded generator, AC, slows down, what will be the effect on the voltage, the current and the frequency?
A. Voltage drops (if there is no voltage regulator), current drops, frequency drops.

Q. What conditions must exist in order that two AC generators will operate in parallel?
A. They must have: same voltage, same frequency, same phase rotation, and be in phase.

Q. Is it necessary that two generators operating in parallel have the same number of poles?
A. No, as long as they have the same regulation characteristics.

Q. Does slip increase or decrease with increase in load on an induction motor?
A. It increases.

Q. Name two main causes that would prevent an AC generator from developing its normal voltage.
A. Insufficient speed and insufficient excitation.

Q. What is meant by power factor? At the same wattage output, will a generator furnish more current to a load drawing unity power factor or to a load drawing 0.8 lagging power factor?
A. Power factor is the ratio of true power to apparent power. It is expressed as a percentage of the apparent power.

$$\text{p.f.} = \frac{\text{Kw}}{\text{KVA}}$$

0.8 p.f. requires greater current for same power.

GENERAL

Q. How does the size of wire affect its resistance?
A. The resistance of a wire is inversely proportional to the square of the diameter, the length remaining the same. In other words, doubling the diameter decreases the resistance 4 times. Quadrupling the diameter decreases the resistance 16 times.

Q. What effect does temperature have on the resistance of a wire?
A. An increase in the temperature of a wire will increase its resistance.

Q. What effect does increasing current have on the temperature of wire?
A. Increases its temperature.

Q. How do electrical losses show themselves?
A. All electrical losses show themselves as heat.

Q. What determines the safe current-carrying capacity of a wire?
A. The diameter or cross-sectional area of the wire.

Q. Why is stranded wire used in some cases instead of solid wire?
A. It is more flexible and less likely to break under bending stresses. Due to greater surface area, heat dissipation is greater.

Q. How does the length of a wire carrying current affect its resistance?
A. The resistance of a wire increases directly in proportion to its length, the diameter remaining the same. In other words, doubling the length of a wire doubles the resistance.

Q. What is a transformer?
A. An electrical device for changing the voltage by means of electromagnetic-induction, with AC current.

Q. What is a step-up transformer? What is a step-down transformer?
A. A step-up transformer is one that raises the voltage and decreases the amperage. A step-down transformer is one that lowers the voltage and raises the amperage. The change in volts is directly proportional as turns in the primary are to the turns in the secondary. The change in amperes is inversely proportional as the turns in the primary are to the turns in the secondary.

Q. What would be the relation of primary volts and amperes to secondary volts and amperes in a 4:1 step-up transformer?
A. The secondary volts would be four times the value of the primary volts, while the secondary amperes would be 1/4 the value of the primary amperes.

Q. Why should insulation resistance measurements be supplemented by a thorough visual inspection of the insulation before arriving at a conclusion relative to the condition of the insulation?
A. Insulation resistance varies with factors such as; humidity, recent use, and ambient temperature, which should be considered when interpreting the reading. Cleaning, drying, repairing and spraying affect insulation readings. Age is one deteriorating factor that compromises its effects with high operating temperatures.

Q. What is a rotary converter? Explain its general construction and operation.

A. An electrical device for converting AC to DC. Also called a synchronous converter. It consists of a synchronous motor and a generator in one machine. The field and armature are similar to a DC generator except that slip rings are located on the end of the shaft opposite from the commutator. AC turns the converter as a synchronous motor. DC is taken from the commutator brushes.

Q. What is an "inverter"?
A. A piece of electrical equipment for converting DC to AC.

Q. All preparations for paralleling AC generators have been made. The synchronizing lamps are going bright and dim in unison at a very slow rate. At what point should the synchronizing switch be closed?
A. When lights are dim.

Q. Give 4 factors that affect resistance in a wire.
A. Length of wire; cross-sectional area of wire; type of material; temperature of wire.

Q. Give 4 advantages of turbo-electric drive ships.
A. 1. Allows high-speed prime mover and low-speed propeller each to operate at its highest efficiency.
2. Allows full power of turbine to be applied to propeller when going astern.
3. Less noise and vibration.
4. Does away with astern turbine, which is of no value under normal operating conditions.
5. Does away with line shafting from turbine to propeller.
6. The electric instruments indicate the power output and can be used to improve plant operation and economy.
7. Main unit may be used for cargo discharge (electric winches).

Q. What effect does increasing current have on the temperature of a wire?
A. Temperature increases.

Q. What effect does temperature have on the resistance of a copper wire?
A. Increase in temperature increases resistance.

Q. How does the size (cross-sectional area) of wire affect its resistance?
A. The resistance of a wire is inversely proportional to the square of the diameter, the length remaining the same. In other words, doubling the diameter decreases the resistance 4 times. Quadrupling the diameter decreases the resistance 16 times.

Q. What determines the safe current-carrying capacity of a wire?

A. The diameter or cross-sectional area, insulation, and use (in free air or a number of conductors in a conduit, etc.).

Q. How would you determine the circular mil area of a 1/4" wire?
A. Change the fraction to thousandths. Drop the decimal point, and square the answer.

1/4" = .250
.250 X .250 = 62,500 circular mils

Q. Describe the electromagnet type "low voltage" or "no field" release used with a manual starting box for a shunt motor.
A. When the line switch is closed, no current can enter the armature or field coils. As the movable arm (lever) touches the first contact point, full line voltage is applied to the field (field develops at once). However, only a limited amount of current can flow through the resistance to the armature. As the armature starts to rotate, the lever is moved to the second point. As each contact point is passed, less resistance is allowed to remain in the armature circuit until the final contact point is made and resistance is cut out entirely. Full line voltage is now applied across the armature. The holding magnet is connected in series with the field. Since field current is flowing through the circuit, its magnetism holds the contact arm (lever) at running position. If field circuit is broken, the magnet is demagnetized and releases the contact arm. Low voltage will also release the arm, Fig. 50.

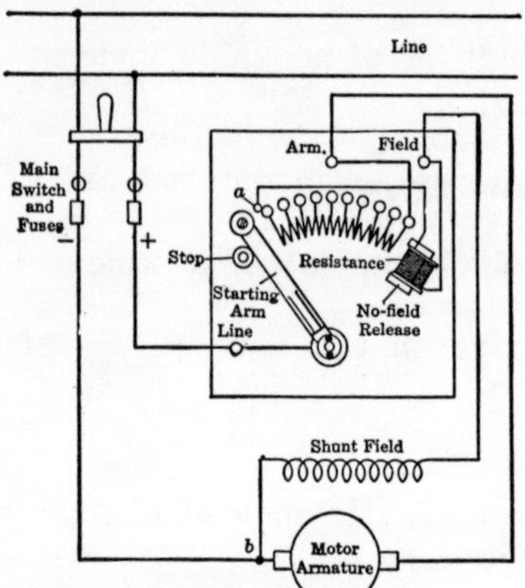

Figure 50.

Q. How does the voltage supplied to a rotary converter compare with the voltage output?
A. The DC voltage is higher than the voltage at the AC slip rings. The ratio is 1 to .707, that is, 70.7 volts at AC end will give 100 volts at DC end.

Q. What would you do if a fire were discovered in the generator coils?
A. Shut down. Use CO_2 or any other approved type of extinguisher.

Q. Are splices permitted in portable cords?
A. No. Portable cords shall be used only in continuous lengths without splices or taps.

Q. What would cause a short circuit in a battery?
A. Collapsed plates touching each other; sludge build-up in bottom of battery casing.

Q. A low power factor in an induction motor with a constant applied voltage: (a) Requires less current for a given power output; (b) Requires more current for a given power output; (c) Would heat the motor dangerously; (d) Would slow the motor down.
A. (b) Requires more current for a given power output.

Q. Synchronous motors are most similar in construction to: (a) D. C. compound motors; (b) Induction motors; (c) Alternators; (d) D. C. shunt motors.
A. (c) Alternators.

Q. A balance relay on a synchronous propulsion motor: (a) Keeps each phase of the motor in balance; (b) Operates when one phase carries approximately 25% more current than any other; (c) Operates when one phase carries approximately 25% less than any other; (d) Operates when there is approximately 120° between phases in a two-phase motor; (e) Is manually controlled by the engineer on watch.
A. (b) Operates when one phase carries approximately 25% more current than any other.

Q. A squirrel-cage induction motor is: (a) A motor that is compound-wound with windings in series; (b) An induction motor in which the primary circuit consists of a squirrel-cage winding suitably disposed in the secondary core; (c) An induction motor in which the secondary circuit consists of a squirrel-cage winding suitably disposed in slots in the secondary core; (d) A squirrel-cage motor consists of a primary conductor that is shunt wound.
A. (c) An induction motor in which the secondary circuit consists of a squirrel-cage winding suitably disposed in slots in the secondary core.

Q. When is a portable tool required to be grounded?
A. 100 volts or over.

Q. Describe a thermocouple.
A. Two dissimilar metals are joined together; when heated, a voltage is produced.

Q. What is the treatment for acid burns?
A. Wash as soon as possible with large quantities of fresh water. Apply boric acid solution, vaseline or zinc ointment.

Q. What is a drip-proof electrical apparatus?
A. A drip-proof machine is one in which the ventilating openings are so constructed that drops of liquid or solid particles falling on the machine at any angle not greater than 15° from the vertical cannot enter the machine either directly or by striking and running along a horizontal or inwardly inclined surface.

PROBLEMS

Q. Calculate the energy supplied in 20 minutes to an electric oven using 30 amps at 115 volts.

A.
$$W = VA = (115)(30) = 3450 = 3.45 \text{ Kw}$$
$$1/3 \text{ of } 3.45 = 1.15 \text{ Kw/20 mins.}$$

Q. With 2 resistances in parallel of 4 ohms and 3 ohms, what current will 110 volts cause to flow?

A.
$$\frac{110}{4} = 27.5 \text{ amps.}$$
$$\frac{110}{3} = 36.6 \text{ amps.}$$
$$64.1 \text{ total amps.}$$

Q. What is the total resistance in the circuit in the preceding question?

A.
$$R = \frac{1}{1/3+1/4} = \frac{1}{\frac{7}{12}} = \frac{12}{7} = 1.71 \text{ ohms}$$

or
$$R = \frac{110}{64.1} = 1.71$$

Q. The panel board instruments indicate an input of 2200 KW to the propulsion motor. What will be the shaft horsepower of motor output if the efficiency of the motor is 98.4%?

A.
$$HP = \frac{V \times A \times 1.73 \times PF \times Eff.}{746}$$
$$= \frac{2200 \times 1000 \times 1.73 \times 1 \times .984}{746} = 5013$$

Q. The resistance of copper wire of a certain size is 2.521 ohms per 1000 feet. What is the resistance of 400 feet?

A.
$$2.521 : x = 1000 : 400$$
$$1000\,x = 1008$$
$$x = 1.008 \text{ ohms}$$

Q. The EMF of a battery is 12 volts on an open and 9 volts on a closed circuit with a current of 6 amps flowing. Find the internal resistance of the cell.

A.
$$R = \frac{E_1 - E_2}{I}$$
$$= \frac{12-9}{6} = \frac{3}{6} = .5 \text{ ohms}$$

REFRIGERATION

AMMONIA

Q. Describe the complete operating cycle of an ammonia refrigerating unit?

A. Ammonia gas is drawn from the system into the cylinders and compressed to approximately 175-200 lbs. pressure. As the pressure increases, the temperature increases and the gas is forced through the separator (or interceptor) to the condenser. The cool circulating water in the condenser changes the ammonia gas to a liquid. The liquid passes to the liquid receiver, then through the main liquid valve and expansion valve to the cooling coils in the refrigerator. On passing through the small orifice in the expansion valve, the liquid drops in pressure and is changed to a gas. As the temperature of ammonia gas at atmospheric pressure is —28° F., it is readily seen that the ammonia gas will begin to boil by extracting the heat in the refrigerator box. The gas is then drawn into the compressor and the cycle commences again.

Q. How would you purge air from the condenser?

A. Attach a hose to the purge cock on top of the condenser and place the other end of the hose in a bucket of water. Open purge cock slowly. Air will bubble into the water but ammonia will start a crackling sound and the odor will be detected.

Q. How would you start an ammonia refrigerating machine?

A. 1. Make sure that gauge glass is half full of oil.

2. Open water supply valves. (Be sure water is flowing freely.)

3. Open wide by-pass valve from discharge to suction side of system.

4. Turn flywheel back to the compression point. (Opposite to direction of rotation.)

5. Turn on power. (Depends on type of drive.)

6. Open discharge valve.

7. Close by-pass valve. (Never close by-pass before opening discharge valve.)

8. Open suction valve. It should be opened only one-quarter turn until system is pumped down and the suction-pressure gauge registers 25 lbs. pressure. Then slowly open the suction valve wide and run the machine until low pressure gauge reads 10 lbs.

9. Open liquid valve.

10. Regulate expansion valve.

11. Proceed to check machine to see that it is operating properly.

Q. How would you recharge an ammonia system?

A. Figure **51.**

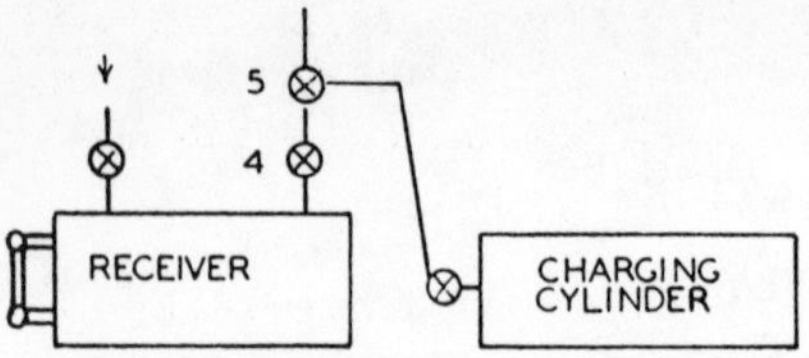

Fig. 51. Recharging an ammonia system.

1. Close liquid valve #4.
2. Remove cap from spare ammonia cylinder.
3. Connect new tank to charging valve #5.
4. Open valve slowly to test all connections.
5. Start machine and open charging valve #5.
6. Regulate suction pressure to about 10 lbs. and charge until high-pressure gauge reaches 160-180 lbs.
7. Close valve on cylinder and charging valve #5.
8. Open liquid valve #4.
9. Liquid should register about half a glass in the receiver.

FREON

Q. How is a Freon-12 system purged?
A. Purge valve located on top of condenser or highest part of discharge line. Procedure same as used with ammonia system.

Q. What types of gaskets are used with a Freon-12 machine? Why?
A. Asbestos or metallic. Freon will dissolve oil and graphite in ordinary gaskets.

Q. What is the purpose of the dehydrator in a Freon-12 unit?
A. To remove the moisture from the refrigerant in the system.

Q. What causes unusually high head pressure?
A. 1. Insufficient cooling water (or cooling water too hot).
 2. Dirty condenser.
 3. Air in condenser.
 4. Choked discharge passages.
 5. Too much refrigerant in evaporating coils.
 6. Operating compressor at excessive speed.

Q. What drying agents are usually used in the dehydrator of the Freon-12 refrigeration system? When is this dehydrator placed in operation?
A. Activated Alumina. Used when adding new refrigerant.

Q. Why is air circulation an important factor in the cooling space of a refrigerating plant?
A. To procure a better heat transfer throughout the box, thus keeping a fairly even temperature in the box.

Q. Explain the purpose of the water regulating valve and the water pressure failure switch which are used in a Freon automatic control system.
A. The water regulating valve should be adjusted so that it maintains a constant condensing pressure regardless of the tempera-

ture of the circulating water. It should also shut off the water supply when the compressor is not operating.

The water failure switch should be set so that it remains closed as long as there is sufficient water flowing to the condenser. The cutout point must be adjusted so that it will open the switch and stop the compressor if the water stops.

Q. What is the proper procedure for shutting down a Freon-12 refrigerating unit for an extended period?

A. Pump down (that is, pump all liquid to receiver); vent thoroughly; drain condenser; lubricate cylinders; drain crankcase and clean it out.

Q. What is the purpose of a suction-pressure regulating valve?

A. When a very heavy load is thrown on a refrigeration system the compressor may become overloaded. To prevent this from occurring, a suction-pressure regulating valve is installed (this overloading would cause excessively high suction temperatures at the compressor). The valve is designed to permit only a predetermined maximum amount of gas to reach the compressor. This prevents the suction pressure in the compressor from going higher than the setting of the valve.

Q. Where are the suction-pressure regulating valves located in a Freon-12 unit?

A. They are used where several compartments are served by the same liquid line. These valves are designed to maintain a constant pressure within the cooling coils higher than the suction pressure and independent of suction-pressure fluctuation. The valve adjustment is obtained by compression of a spring, and a diaphragm or bellows balances the flow against minor fluctuations.

Q. Describe a solenoid valve. What is its function?

A. The solenoid valve is located ahead of the thermo-expansion valve. When the current is on, the magnetic coil of the valve is energized, causing the plunger to lift the valve off its seat and permitting the refrigerant to flow through. When any of the other control devices breaks the electrical circuit, the magnetic coil releases the plunger, closing the valve and stopping the flow of refrigerant.

Q. What is the purpose of the thermo-expansion valve?

A. This is an automatic control for the temperature of the gaseous refrigerant leaving the compartment. It controls the quantity of liquid refrigerant that passes into the evaporator coils and is designed to maintain a constant degree of superheat of the gas leaving the coils regardless of the suction pressure. Thus it has the dual purpose of giving automatic expansion control and insures that all the liquid refrigerant has been evaporated before it is returned to the compressor.

GENERAL

Q. What are the characteristics of Freon-12?
A. Colorless; odorless; boiling point —21.7°F. at atmospheric pressure; non-irritating; non-poisonous; non-corrosive to metals ordinarily used in refrigerating units; non-inflammable; non-explosive; an excellent solvent.

Q. How would you add lubricating oil to a Freon compressor preventing air from entering system?
A. 1. Wipe out and clean oil pump thoroughly.
2. Insert pump into oil can.
3. Connect pump discharge hose to the compressor crankcase oil charging valve — do not tighten.
4. Displace the air in the pump and connections by several strokes of the pump handle until oil appears at the oil charging valve. Tighten connection.
5. Pump oil into compressor. Fill to about 3/4 of the bull's-eye or gauge glass sight glass.
6. Close charging valve tightly, replace valve seal cap, disconnect hose and replace the seal cap on the oil charging valve outlet.

Q. What is meant when a refrigeration machine is said to have a capacity of 5 tons?
A. This means that the machine can remove 2000 X 5 X 144 = 1,440,000 BTU's per hour from the foods in the icebox. (A ton of refrigeration is equal to the melting of one ton of ice in 24 hours. This is 2000 X 144.)

Q. Name 3 requirements of substances to be used as refrigerants.
A. Low boiling point; high latent heat; should not be injurious to foods in case of leakage; operate within reasonable pressures.

Q. What is the principle of mechanical refrigeration?
A. The absorption of heat under temperature, pressure, compression and expansion.

Q. What is the function of the expansion valve?
A. Regulates amount of refrigerant passing to the expansion coils.

Q. What effect has liquid ammonia on petroleum lubricating oil?
A. Very little, unless water is present.

Q. What is the function of the compressor?
A. To pull the gas through the system, compress it to a higher pressure and temperature and discharge it to the condenser.

FIRE, EMERGENCY, ORGANIZATION, ETC.

Q. What is the minimum temperature for boiler feedwater entering a boiler?

A. The minimum temperature of boiler feedwater allowed to enter any marine boiler is 100°F.

Q. What are the regulations on dampers?

A. Where dampers are installed in the uptakes or funnels, the arrangement shall be such that it will not be possible to shut off the gas passages from the operating boilers.

Q. How long a crack in a staybolted surface may be repaired by fusion welding?

A. Cracks in staybolted plates may be welded provided they are located entirely within staybolted areas and the total length of any crack or series of consecutive cracks does not exceed two staybolt pitches.

Q. How large a crack may be welded on the shell or drum of a boiler?

A. Cracks may be welded between tube holes in the shell of water-tube boiler drums, provided they are not more than two cracks in any one row in any direction, not more than a total of four cracks in a drum, and further provided the welding meets the requirements for Class I welded pressure vessels and is approved by the Commandant.

Q. How may fusion-welding repairs be made on boilers?

A. By a certified, qualified welder, upon approval of the inspector.

Q. What are the requirements for pressure gages on boilers?

A. Each boiler must be fitted with a reliable pressure gage graduated to approximately twice the pressure to which the safety valve is set but in no case less than 1-1/2 times this pressure. Each double-ended boiler must have two, one for each end, installed where they are easily seen. Error in reading allowed is 1-1/2% of maximum reading of gage.

Q. What are the G. R. &R. on fuel oil pumps, heaters and strainers?

A. All vessels having oil-fired power boilers shall have at least two fuel oil service pumps, each of sufficient capacity to supply all the boilers at full power, and arranged so that one may be overhauled while the other is in service.

At least two fuel oil heaters of approximately equal capacity shall be installed and so arranged that any heater may be overhauled while the others are in operation.

Suction and discharge strainers shall be of the duplex or other type capable of being cleaned without interrupting the oil supply.

Q. What are the duties of a chief engineer regarding the bunkering of fuel oil?
A. Upon receiving fuel oil, the chief engineer must obtain and seal a half pint sample and retain until that lot of fuel oil is used. The chief engineer must also log the quantity received, the vendor's name, the producer's name and the flash point as certified by the producer.

Q. What are the requirements regarding pumping systems to supply fuel oil to oil burners?
A. Two separate pumping units. Relief valve in closed circuit on discharge piping, discharging into the suction line or back into the tank; service oil pump equipped with control means from readily accessible position outside of boiler room in the event of fire in the fireroom. The return line from the burners shall be so arranged that the suction piping cannot be subjected to discharge pressure.

Q. What are the rules regarding interchanging of service of fuel oil piping?
A. Oil piping systems for the transfer or discharge of cargo or fuel oil shall be separate from other piping systems as far as practicable, and positive means shall be provided to prevent dangerous interconnection in service.

Q. Describe a standard type soda-acid fire extinguisher.
A. This consists of a 2-1/2-gal. cylindrical copper tank containing 2 gals. of water and 1-1/4 to 1-1/2 lbs. of bicarbonate of soda. In the top of the tank is a bottle with porcelain stopper containing 4 ozs. of sulphuric acid. When extinguisher is turned bottom-up, the stopper falls out, allowing the acid to mix with the soda mixture. This creates a gas that forces the liquid out of the hose.

The soda-acid extinguisher must not be used on electrical or oil fires.

This extinguisher must be discharged at annual inspection, recharged, and a tag attached showing the date of recharge.

Q. What is electric bonding?
A. All tank vessels must be electrically bonded to the shore piping previous to the connection of cargo hose and this connection maintained until all cargo is handled and spillage is removed.

Q. What are the regulations regarding the piping system between the fuel oil service pump and the burners.
A. Piping between service pumps and burners shall be located so as to be readily observable. The relief valve located at the pumps and the relief valves fitted to the fuel oil heaters may discharge

back into the settling tank or the suction side of the pump. The return line from the burners shall be so arranged that the suction piping cannot be subjected to discharge pressure. Extra heavy seamless steel piping must be used where the temperature exceeds 450°F. or the pressure exceeds 350 p.s.i.

Extra strong valves and fittings must be used. Cast iron valves and fittings are not permissible. Bushings and street ells are not permissible.

If screwed bonnet valves are employed, they shall be of the union bonnet type capable of being packed under pressure.

Unions shall not be used for pipe diameters of one inch and above.

Q. What are the rules regarding vent pipes on fuel oil tanks?

A. All tanks must be fitted with vent pipes. Tanks having a comparatively large surface must have at least two, one of which is located at the highest part of the tank. The area of the vent pipe must be at least equal to the area of the filling pipe, but in no case less than 2-1/2" in diameter. They must terminate in a U-bend and be fitted with flame arrestors or flame screens.

Q. Why are oil tank vent pipes covered with flame screens and what attention must be given them?

A. To keep tank free of sparks and dirt. Keep screen clean for air and gas passage from the tank; do not paint screens.

Q. Why must the area of a tank vent pipe be equal to that of the filling pipe?

A. To prevent a pressure on the tank when filling.

Q. What are the requirements for fuel oil manifolds?

A. Manifolds must be located in accessible places and distinctly marked to indicate the tank or compartment to which they are attached.

Q. What are the requirements for fuel oil tank filling lines?

A. Filling pipes may be led directly from the deck into the tanks or to a manifold in an accessible location permanently marked to indicate tanks to which they are connected and from which the oil can be distributed to the various tanks through the piping system.

The end of filling pipes shall be fitted with a shutoff valve which shall be kept closed when not in use. Alternate arrangements where gravity systems for filling are employed will be given special consideration by the commandant.

Piping subject to head pressure from oil in the tank shall be fitted with shutoff valves at the tank which shall be remotely controlled from a readily accessible and safe location outside of the compartment in which the valves are located.

Q. Where must overflow pipes on fuel oil tanks be located?

A. Overflow pipes, equipped to discharge through the ship's side, must be located as far above the load line as practicable and be provided with a non-return valve located on the ship's side. Where the overflows do not extend above the freeboard deck before discharging overboard, they shall be provided with one automatic non-return valve fitted with a positive means of closing from above freeboard deck or with 2 automatic non-return valves without positive means of closing, provided the upper valve is so located above the deepest load line, as to be always accessible for examination under service conditions, and is of a type which is normally closed.

Two non-return valves shall always be used unless it is impractical to locate the upper valve in an accessible position, in which case the one non-return valve with positive means of closing will be acceptable.

Q. What are the rules regarding sounding pipes?

A. Oil and water tanks shall be provided with manual means of sounding. All other tanks and hold compartments with or without bilge drain wells which are not at all times accessible under service conditions shall be fitted with sounding pipes of not less than 1-1/2" pipe size which shall be led as straight as possible to within 2" of the lowest part of the tank or compartment to a position which is always accessible. (In addition, tank liquid level indicators may be fitted as an auxiliary means of indicating the liquid level.)

Where sounding pipes terminate below the freeboard deck on cargo vessels, same shall be fitted with gate valves, and on passenger vessels, where the valves terminate below the bulkhead deck, same shall be fitted with self-closing gate valves.

The upper ends of sounding pipes terminating at a deck shall be protected by a screw cap or plug. No perforations or openings will be permitted throughout the length of a sounding pipe where fitted to oil tanks. Provision shall be made to prevent damage to the vessel's plating by the striking of the sounding rod.

Q. How are safety valves installed?

A. Safety valves are installed in an upright position with stems vertical. They are to be connected directly to the boiler or as close thereto as possible.

Except in the case of superheater valves, they are not to be connected to any fitting with outlets for any purpose other than the escape of steam through the safety valve.

They may be fitted with internal dry pipes.

They must be fitted with an escape pipe, the area of which must equal the combined area of all valves discharging into it. Pipes must be supported so that the weight of the pipe is not transmitted to the body of the valve.

Q. What are the minimum and maximum sizes of safety valves?

A. Not less than 1-1/2" nor more than 4". Safety valves of 4-1/2" may be used for replacements on existing boilers. These diameters apply to the valve inlet.

Q. What type safety valve is used with superheated steam? Why? At what pressure is it set?
A. Exposed spring type, which is used to keep the spring clear of excessive temperature, as would be the case in the enclosed spring type. To prevent damage to the superheater, the drum safety valves shall be set at a pressure not less than that of the superheater safety valve setting, plus 5 pounds minimum, plus approximately the normal load pressure drop through the superheater.

Q. What are the regulations regarding repair of a cracked furnace?
A. Cracks in a corrugated furnace may be repaired by fusion welding, provided the length of crack does not exceed 20". Where repairs are extensive, the defective portion is cut away and patch riveted on the water side.

Q. When are boiler stop valves opened for inspection?
A. All valves and other boiler mountings and attachments shall be opened every fourth year at the annual inspection. They shall be removed after each eight years of service for examination of the studs or bolts connecting them to the boiler.

Q. What are the requirements for by-pass valves?
A. All steam stop valves in sizes exceeding six inches shall be fitted with by-passes for heating the line and equalizing the pressure before the valve is opened.

Q. How is adequate size of a safety valve demonstrated?
A. By accumulation test. Steam outlets are shut and fires are forced to their maximum capacity for 15 minutes for fire-tube boilers and 7 minutes for watertube boilers. During this period the pressure must not rise more than 6% above the maximum allowable working pressure.

Q. What information is stamped on the body of a safety valve?
A.
1. Name or registered trade-mark of manufacturer
2. Serial number
3. Inlet diameter
4. Operating pressure
5. Discharge capacity
6. Safe working pressure
7. Blowdown in psi

Q. If the inspector is in any doubt as to the safety of a boiler, what is done?
A. He may require the boiler to be drilled to determine the actual thickness.

Q. What information is found on the name plate of a boiler?

A. 1. Inspection Bureau Number
2. Tensile Strength ( ) lbs. per sq. in. (p. s. i.)
3. Hydrostatic test ( ) p. s. i.
4. Original Working Pressure ( ) p. s. i.
5. Built By......................................
6. Steel By......................................
7. Date.......................
8. Inspector's Initial ....

Q. What are the requirements for stop valves?
A. Stop valves must close with a right-hand motion when facing the stem. They must be installed to seat against the boiler pressure and means must be provided to indicate at a glance whether they are open or closed.

Q. What precautions should be observed when wearing a gas mask?
A. 1. Should not be used in space where there is less than 16% oxygen.
2. Be sure canister is not more than one year old (stamped on canister).
3. Tighten straps sufficiently to keep rubber facepiece tight against face.
4. Carry a flame safety lamp to warn of low oxygen content in space where working.

Q. By percentage, how much oxygen does normal air contain?
A. 21%

Q. At what percentage of oxygen will a flame safety lamp cease to burn?
A. 16%

Q. At what percentage of oxygen does unconsciousness occur in human beings?
A. 10%

Q. How may oil fumes be dangerous to humans? What precautions are necessary when entering a tank which had contained fuel oil?
A. The fumes contaminate the air and get into the lungs. Gases liberated by fuel oil are heavier than air and stay near the bottom of the tank, under beams, etc. U. S. Coast Guard regulations state that fuel oil tanks must be checked by an authorized chemist before entering.

Q. Explain the combination letter and number system used to classify hand-portable and semiportable fire extinguishers required aboard inspected vessels.
A. The letter indicates the type of fire on which the extinguisher is to be used. The number indicates the size. Sizes 1 and 2 are considered portable. Sizes 3, 4, and 5 are semiportable.

Q. What are the general forms of fire-extinguishing apparatus used on board vessels? For what purpose is each type used?

A. 1. Soda-acid: Fires in ordinary combustible materials where quenching and cooling effects of quantities of water are needed.

2. Foam: Fires in flammable liquids, greases, etc., where a blanketing effect is essential. Also same as 1.

3. Carbon dioxide: Fires in electrical equipment where use of a nonconducting extinguishing agent is of prime importance. Also same as 2.

4. Dry chemicals: Same as 2 and 3.

5. Sand: In fireroom; used on small oil fires to prevent them from spreading.

Q. What is meant by Class A, B and C fires?

A. Class A fire: Ordinary combustible materials where quenching and cooling effects of large quantities of water are needed.

Class B fire: In flammable liquids, greases, etc., where a blanketing effect is essential.

Class C fire: In electrical equipment where the use of non-conducting agents are of first importance.

Q. What are the requirements for the distribution lines of the steam-smothering system to cargo holds? How is the minimum diameter of pipe to each space determined?

A. The steam supply (100 psi) shall emanate from not more than three stations in easily accessible locations on the weather deck. Lines to paint lockers and similar spaces may be taken from the nearest steam supply line. Each distribution line to each compartment shall be fitted with a shutoff valve. The valve shall be marked. The manifold steam supply shall be fitted with a master valve. Provisions shall be made for draining. If located on open deck, the manifolds shall be enclosed in a cabinet and marked. Steam piping shall not be led in or through spaces accessible to the people on board (except machinery spaces and corridors). The size of the pipe and number of branch lines is determined by the volume of the space to be supplied.

Q. What tests are required for emergency lighting and power systems?

A. Operated and checked once a week. Tested under load for at least 2 hours once a month. Storage batteries for emergency lighting and power systems tested at least once every 6 months vessel is in operation.

Q. What installation test shall be made on new fuel oil discharge piping?

A. Fuel oil discharge piping between the pump and the burners shall be hydrostatically tested in the presence of an inspector at a pressure of 1-1/2 times the maximum pressure but not less than 500 pounds per square inch.

Q. During the vessel's annual inspection, what are the duties of all engineering officers in regards to defects or imperfections in the machinery?
A. Inform the Coast Guard inspector and assist in every way possible.

Q. What care is required for portable fire equipment?
A. Checked yearly. Dumped, refilled and date labeled. CO_2 weighed and filled if below required specifications. Kept in easily accessible places. Hose connections and hoses kept in good condition.

Q. Describe a fog nozzle.
A. A combination nozzle which can supply a straight stream of water or a fog spray. The nozzle lever has three positions. In the forward position the valve is closed. In the vertical position it produces a spray or fog. In the third position (back) it gives a solid stream of water.

Q. Describe the alarms and indicators used on watertight doors.
A. Each door is fitted with an electric alarm bell. This bell starts to ring 7 to 10 seconds before the door starts closing if operated from the bridge or any remote station. There is an indicating board on the bridge fitted with individual lights for each door to show whether the door is open or closed.

Q. Describe the signals displayed when loading or discharging fuel oil.
A. By day, a red flag (International code flag "B") flown from the triatic stay. By night, a red light.

Q. What attention should watertight doors receive?
A. (From G. R. & R.) It shall be the duty of the master to see that all watertight doors in subdivision bulkheads that may be opened at sea, and all mechanisms, remote controls, and indicators connected therewith, shall be periodically inspected at least once in each week that the vessel is navigated to be assured that they are in proper operating condition. On vessels in which the voyage exceeds one week in duration, these doors shall be operated before the vessel leaves port. All such doors shall be operated daily. The date of the test and the condition of the equipment shall be noted in the official logbook.

Q. What electrical equipment should be provided with remote control shut-offs?
A. Fuel oil service pumps. Emergency bilge pumps. Submersible pumps. Motor-operated watertight doors.

Q. Describe the procedure to be followed in the event a main condenser overboard line ruptures.

A. Shut down main propulsion unit. Shut down main circulator. Close sea suction and overboard discharge valves. Make necessary repairs.

Q. Discuss the proper procedure to be followed if an oil spill occurs while vessel is in port.
A. Reduce any possible fire hazards. Notify proper fire officials so that remedial action can be taken.

Q. What precautions are taken when pumping an inflammable liquid with a centrifugal pump?
A. Pump must be in good mechanical condition; shaft seals, wearing rings, bearings, couplings and the shaft true so as to preclude any heat being generated within the pump itself. When starting the pump it must be well vented and primed, the discharge valves opened to prevent reaching the shut-off head with its attendant heat generation. Gland seals or packing must be lubricated; where water cooling is required or smothering glands are fitted, a steady supply should be maintained. Tanks should not be allowed to drop to a level whereby the pump will lose suction. When nearing the low limit, a stripping pump should be started and the main pump secured. A strict watch schedule should be maintained at all times when pumping inflammable liquids.

GENERAL QUESTIONS

Q. If the volume of a unit mass of a perfect gas is doubled and the temperature is kept the same, what will happen to the pressure of the gas? What law is this?
A. If the volume of a unit mass of a perfect gas is doubled and the temperature is kept the same, the pressure will be decreased by one-half. This is known as an example of Boyle's law.

Q. Describe the original tests made upon new arc- or gas-welded pressure vessels.
A. Arc- or gas-welded vessels which have been both stress-relieved and radiographed shall be hydrostatically tested to not less than 1-1/2 times the maximum allowable pressure for a sufficient time to permit an inspection of all joints and connections. Welded vessels which have not been stress-relieved and radiographed shall be given a thorough hammer or impact test and, following the hammer test, the vessels shall be hydrostatically tested to 1-1/2 times the maximum allowable pressure.

Q. What welders are allowed to make repairs to a boiler?
A. Only welders who have been examined and certified as to their qualifications by an Inspector of the U.S. Coast Guard, American Bureau of Shipping, or the Bureau of Ships of the Navy Department, and hold a certificate which is still in force.

Q. Upon what does the size of a safety valve depend?
A. The size is determined by the capacity and the steam pressure of the boiler.

Q. What fittings are usually found on condensers?
A. Exhaust-steam connection, circulating-water inlet and outlet connection, condensate connection, air-suction connection, relief-valve connection, manholes or handholes for tube access, connections for thermometer, make-up feed water, drains.

Q. How would you clean the steam side of a condenser?
A. By boiling out with boiler compound, sal soda, or a similar solution.

1. Secure discharge from condenser to hot well by closing valve or inserting a blank flange.
2. For each 1000 gals. of water the condenser will hold, make a mixture of 200 gals. fresh water with 100 lbs. boiler compound (approximately 25 lbs. to a barrel—this allows for easier mixing). Sometimes 5 gals. of kerosene are added for each 200 gals. of mixture.
3. Fill the condenser with fresh water.
4. Turn live steam into the condenser through the boiling-out connection and bring contents to boiling point at atmospheric pressure.
5. Boil about 12 hrs.; drain contents.
6. Wash out several times with fresh water.

Q. How would you clean the water side of a condenser?
A. Any of the following methods may be used: 1). Brush out tubes with wire brush; 2). Blow out tubes with air lance.

Water boxes and tube sheets should be wire-brushed and all exposed surfaces painted with anticorrosive paint.

Q. How often should the condition of the zinc plates be examined; what should be done when examining; what is the sign that the zinc is not working properly; what should be done about it?
A. At least every two months, zinc should be thoroughly scaled with scraper and wire brush, and if more than half deteriorated, should be renewed. A zinc not deteriorating is not doing its work. Metallic contact between such a zinc and the metal of the condenser should be made adequate.

Q. Sketch 2 methods of installing condenser tubes to allow for expansion and contraction.
A. Figure 52.

Q. What are some of the principal causes for inadequate vacuum?
A. 1. Air leaks into the system.
2. Division plate carried away.

3. Dirty condenser.
4. Insufficient circulating water.
5. Inefficient or improperly operated pumps.
6. High injection water temperature.

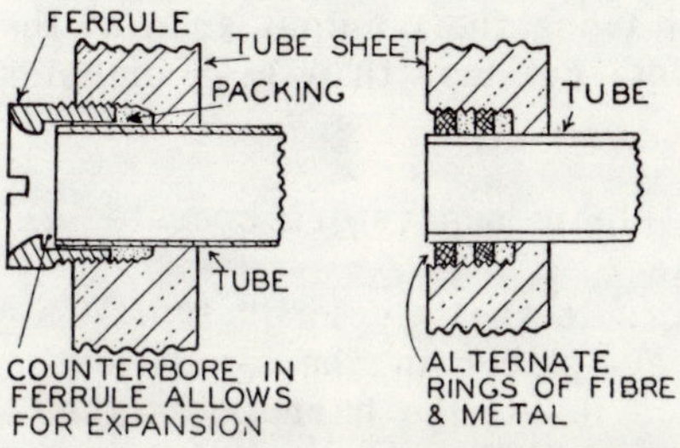

Fig. 52. Condenser tube installations.

Q. How are small leaks in a condenser frequently stopped?
A. By adding a bucket or two of oatmeal or sawdust to the inlet water. In most cases this is added in the fire or bilge-pump discharge manifold and pumped into the condenser. The oatmeal or sawdust, on passing through the tubes, sticks in the fracture, swells up and stops the leak temporarily.

Q. How does a surface condenser operate?
A. Sea water is pumped through tubes in the condenser shell and the exhaust steam passes around these tubes. On coming in contact with the cooler tube surfaces, the steam condenses and falls to the bottom of the condenser.

Q. How are salt-water leaks detected?
A. A salinity test of the condensate is made at least once a day. A sudden increase in water in the boilers may denote a tube carried away completely. Most plants today are fitted with salinity indicators which show the leak immediately.

Q. How are condenser leaks found?
A. 1. Fill steam space with warm water, and note if water comes out of any tube. Remember to blank off main air-pump discharge to hot well to prevent testing water from getting into feed system.

A slight pressure is sometimes added by either using air pressure or installing a temporary standpipe and filling with water to give a head.

In certain types of turbine installations, steam is put on the L. P. gland-sealing system to prevent the testing water from running out.

In the above method, Spectro powder may be added to the water to make the leak easier to detect. It turns the water a brilliant green color.

2. Another method is to start the air ejector and create a vacuum. Passing a candle in front of the tubes, a leaky tube will draw the flame in.

3. Using the same method of producing a vacuum, a number of sheets of plastic (thin) may be placed over the tube ends. The vacuum will pull the plastic against the leaky tube.

Q. Name three tests usually given boiler water daily.
A. Hardness; chlorinity; alkalinity.

Q. What are the G. R. &R. regarding the required area of a dry pipe and openings in relation to the openings to which they are attached?
A. Area of openings must be at least twice the nominal area of the dry pipe. Openings in pipe to be slots not less than 1/4" or holes not less than 3/8" diameter.

Q. Explain the principle under which steam separators operate in removing condensate from the steam.
A. One of the most common separators is the "Cyclone" type. This consists of a number of round barrels placed in the steam and water drum. There are curved vanes inside the barrel. As the steam swirls around through these vanes, the water is thrown to the outside by centrifugal force and drips back to the water in the drum. The steam continues upward to a collector pipe and to the valves, Fig. **53.**

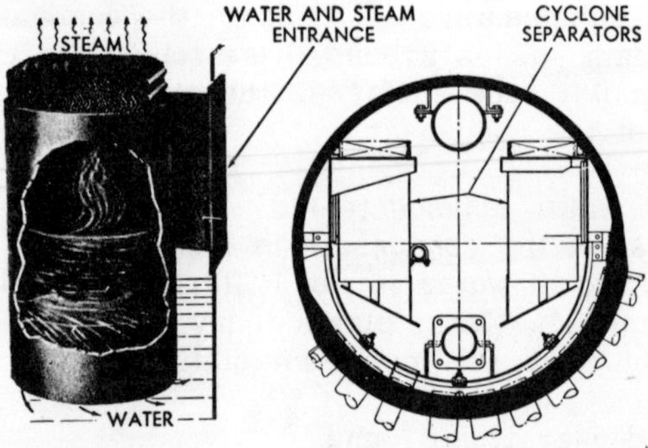

Figure 53.

Q. How may the vacuum in a condenser be determined by temperature? Under what turbine operating conditions will this method be inaccurate?
A. By referring to the corresponding pressures in the steam tables. When running at slower speeds.

Q. What is the theory of deaeration?
A. Any gas, such as oxygen, is soluble in water in proportion to its own pressure. When water is heated to the boiling point, the pressure of the water vapor will be equal to the total pressure of the atmosphere pressing on the water—consequently the solubility of gases will become zero. The first step in deaeration is therefor to raise the water to the boiling point. As the bubbles pass through the surface they must be carried away immediately so that they will not build up a partial gas pressure causing the gas to go back into solution.

Q. Why is the thermal efficiency of an injector high?
A. The heat of the steam used in the injector is practically all regained in heating the feed water.

Q. Describe the feed-water deaerating process used on modern installations.
A. The feedwater is sprayed into the deaerating tank through an atomizer into an atmosphere of steam. This spraying heats the water almost to the steam temperature and at the same time removes much of the dissolved oxygen. The mass of mist created by the water and steam results in a large liquid surface. The steam agitates and scrubs the water surfaces and carries away the liberated oxygen. The deaerated water falls to the bottom of the tank and the steam and liberated gases are led to the preheater thence to the vent condenser.

Q. Describe a thermoelectric pyrometer.
A. This consists of two wires of dissimilar metals connected at both ends. When they are subjected to different temperatures, an electrical potential is set up between them. This voltage is almost directly proportional to the temperature difference, therefore, a voltage measuring instrument placed in the circuit will measure temperature.

Q. In making boiler repairs, it is found that a number of new tubes are too short. May new ends be welded on?
A. Tubes may be lengthened by forge, electric-resistance, arc or gas butt-welding methods. In each instance, the type of repair must meet the requirements as set up by U. S. Coast Guard standards.

Q. What effect does temperature have on salinity readings?
A. If the sample of water being tested with a salinometer drops in temperature below that marked on the salinometer, a false reading will be obtained. This is due to the fact that water contracts when cooling, down to a limit of 39°F.

For every 10° less temperature, the salinometer will show 3/4 oz. more than the actual density. That is, if the sample cools to 160°, it would be 40° lower than it should be (if reading on the 200 scale). This would mean

$\frac{40}{10} = 4 \times 3/4 = 3$ oz. in excess of the real density.

Q. What effect will magnesium chloride have on boiler water?
A. Magnesium chloride reacts with water to form hydrochloric acid and magnesium hydroxide. $MgCl + H_2O = HCl + MgO$

The hydrochloric acid will show up in boiler water testing.

This acid reacts with iron forming ferrous chloride and free hydrogen. This results in a slow dissolving process of the boiler metal. $HCl + Fe = FeCl + H$

Q. How do you read a "U"-tube type manometer?

A. By reading the difference between the two legs of the tube. In Fig. **54.** the reading is 1". Draft is measured in inches of water. 1" = .036 pounds pressure.

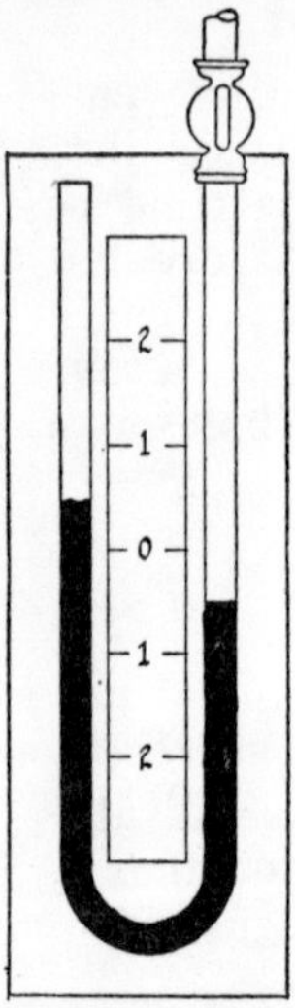

Fig. 54. Manometer or draft gage.

Q. How is a manometer connected?

A. One end is connected to the air duct leading to the boiler. The other end is open to the atmosphere.

Q. Explain how the following conditions will be indicated and how they will affect the combustion of the fuel oil in the furnace: 1). Excessive air; 2). Insufficient fuel oil pressure; 3). Water in the oil; 4). Low fuel oil temperature.

A. 1. White smoke at stack; incandescent white furnace flame with jagged edges; will record a low CO_2 reading and a high O_2 reading; will cause excessive cooling of furnace.

2. Drop in steam pressure; possible carbonization of burner tips and oil dripping into furnace rather than burning.

3. Sputtering burners; drop in steam pressure.

4. Drop in steam pressure will probably be noticed first; fires will gradually go out.

Q. Define specific gravity.

A. Specific gravity is the ratio of the weight of a given volume of a substance with the weight of an equal volume of water at the same temperature and pressure.

Q. What is the principle of the Hays Orsat gas analyzer?

A. This instrument shows the amount of carbon dioxide (CO_2) in flue gases. A sample of gas is taken and accurately measured. This sample is then passed through an absorbing medium which removes the CO_2. The sample is measured again and its volume compared with the original sample. The difference is read in an exact percentage on a scale. If it is a Triple unit, it also measures the amount of oxygen (O) and carbon monoxide (CO).

Q. If a boiler feed pump in operation fails to maintain the water level in the boiler, what would you do? What are possible reasons for its failure?

A. Start the other feed pump. If this also fails, change to auxiliary feed line (main feed check may be inoperative). If neither pump, nor both pumps together, can maintain water level, try the steam injector. If this works, the feed pumps need overhauling.

Q. What is the purpose of a manometer?

A. To measure air pressure to the furnaces. This reading along with that shown on the CO_2 recorder can be interpreted to make proper adjustments to increase boiler efficiency.

Q. What are the advantages and disadvantages of an injector?

A. Advantages: Cheap; occupies small space; easy to repair: no moving parts.

Disadvantages: Will not start with steam under designed pressure; very touchy, must be handled carefully.

Q. What are the absorption chemicals used in the Orsat gas analyzer?

A. For CO_2: Caustic potash
For CO: Cuprous oxide
For O: Pyrogallic acid

Q. How is approximate amount of excess air figured from the CO_2 reading?

A. Subtract % of CO_2 reading from 14 and multiply by 2.

Example:. CO_2 recorder registers 7%
14-7 = 7   7 X 2 = 14 lbs. excess air

Q. Can acids be used for cleaning heat exchangers?

A. Yes, but certain safety precautions should be used. Hydrochloric acid can be used but, being a liquid, there is the problem of safe storage. Sulfamic acid can also be used. This comes in powder form and is cleaner and safer to store. Regardless of which type of acid is used, take all safety precautions possible. Use goggles, a respirator or mask, and rubber gloves, Have plenty of ventilation. If there are any thermocouples, zinc plates, or other equipment that might be damaged by the acid, remove them from the system. When finished with acid circulation, do not dump it into the bilge.

Q. What are the desirable qualities of a good lubricating oil?

A. Good adhesion and cohesion properties, physical stability in regards to temperature and pressure, chemical stability against oxidation.

Q. What is the principle of a centrifuge?

A. The centrifuge purifier consists of a number of hollow cylinders or bowls which are rotated at a high speed. As the oil spins around, the heaviest impurities are deposited on the wall of the vessel, the water forms the next whirling layer and oil, being lighter, separates into a layer closest to the axis. These three substances are removed through spouts on the side of the centrifuge.

Q. What advantages are gained with the use of hollow shafting? How does the strength of a hollow shaft compare with a solid shaft?

A. First of all, the weight of the shaft is reduced. The I. D. is usually 60% of the O. D. This figures out that the hollow shaft will weigh about 64% as much as the solid shaft. At this percentage, the hollow shaft is approximately 87% as strong as a solid shaft but will operate at approximately a 15% higher torsional stress. Another comparison is that, for the same weight, a given length of hollow shaft with I. D. 60% that of its O. D. will have an O. D. 1-1/4 times the O. D. of a solid section of shaft the same length, but will be 70% stronger.

Q. How should packing be installed in stuffing boxes of moving pump rods?
A. With cuts staggered to cut down leakage as much as possible. Use slant cuts and leave approximately 1/4" opening (depending on size of rod).

Q. How is the condensate pump protected while maneuvering?
A. A small priming line is connected from the hot well to supply a sufficient amount of water to keep it from running dry.

Q. What conditions may cause groaning in the steam end of a reciprocating pump?
A. Cracked or broken rings. Another common cause is the fact that the steam cylinder and liquid cylinder are not in a straight line after the pump becomes heated. This is due to the fact that the two cylinders are so far apart and connected by small rods. The steam cylinder expands more than the liquid cylinder and forces it further away from its common supporting column. This throws the center line of the steam cylinder farther from the column than the center line of the liquid cylinder. This will cause a groaning noise. To remedy, place shims behind the foot of the liquid cylinder when it is cold.

Q. What is the reason for installing a recirculating line on most centrifugal boiler feed pumps?
A. To prevent a pressure build-up in the event the boiler checks are closed.

Q. Explain "suction head" in reference to pumps.
A. Suction head is the distance between the surface of the liquid to be pumped and the suction of the pump. However, the height is limited by the effects of atmospheric pressure: 14.7 psi. Since a 1-ft. head of water exerts a pressure of approximately 1/2 lb., the head of water would theoretically be limited to about 30 ft. Friction cuts this to about 26 ft. Viscosity will also have an effect on the suction head. For example, it would be impossible to lift cold fuel oil a height of 26 ft.

Q. What effect does the speed of the engine have on a magneto type ignition system on a gas engine?
A. Since the magneto is driven by the engine, the spark is very weak when first turning over. As the speed increases, the intensity of spark naturally increases.

Q. Explain viscosity vs. temperature of liquids.
A. As temperature decreases, viscosity increases (liquid gets thicker). As temperature increases, viscosity decreases (liquid gets thinner).

Q. Discuss the advantages and disadvantages of the gate valve with the globe valve.
A. Probably the greatest advantage of the gate valve is the fact that when it is wide open there is no restriction to the flow through it. They are operated wide open or closed. The fact that the valve closes across the orifice instead of being pushed on or into it is a disadvantage. If the valve is throttled, the liquid will wire-draw around the edges and ruin the disc, making it unusable. Gate valves should never be used for steam. The globe valve has the disadvantage of having to let the liquid or gas pass in a roundabout way through the seat. This causes friction. The globe valve is easier to grind-in and keep in good condition.

Q. What distance must a globe valve be raised from its seat in order to give an area of opening which is equal to the area of the seat?
A. 1/4 of its diameter

Example: How much will a 4" valve have to be lifted from its seat to obtain full opening?

1/4 of 4" = 1"

Proof: Area of opening $= 2\pi rh$

Area of valve $= \pi r^2$

We want these two to be equal, therefore,

$$2\pi rh = \pi r^2$$

$$h = \frac{\pi r^2}{2\pi r}$$

$$h = \frac{r}{2}$$

$$h = \frac{2r}{4}$$

$$h = \frac{D}{4}$$

Q. What is meant by bilge injection?
A. A valve placed in the suction line to the main circulator.

Q. What is the purpose of the bilge injection?
A. To pump out the bilges in case of emergency.

Q. What precaution is taken in installing the bilge injection to prevent accidental flooding of the ship?
A. It is a nonreturn valve. Be sure it closes with pressure above the disc.

Q. What are the advantages of oil over coal as a fuel?

A. 1. Amount of heat supplied to furnace can be controlled more quickly and over wider ranges.
2. Less storage space needed.
3. Smaller fireroom crew required.
4. Reduced fuel-handling time.
5. Cleaner to handle.

Q. In what position and location should reducing valves be installed?

A. Reducing valves should be installed in a vertical position in the highest point in the pipe line. They should be located so that the bonnet and working parts can be removed without breaking pipe joints.

Q. Explain the construction of a steam gauge.

A. Consists of an elliptical-shaped tube called a "Bourdon tube," mounted with one end free to move, Fig. **55.** Steam pressure is admitted to the secured end, tending to straighten the tube out. The free end moves accordingly and a pointer attached to it travels across the face of the gauge, showing registered pressure.

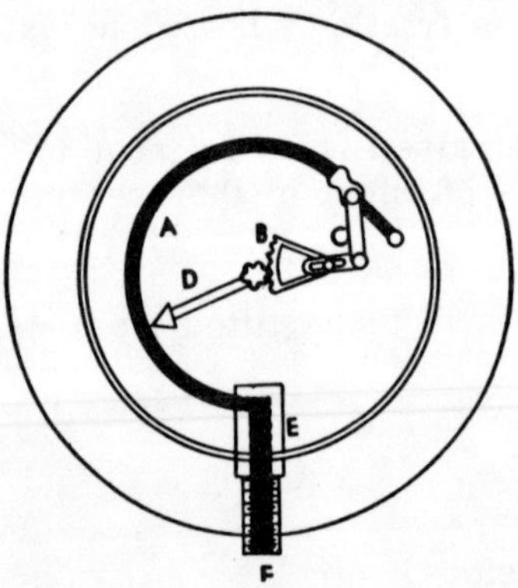

Fig. 55. Bourdon pressure gage.

Q. Does steam act directly on a steam-pressure gauge? Why?

A. No. There is a U-bend or loop in the connecting pipe between the boiler and the gauge. This fills with water from condensation and protects the gauge from direct contact with the steam. Uninformed operators frequently discover this water and attempt to keep it drained out in the mistaken belief that the gauge must be clear of water in order to register correctly.

Q. What makes a pop-type safety valve pop?

A. Extra area added to the valve area by a lip, an adjustable ring, or similar device. Slight lifting of seat admits steam to an outer valve area much larger than the seat, thus subjecting the spring to a pressure greater than that exerted by the seat alone, so the spring will compress quickly. When pressure is about 3 lbs. below allowable working pressure, this process reverses itself and the spring applies its tension and closes the valve sharply.

Q. Which valve is located next to the boiler, the feed stop or check? Why?

A. The feed stop valve. The feed is changed over to the auxiliary feed line. The main feed stop is then closed. This prevents steam and water from blowing back out of the drum. The discharge from the feed pump to the main feed line is closed. The main feed check valve is now isolated and can be worked on.

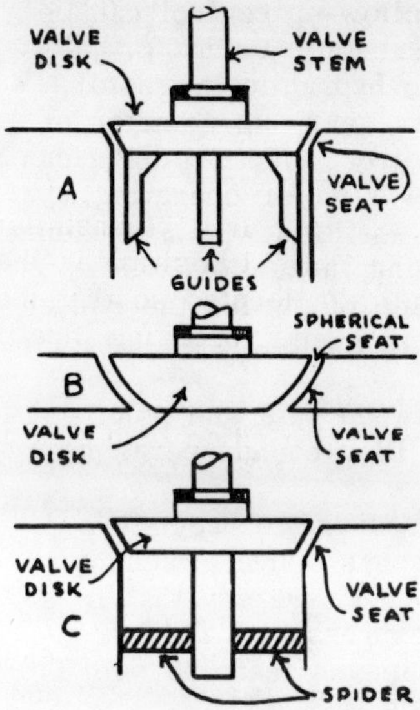

Figure 56.

Q. How are most valves constructed so that the discs will not tilt, bind or jam in an open position?

A. 1. Guides are cast on the lower end of the valve disc. These guides fit into the chamber which extends below the valve seat, Fig. **56A.**

2. Valves with spherically ground seats and discs will not jam or bind, Fig. **56B.**

3. An extended stem on the bottom of the valve disc travels in a spider in the body of the valve, Fig. **56C.**

Q. What are the indications of air in the hydraulic telemotor system? If air is present, how is it removed?

A. With steam on, put helm hard over several times in each direction. If air is present, the rudder will follow the wheel very sluggishly and will not run to the hard-over position unless the wheel is turned past the usual angle.

To remove air, pump up telemotor with hand by-pass valve open, circulating the liquid briskly until it is observed to flow for each stroke of the pump, and not in a steady stream. Close the after telemotor discharge valve and maintain pressure on the pump, at the same time bleeding the air from the highest part of the forward telemotor by slacking back the relief nuts or opening air cocks. Try the after telemotor air cocks also. Hold the air cocks open until a rush of liquid displaces the air. Close all valves and close pump discharge valve last.

Q. Describe the hydraulic type steering gear. What provisions are made, if any, for absorbing excessive rudder shock on this type gear?

Fig. 57. Hydraulic type steering gear.

A. This is probably the most efficient method of operating the rudder, Fig. **57**. The installation cost is higher than others but upkeep is generally lower. An automatic follow-up control can be used with mechanical means of installation from bridge or pilot-house, which is dependable, such as the hydraulic telemotor, shafting or wire rope. It is very efficient gear, as a motor of about half the horsepower of that for a direct electric drive can be used. The pumping unit, consisting of a motor operating a variable-stroke pump at constant speed, is fitted with a pressure regulator to prevent overloading or stalling the motor.

Under test, this gear has proved capable of putting the rudder hard over to hard over in 8 seconds at 15 knots.

Q. Explain the meaning and difference between psia and psig.

A. Psia-pounds per square inch absolute; Psig-pounds per square inch gage.

When a gage is open to atmospheric pressure, it registers zero. In other words, it registers pressures above atmospheric pressure. When a gage registers 250 pounds, we say that it is 250 psig. If one wants to obtain absolute pressure, add atmospheric pressure to gage pressure. This means that 250 psig is 250 plus 15 (14.7 to be exact) or 265 psia.

Q. What is a compound gage?

A. One that registers both pressure and vacuum.

Q. What causes a safety valve to chatter?

A. Defective seat or disc. Loose blowdown ring. Blowdown ring out of adjustment.

Q. What is the main difference between a safety valve and a relief valve?

A. A safety valve opens when pressure set for is exceeded. It has a blowdown adjustment which holds it open until pressure has dropped 10 to 15 pounds below that for which it was set to open. A relief valve opens when pressure it is set for is exceeded and closes as the set pressure is reached again.

Q. Describe a stern tube bearing for a vessel and tell how it is lubricated and how water is prevented from entering the vessel.

A. In most installations, the tail shaft is fitted with a bronze liner and the stern tube is fitted with hard rubber strips. The latter are installed in sections and have rounded edges to allow water to run in the grooves and supply lubrication, Fig. **58**.

Water is prevented from entering by means of a watertight bulkhead fitting, Fig. **59**.

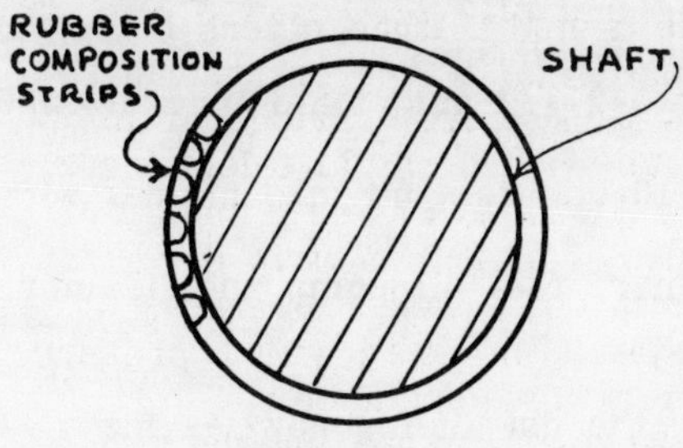

Figure 58.

Q. State the procedure used in case of a fire in generating coils.
A. Shut down; use CO_2.

Q. What different materials are used for linings of outboard shaft bearings and what are their relative advantages?
A. Lignum vitae has been used extensively for years. This is one of the hardest woods known and is self-lubricated by allowing a small amount of sea water to run through the stern tube. In recent years, a composition rubber has come into wide use. One of the outstanding advantages of this latter type is that mud, sand, etc., have practically no cutting or erosion effects on the bearing surfaces.

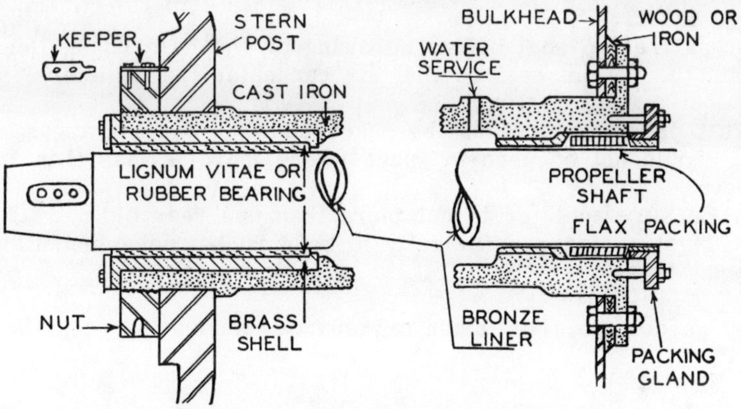

Fig. 59. Stern tube and propeller shaft.

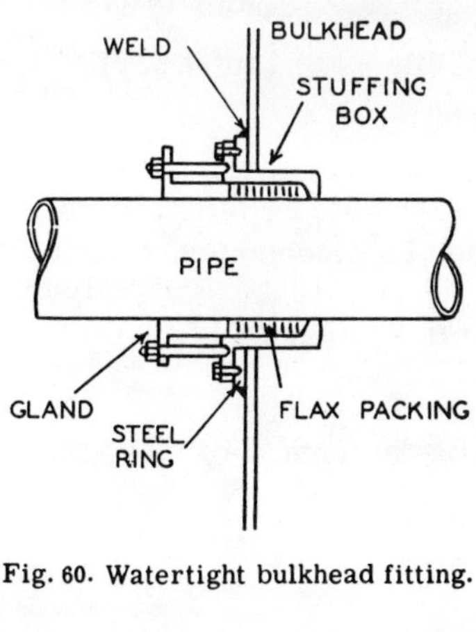

Fig. 60. Watertight bulkhead fitting.

Q. Describe one type of stern tube bearing.
A. Composition rubber strips are inserted into grooved channels in the stern tube. These strips have a semicircular outer edge. The shaft rests on the upper edges of these strips and mud, sand, etc., have free access to run along the grooves and wash out through the stern (see Fig. 58).

Q. Sketch a watertight bulkhead fitting.
A. Figure 60.

Q. What fire-fighting equipment is required in machinery spaces on steam-propelled vessels burning fuel oil?

A. 1. Fixed system of CO_2, foam or water spray.

2. Portable fire extinguishers: two-2-1/2 gal. foam-type in each boiler room.

3. On vessels over 1000 G. T., 10 cu. ft. sand in container with scoop.

4. Drip pans under each oil burner, fuel oil pump and strainer, to prevent oil from reaching bilges.

Q. What advantages are provided by the use of fog nozzles for fighting fire aboard ship in comparison to a nozzle which produces a direct high-pressure stream of water?

A. The direct-type nozzle usually hits the fire with such force that it spreads the fire. The fog nozzle forms an umbrella of spray (cutting off the air supply) which blankets the fire.

Q. Explain the method or procedure in attaching a new canister to a gas mask.

A. Break the top seal before screwing on the hose connections. Break the bottom seal so that air can enter the canister

Q. What precautions should be observed when wearing a gas mask?

A. 1. Should not be used in space where there is less than 16% oxygen.

2. Be sure canister is not more than one year old.

3. Tighten straps sufficiently to keep rubber facepiece tight against face.

Q. By percentage, how much oxygen does normal air contain?

A. 21%.

Q. Sketch a single-pass counterflow feed-water heater.

A. Figure 61.

Q. Sketch a fire-side and a water-side fusible plug (cut-away).

A. Figure 62.

Q. Who sets safety valves and adjusts them to keep the boiler within safe working pressure?

A. Duly authorized inspectors only, except in emergency.

Q. Sketch an apparatus for cleaning fuel-oil heaters.

A. Figure 63.

Q. Sketch a simple diagram showing the fundamentals of the Bailey Combustion Control System.

A. Figure 64.

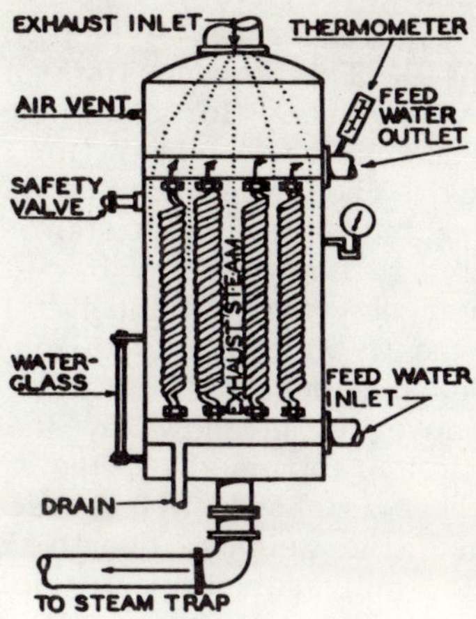

Fig. 61. Single-pass counterflow feed-water heater.

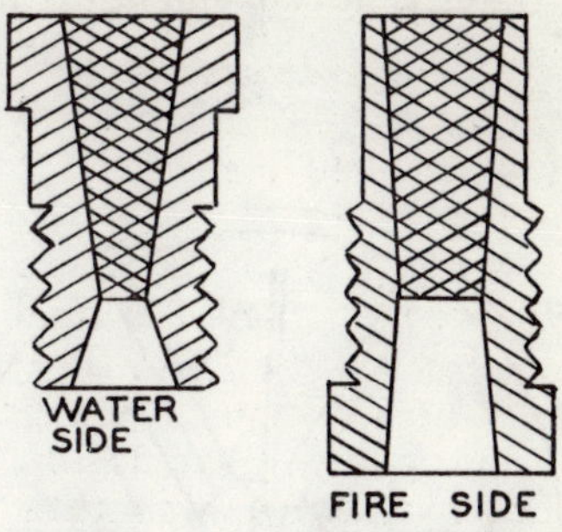

Fig. 62. Fusible plugs.

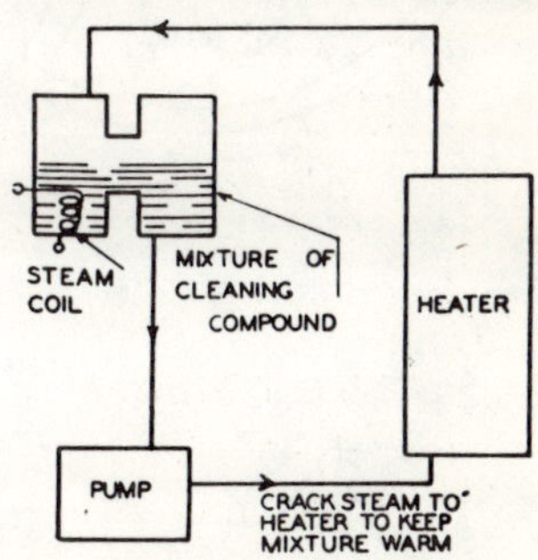

Fig. 63. Cleaner for fuel-oil heaters.

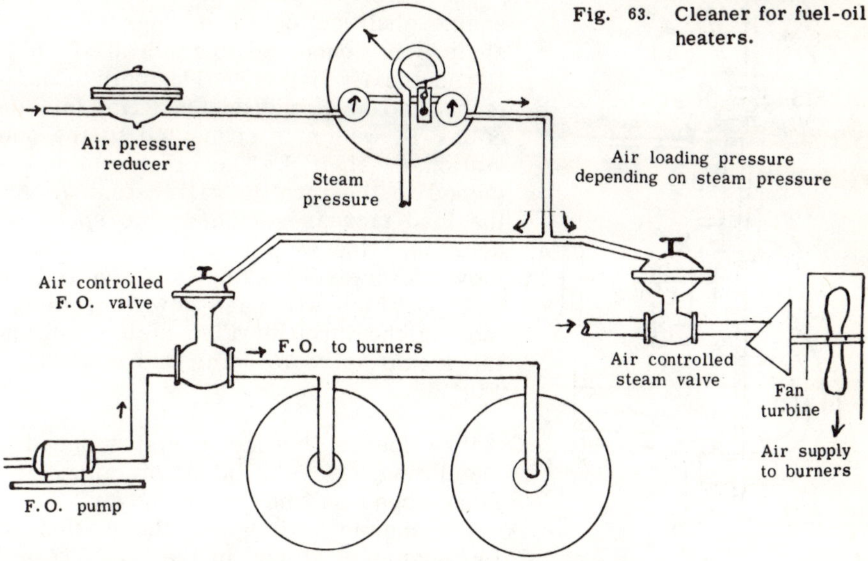

Fig. 64. Bailey Combustion Control System.

Q. By using a diagram, explain the operation of a combination centrifugal force and inertia type speed governor.

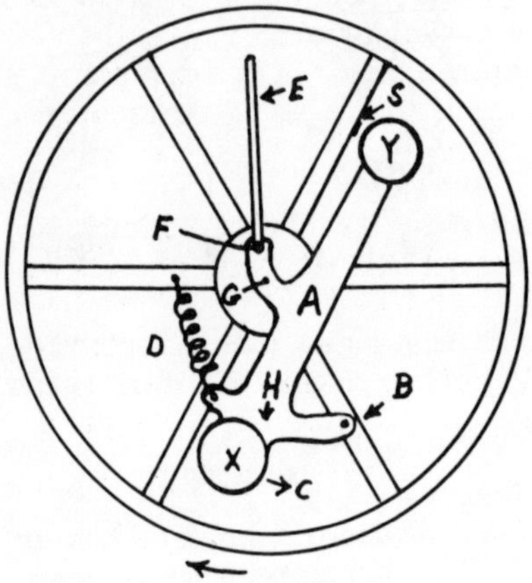

Figure 65.

A. The governor bar A is pivoted at point B. This bar consists of a heavy weight X (for centrifugal force) and a smaller weight Y (for inertia force). The valve rod E, which is connected to the valve stem, is connected to the governor bar at F. The center of gravity of the bar is at H. As the flywheel starts turning clockwise, this center of gravity wants to move out as shown by the arrow C. It is restrained by the spring D. If there is a sudden increase in speed, the weights, due to their inertia, pull against the tension of the spring. This movement will bring the valve rod pin F closer to the center G of the flywheel. This reduces the valve travel and thereby slows down the engine. The bar is kept from swinging too far by a metal stop S secured to the spoke on the flywheel, Fig. **65.**

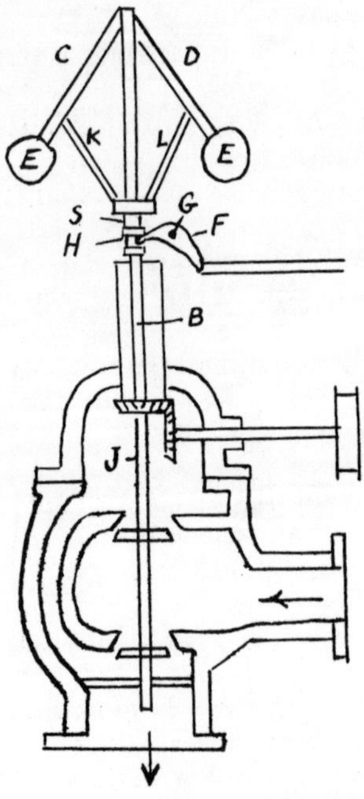

Figure 66.

Q. By using a diagram explain the operation of a fly-ball type governor.

A. The governor pulley A is belted to the engine shaft pulley. The inner turning shaft B is connected to the rods C and D and the fly balls E. The bell crank F is pivoted at G and one end fits into the collar H, which is connected to the outer stationary shaft S. The other end is attached to the throttle valve stem J. As the load increases, the engine speed will decrease, the weight of the fly balls will move downward pushing down on rods K and L, which will push down on collar H, moving the throttle valve stem, and open the steam valve to admit more steam, Fig. **66.**

Q. How does a flywheel assist a speed regulating governor in maintaining a constant speed on the prime mover?

A. The inertia built up by the motion of the engine is stored in the flywheel and keeps the machine running smoother.

Q. Why should the water level or its temperature in the direct contact, deaerating feed-water heater not be permitted to rise or fall any appreciable amount?

A. If water level rises, it will cut down the efficiency of the unit. If the temperature falls, it will not process (deaerate) properly. Also, the pressure and temperature at the feed-water suction is very near the boiling point. Any abrupt change may cause the feed water to vaporize and become vapor-bound. The feed pump may burn up any water-lubricated rings or packing if it becomes vapor-bound.

Q. What drains are usually led or pumped to the direct contact, deaerating feed heater on a modern vessel? What care should be taken with these drains?

A. Whistle drain, HP reducing station drains, feed pump recirculating line, vent condenser drain, HP heater drain. Be sure traps are in perfect working order at all times. Do not leave by-pass valves open.

Q. During what operating conditions does the deaerating feed heater frequently tend to overheat? How should this overheating be kept to a minimum?

A. If the back pressure regulating valve is not operating properly. The whistle drain trap or HP heater drain trap by-pass valves may be open or the traps may not be functioning properly.

Q. What is meant by the displacement of a pump?

A. The amount of liquid that it discharges. In a reciprocating pump this would be the area of the piston times the length of the stroke (volume). This would then be converted to gals. and finally multiplied by strokes per minute. This would give displacement in g. p. m. (gallons per minute). In a rotary pump, the manufacturer indicates the g. p. m. on the name plate.

Q. What is the difference between displacement and the actual discharge of a pump?

A. The displacement is figured from actual dimensions. The actual discharge will be less, due to losses caused by leakage past the piston, gland leakage, friction losses, etc.

Q. What is the advantage of piston-type valve taking steam on the inside rather than on the ends?

A. The valve stem packing will be under a lower pressure.

Q. When bunkering, what precautions would you take to prevent spills?

A. Indicators and alarms in perfect condition; sufficient help standing by F. O. valves and sounding tubes; know how much oil is needed in each tank; have a slack tank ready for any overflow condition; know discharge capacity of pump supplying you; have pump slowed down when topping off.

Q. What are the advantages of a centrifugal pump in comparison with a vertical reciprocating pump?

A. A smoother, more powerful discharge; better adapted to motor or turbine drive; fewer lubrication and wear problems; closed discharge valve does not cause a high discharge pressure.

Q. During constant flow conditions, when water flows horizontally through a pipe that has a narrow constriction, how is the speed (velocity) and pressure of the liquid affected in passing through this restriction?

A. Pressure drops and velocity increases.

Q. What is the difference between a deaerator and a DC heater?

A. A DC (direct contact) heater is one in which the steam and feed water are brought into direct contact with each other.

A deaerator removes dissolved gases from the feed water. This is accomplished by heating the feed water to a point where the gases are no longer soluble and then scrubbing the steam by spraying it through fresh water.

Q. Why is it necessary for a steam pump to have steam admitted during full stroke?

A. In order to maintain a full discharge pressure from the liquid end by having a full power stroke on the steam end.

Q. What prevents a duplex steam pump piston from striking at the ends of the stroke?

A. By properly setting the steam valves. If the clearance volume is very small, cushion valves are usually fitted to the pump. These valves are manually controlled and allow steam to by-pass directly to the exhaust ports.

Q. If one of the steam pistons in a duplex pump repeatedly strikes the head on each stroke, what fault is indicated? What may be done to remedy it?

A. Too long a stroke. Shut down pump, remove valve stem pin and turn stem in proper direction about one turn. The direction will depend on the operator deciding which piston is striking which end. By turning the stem, which turns in the nut in the valve, the valve position can be changed without removing the steam chest cover. (Remember that the crosshead from one rod works the valve on the other side.)

Q. What is a pneumercator?

A. A pneumercator, or tank level indicator, is used for measuring the height of a liquid in a concealed or remote tank, such as fuel oil settling tanks, deep tanks, etc. A balance chamber is placed at the bottom of the tank and a mercury gauge is placed at a convenient place outside the tank. The balance chamber has a small hole located near the bottom. As the liquid rises above this hole, it traps the air in the balance chamber and compresses it. This compression of the air forces the mercury to rise in a tube in proportion to the head of pressure of the oil that caused the air in the balance chamber to be compressed. A certain amount of oil, on entering the balance chamber, will rise above the zero level and the volume of compressed air will be too small to give an accurate reading. To compensate for this loss, a hand air pump is used to force air into the balance chamber, bringing the level of the oil below the top of the hole in the balance chamber. Any pressure that might build up in the tank is compensated for by connecting the top of the indicator tube with the air space in the tank.

Q. What is a rotary regenerative air heater? How is it cleaned?

A. This type of air heater is shown in Fig. 67. It consists of slightly separated metal plates, supported in a frame attached to a slowly moving rotor shaft, and arranged edge-on to the gas and air flow. As these plates pass progressively through the gas stream, they are heated. Then, in passing through the air stream, they give up heat to the air before again entering the gas stream. This is called the regenerative cycle. The plates are placed in segmental groups to fill a complete circle around the shaft. Seals are provided to reduce air filtration into the gas. Gas and air ducts and the casing around the heater confine the gas and air to the desired flow paths. Soot blowers are used to clean heating elements.

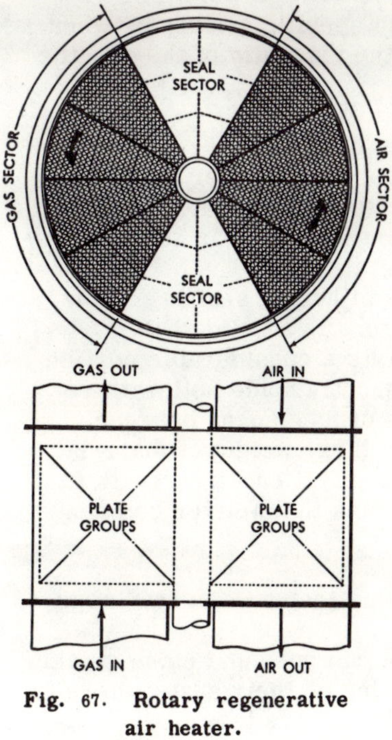

Fig. 67. Rotary regenerative air heater.

Q. Describe the original tests made upon new arc- or gas-welded pressure vessels.

A. Arc- or gas-welded vessels which have been both stress-relieved and radiographed shall be hydrostatically tested to not less than 1-1/2 times the maximum allowable pressure for a sufficient time to permit an inspection of all joints and connections. Welded vessels which have not been stress-relieved and radiographed shall be given a thorough hammer or impact test and, following the hammer test, the vessels shall be hydrostatically tested to 1-1/2 times the maximum allowable pressure.

Q. Explain the significance of the terms SSU and SSF in connection with fuel or lubricating oils.

A. The terms SSU and SSF are abbreviations for seconds Saybolt Universal and seconds Saybolt Furol, respectively. These abbreviations refer to the viscosity of the oil. The units of these two standards of viscosity are units of time, namely, seconds. The viscosity is determined by finding how long it takes a fixed quantity of oil to flow through an orifice of a viscosimeter at a standard temperature. The SSU units are generally used for the lighter bodied oils, such as Diesel and lubricating oils; the SSF units are used for the heavier oils such as Bunker C. Over the usual ranges of viscosity the SSU viscosity is approximately 10 times the SSF viscosity.

Q. How does the specific gravity and the viscosity of the liquid to be purified affect the rate of separation in a centrifugal purifier?

A. In general, the greater the difference in specific gravity between the liquids to be separated and the lower the viscosity of the oil, the greater will be the rate of separation.

Q. What is meant by normalizing steel after forging?

A. Normalizing consists of allowing the steel to cool after forging to a temperature below the critical range, then reheating it to a proper temperature to refine the grain and allowing it to cool in still air.

Q. What advantages are gained with the use of hollow shafting? How does the strength of a hollow shaft compare with a solid shaft?

A. The shafts of high-speed vessels constitute a considerable portion of the total machinery weight and are frequently made hollow to reduce weight. Hollow shafts permit the use of long spans between bearings. The removal of a comparatively large weight of material from the bore of the shaft has only a small effect on its strength in torsion and bending and a very small increase in diameter compensates for this loss.

Q. Why are many small valves used in a pump instead of a few large ones?

A. Small valves have less lift and therefore can open and close quicker and are subjected to less wear from pounding on their seats; there is less friction, and the shorter studs or spindles are less likely to become broken from shocks or unequal distribution of pressure. A number of valves are not likely to be seated at the same time and less shock is thereby imparted to the valve deck than when a large valve comes down on its seat; small valves are cheaper to construct and repair, and the valve seats are easier to replace than those of larger valves.

Q. Given a 6-inch diameter nut, pitched four threads to the inch, how would you mark the nut so that each graduation would represent 0.010 inch?

A. Since one turn of the nut represents 0.250 inch along the bolt, graduate the nut into 0.250/0.010 or 25 equidistant markings around the top surface of the nut.

Q. Describe the various provisions for relief of expansion stresses of steam piping.
A. Various shapes of expansion joints (loops) in the pipe. Packed slip joints.

Q. What is a scantling?
A. The dimensions of all structural parts, such as frames, girders, plating, used in building a vessel.

Q. What is a strake?
A. A range of planks or plates abutting each other and extending the whole length of the ship.

Q. A box girder compares with what part of a vessel?
A. A hollow steel girder of rectangular cross section, made of riveted or welded plates and angles.

Q. Explain how a flywheel assists a speed-regulating governor in maintaining a constant speed on the prime mover.
A. It maintains the crank at a uniform speed thereby reducing the hunting of the governor that would otherwise result from oscillations.

Q. What is a pressure-vacuum relief valve? Where is it used?
A. A special type valve placed on vent lines on cargo tankers. The newer types are weight-loaded and usually open at a pressure of from 0.5 to 1.5 p.s.i. and the vacuum valve opens at a vacuum up to 0.75 pounds. The pressure side prevents a pressure from being built up in the tank as it is being filled. The vacuum side prevents a vacuum from being created in the tank as it is being pumped out.

Q. Describe a fuel oil transfer system.
A. The transfer system consists of the transfer pump, its suction and discharge manifolds and the pipelines leading to and from (suction and filling) the double bottom and settling tanks.

Q. What is meant by the thermal efficiency of an engine?
A. It is a ratio between the heat added and the heat lost.

$$T.E. = \frac{T_1 - T_2}{T_1}$$

T_1 = heat added $\qquad$ T_2 = heat lost

Q. What are the desirable qualities of a good lubricating oil?
A. Two characteristics of a good lubricant are cohesion and adhesion. Cohesion is the attraction that particles in a mass have for each other. Adhesion is the property by which one mass will cling to another. Two other important characteristics are viscosity and emulsification properties.

Q. If the volume of a unit mass of a perfect gas is doubled and the temperature is kept the same, what will happen to the pressure of the gas? What law is this?

A. If the volume of a unit mass of a perfect gas is doubled and the temperature is kept the same, the pressure will be decreased by one-half. This is known as an example of Boyle's law.

Q. What welders are allowed to make repairs to a boiler?

A. Only welders who have been examined and certified as to their qualifications by an Inspector of the U.S. Coast Guard, American Bureau of Shipping, or the Bureau of Ships of the Navy Department, and hold a certificate which is still in force.

Q. (a) Where is the back pressure valve installed? (b) Why is this valve so important?

A. (a) It is a spring-loaded valve fitted in the exhaust line. (b) To provide constant pressure in the exhaust line, to provide a cushion to all reciprocating auxiliaries and to maintain a set pressure of steam on the feedwater heater.

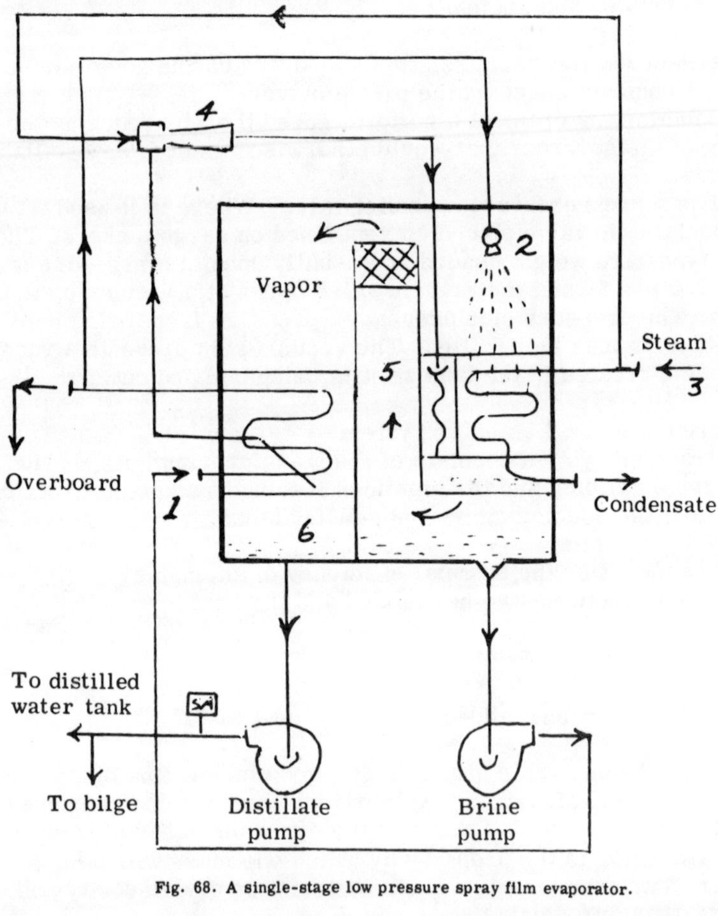

Fig. 68. A single-stage low pressure spray film evaporator.

Q. Sketch a simple single-stage low pressure spray film evaporator and explain operation.

A. Refer to Fig. 68.

1. Sea water is pumped through the distillate condenser (1). A portion is sent in as feed water for the evaporator the rest goes overboard.

2. Feedwater enters the evaporator shell through spray nozzles (2) spraying the tube bundles.

3. Heat for vaporization of the sea water is provided by steam from the auxiliary system (3).

4. The steam air ejector (4) maintains low pressure, low temperature operation.

5. The feedwater is heated and flashed (5). Vapors generated are drawn to the condenser after passing through the separator to insure distillate purity.

6. The distillate is collected in the distillate trough (6) and flows to the distillate pump.

7. The salinity indicator (SA) is set to sound an alarm and dump the distillate to the bilge if the indicator shows more than the prescribed amount of grains per gallon (gpg).

Q. Explain the operation of the Nash air pump.

A. Figure 69. A rotor (1), in dynamic balance, revolves freely, without metallic contact, in a circular casing (3) containing a liquid, usually water (4). This rotor (1) is a circular casting consisting of a series of blades projecting from a hollow cylindrical hub through which the shaft has been pressed. These blades are shrouded at the sides and form a series of chambers. Note that the curvature of the blades is in the direction of rotation.

Starting at point "A", the chambers of the rotor are full of water. This water rotates with the rotor, but follows the contour of the casing (3), due to centrifugal force. The water (4), which entirely fills the rotor chamber at "A," recedes as the rotor advances, until at (5), the rotor chamber is empty. The converging casing forces the water back into the rotor chamber, until at (6), it is again full. This cycle occurs once in each revolution. As the water is caused to recede from the rotor chamber at (7) it is replaced by air entering through "Inlet Port" in conical casting (2), connected with the "Pump Inlet". As the rotor turns 360 degrees, and the water is forced by the casing back into the rotor chamber, the air that has filled the chamber is forced through the "Discharge Ports" provided in the conical casting (2), to the Pump Discharge.

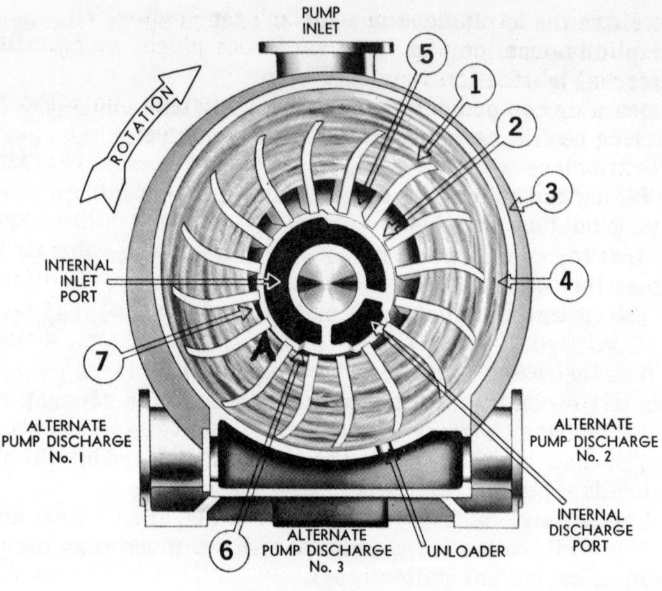

Fig. 69. Nash air pump.

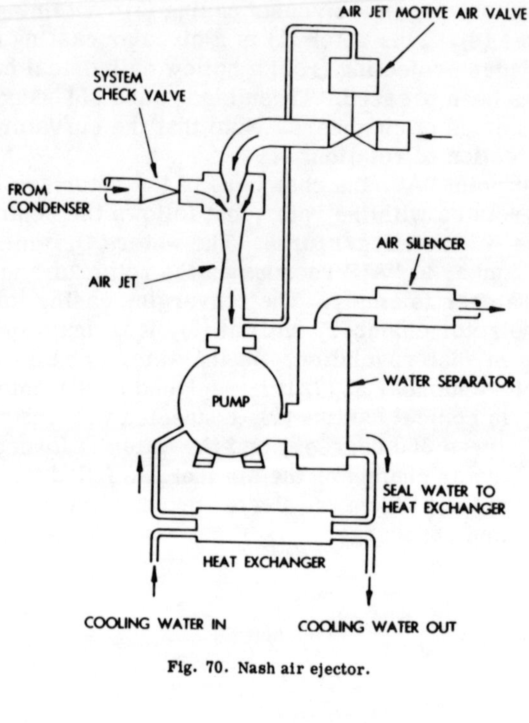

Fig. 70. Nash air ejector.

Q. What are the advantages of the Nash air ejector.

A. Simplicity—one moving part cast in one piece, no metallic contact, no internal lubrication required.

Vacuum or compressed air produced is clean and without pulsation, requiring no filters or other accessory devices.

Vibrationless operation—does not require special foundation, expensive piping and complicated installation procedures.

Will handle quantity of liquids with the air without damage.

Increased capacity when handling hot gases and saturated vapors because the cool liquid compressant has a condensing effect.

Requires less maintenance than electric motors.

Low weight and small size eliminates need for special foundations and requires less space.

The air ejector uses atmospheric air as the motivating medium instead of the conventional compressed air or steam. Air is introduced into the ejector nozzle from atmosphere. This air passes through a specially designed superaccoustic nozzle into the suction chamber or body of the ejector.

The motive air is then moving at a velocity several times that of sound. This high velocity air entrains the suction gases entering the ejector body from the system being evacuated, and compresses the gas and motivating air during its passage through a diffuser tube, to a higher pressure. The Nash pump takes suction at the discharge of the Nash air ejector and carries on evacuation in the usual manner, Fig. 70.

Q. What is the heating surface in sq. ft. available in a nest of 120 boiler tubes, 2" in diameter and 10' long?

A. Referring to Fig. **71,** the dotted circle represents the tube 2" in diameter. If the tube were split and rolled out flat, it would appear as ABCD, AB being equal to the circumference of the tube.

1. Circumference of tube = πD
   = 3.1416 X 2 = 6.2832 ins.
2. Changing the length of the tube to inches,
   10 X 12 = 120"
3. The total area is now the L X W
   120 X 6.2832 = 753.984 sq. ins.
4. There are 144 sq. ins. in 1 sq. ft.
   Therefore,
   753.984 ÷ 144 = 5.236 sq. ft. of heating surface in one tube.
   5.236 X 120 (tubes) = 628.32 sq. ft. in 120 tubes.

Q. Find the maximum allowable pressure on a dished head with the pressure on the concave side. Thickness is 5/8"; the head is dished to a radius of 52 inches; the maximum allowable stress is 17,000 psi and the joint efficiency is 88%.

A.

$$P = \frac{STE}{.833\ R}$$

$$= \frac{(17000)(.625)(.88)}{(.833)(52)}$$

$$= \frac{9350}{43.316}$$

$$= 216$$

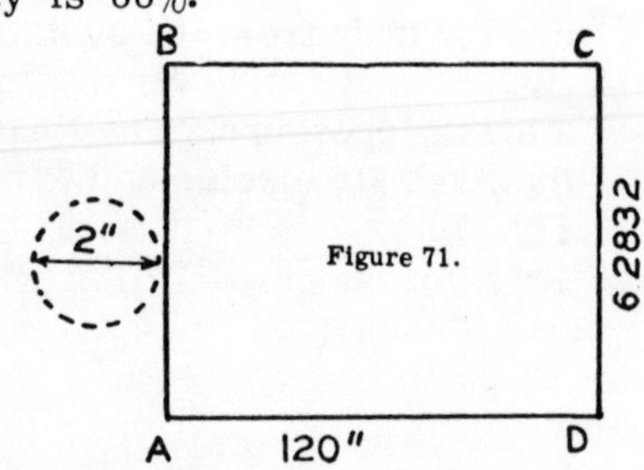

Figure 71.

Q. Calculate the maximum allowable pressure for the shell of a boiler using the following data:

Maximum allowable stress . . . . 17,500 psi
Minimum thickness of shell . . . . . . 0.9"
Inside radius of shell . . . . 5'
Minimum efficiency of joint 88%

A.

$$P = \frac{STE}{R + 0.6T}$$

$$= \frac{(17500)(.9)(.88)}{60 + (0.6)(.9)}$$

$$= \frac{13860}{60.54} = 228.9 \text{ or } 229$$

Q. What is the weight of a 55 gallon drum of oil? Drum weighs 14.5 lbs. S.G. of oil .86.

A. Total weight of oil = (8.33)(.86)(55) = 393.8 lbs
393.8 + 14.5 = 408.3 lbs. total weight

Q. What is the horsepower of a boiler evaporating 12,000 lbs. of feed water per hour from a feedwater temperature of 170° F. into steam at 150 lbs. gauge? (Total heat of 1 lb. of steam at 150 lbs. gauge is 1195 B. T. U.)

A. 170° — 32° = 138 B. T. U. required to raise 1 lb. from 32° to 170°

1195 — 138 = 1057 B. T. U. added to each pound of steam generated

1057 X 12,000 = 12,684,000 B. T. U. added to 12,000 lbs. of steam per hour

One (1) boiler horsepower = 34.5 lbs. of water at 212° F. converted to steam at 212° F.

970 B. T. U. = B. T. U. required to convert 1 lb. of water at 212° F. to steam at 212° F.

34.5 X 970 = 33,465 B. T. U. (one boiler horsepower)

$$\frac{12,684,000}{33,465} = 379 \text{ horsepower}$$

Q. How many 2-1/2" steel stays will be required to stay an area of 3700 sq. in.? Boiler pressure 200 lbs. Stress allowed on stays 9000 lbs. per sq. in.

A. Strength of 1 stay = .7854 X 2-1/2 X 2-1/2 X 9000 = 44178.75

Pressure on plate = 200 X 3700 = 740,000

740,000 ÷ 44178.75 = 17 stays

Q. A boiler supplies a pump of the dimensions 6" X 4" X 8" with 60 lbs. of steam. Against what pressure will the pump work?

A.
$$P = \frac{\text{Area steam piston X Steam pressure}}{\text{Area water piston}}$$

$$= \frac{.7854\,(6)^2 \times 60}{.7854\,(4)^2}$$

$$= 135 \text{ lbs.}$$

Q. What steam pressure would be required to test a boiler at 140 lbs. with a pump of the dimensions 8" X 4" X 10"?

A.
$$P = \frac{(\text{Area water piston})^2 \text{ X Steam pressure}}{(\text{Area steam piston})^2}$$

$$P = \frac{W^2 \times P}{S^2} = \frac{(4)^2 \times 140}{(8)^2} = 35 \text{ lbs.}$$

Q. If 5000 bbls. of oil were received at 100° F., what amount would you have at 60° F.?

A. For each degree in change of temperature, the volume of oil per unit changes .0004.

Therefore: $100° - 60° = 40°$ temp. difference.
$40° \times .0004 = .016$ change in volume per bbl.
$.016 \times 5000 = 80$ bbls., the change in volume in 5000 bbls.
$5000 - 80 = 4920$ bbls., amount on hand at 60°.

Q. What is the maximum allowable working pressure on the shell of a boiler with the following dimensions?

D = 11' 6"
T.S. = 60,000 lbs. per sq. in.
Thickness of plate = 1-1/2"
Efficiency of joint = 80%
Factor of safety = 4.5

A.

$$W = \frac{STE}{RF}$$

$$= \frac{\overset{20,000}{60,000} \times 1.5 \times .80}{\underset{23}{69} \times \underset{3}{4.5}}$$

$$= \frac{16000}{69}$$

$$= 231.883$$

W = Maximum allowable working pressure in lbs. per sq. in.

S = Tensile strength

T = Minimum thickness of plate in inches

E = Efficiency of joint

R = Radius of boiler in inches

F = Factor of safety

Q. A pipe 12" in diameter is used as a dry pipe in a boiler. The slots in the pipe are 6" long and 1/4" wide. How many slots must there be to make an area equal to double the sectional area of the pipe?

A. Area of pipe $= .7854\ D^2$
$= .7854\ (12)^2$
$= .7854\ (144)$
$= 113$ sq. ins.

Double this $= 113 \times 2 = 226$ sq. ins.

Area of one slot $= (6)(1/4) = 1.5$ sq. ins.

$226 \div 1.5 = 150.7 = 151$ slots required

Q. If an engine has a 6 ft. stroke and a connecting rod 8 ft. long, what distance downward would the piston have traveled when the crank is on the quarter?

A. In Fig. **72a,** AB represents the connecting rod and BC the web of the crank. The connecting rod being 8 ft. long and the crank 3 ft. long, the total distance from A to C, when the crank is on the top dead center, is 11 ft. In Fig. **72b,** the crank has moved

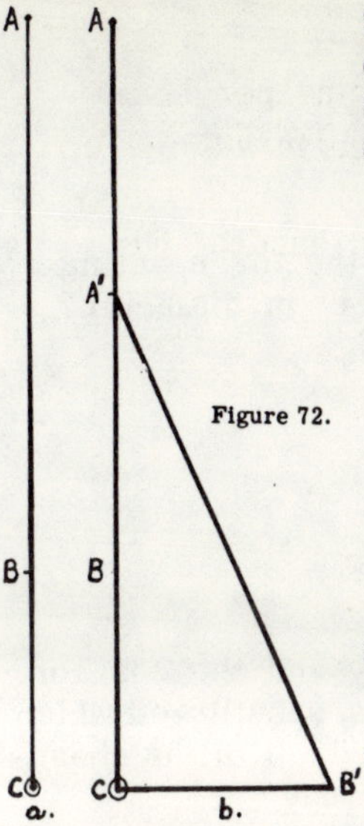

Figure 72.

90° to its quarter position and the piston has dropped to $\underline{A'}$. The connecting rod is now represented by $\underline{A'B'}$ and the crank web by $\underline{B'C}$. Triangle A'B'C is a right triangle.

Using the right triangle (Pythagorean) theorum:

$$(A'B')^2 = (CB')^2 + (A'C)^2$$
$$(8)^2 = (3)^2 + (A'C)^2$$
$$64 = 9 + (A'C)^2$$
$$64-9 = (A'C)^2$$
$$55 = (A'C)^2$$
$$\sqrt{55} = A'C$$
$$7.42 = A'C$$

Since AC = 11'
and A'C = 7.42'
then AC−A'C = AA' or
11'−7.42' = 3.58'

Therefore, the piston has moved down 3.58' in its stroke when the crank is on the quarter.

Q. How many studs are required for a 30" diameter cylinder with 160 psi, studs 1-1/8" diameter, allowable stress per sq. in. of stud, 4800 lbs.?

A. Area cylinder $= \pi r^2$

$$= \pi(15)^2$$
$$= \pi(225)$$
$$= 706.86 \text{ sq. ins.}$$

Total load under head = (706.86)(160) = 113097.6 lbs.
Load each stud supports per square inch =
113097.6 ÷ 4800 = 23.56 psi
Cross-sectional area one stud $= (.7854)(1.125)^2$
= .99 or 1
Number studs = 23.56 ÷ 1 = 23.56 or 24 studs

Q. A compound engine has cylinders 26" and 69" diameter. The mean effective pressures are 67 and 14.75 lbs., respectively. The stroke is 4 ft. and the revolutions 60 per min. Find the combined horsepower.

A.
$$HP = \frac{PLAN}{33000}$$
$$= \frac{(67)(4)(.7854)(26)^2(60 \times 2)}{33000}$$
$$= 517.5$$
$$HP = \frac{PLAN}{33000}$$

$$= \frac{(14.75)(4)(.7854)(69)^2\ (60 \times 2)}{33000}$$

$$= 802$$

Combined HP = 1319

Q. What is the pressure in an engine cylinder if the force acting on the piston is 75,000 lbs. and the piston is 14" in diameter?

A. Area of piston $= .7854\ D^2$

$= .7854\ (14)^2$

$= .7854\ (196)$

$=$ 154 sq. ins.

Pressure in cylinder $= \frac{75000}{154} = 487$ psi

Q. The consumption of lube oil is 9 gals. per day and the cost of the oil is .58¢ per gal. Another quality of oil is substituted which cost 25% less but of which 20% more is used. Which oil is cheaper to use and what is the difference in cost per day?

A. First grade oil, 9 gals. at .58¢ gal. = $5.22 per day
Second grade oil, 58¢ — 25% = .435¢ per gal.
9 gals. + 20% = 10.8 gals. per day used of second grade oil
Cost second grade oil = (.435)(10.8) = $4.698
= $4.70 per day

Second grade oil is cheaper by

$5.22
4.70
$ .52 per day

Q. How long will it take a vessel having a wheel with a 16' pitch, turning 72 RPM, to traverse 580 nautical miles; slip 10%?

A.

$$K = \frac{PN\ (1\text{–slip})}{101.3}$$

$$= \frac{(16)(72)(1-.10)}{101.3}$$

$$= \frac{1036.8}{101.3}$$

$$= 10.2 \text{ knots}$$

$580 \div 10.2 = 56.8$ hrs.

Q. If an indirect valve measures 34" between the steam (admission) edges and the distance between the inside edges of the ports is 37-1/4", what is the steam lap of the valve?

A.

37-1/4
− 34
3-1/4

3-1/4 ÷ 2 = 1.625 or 1-5/8"

Q. There are 150 40-watt lamps in an installation. The voltage is 220. What is the horsepower required to drive the generator if its efficiency is 92.5%?

A. Total watts = 150 X 40 = 6000

746 watts = 1 horsepower

$$\frac{6000}{740} = 8.04 \text{ horsepower}$$

$$\text{Efficiency} = \frac{\text{electrical horsepower}}{\text{shaft horsepower}} \left[\frac{\text{Output}}{\text{Input}}\right]$$

$$\text{Shaft horsepower} = \frac{\text{electrical horsepower}}{\text{efficiency}}$$

$$= \frac{8.04 \times 100}{92.5}$$

$$= 8.69$$

PUMPS

The following information on pump problems should be noted carefully:

1. A stroke should be considered as a movement of one piston or plunger over its maximum travel in one direction.

2. A double stroke or revolution should be considered as a movement of one piston or plunger through a maximum travel in both directions, or the equivalent of two single strokes.

3. Regardless of the type of pump, a stroke refers to the movement of only one piston or plunger over its maximum travel in one direction.

4. The strokes, as given in the problem involving the duplex double-acting pump, should be multiplied by 2 in order to obtain N. Where double strokes of a duplex-acting pump are given, they should be multiplied by 4 in order to obtain the total strokes.

5. In a simplex double-acting pump where strokes are given, they should be considered the total power strokes to be used as N. Where double strokes are given, they should be multiplied by 2 in order to obtain N.

6. In regard to simplex single-acting pumps, the strokes given should not be considered as the strokes used by N, but should be divided by 2 in order to obtain the effective working strokes. Where double strokes are given, the given number should be used as N.

The following information regarding the figures on pump name plates should be memorized:

10" X 6" X 8"

1. The first figure represents the diameter of the steam piston: 10".

2. The second figure represents the diameter of the water or liquid piston: 6".

3. The third figure represents the stroke: 8".

Q. A duplex double-acting pump 6" X 8" X 6" makes 45 S. P. M. Slip 4%. What is its actual discharge in gallons per minute (G. P. M.)?

A. G. P. M.

$$= \frac{\text{Vol. of cylinder X number working strokes X efficiency}}{231}$$

$$= \frac{.7854 \times (8)^2 \times 6 \times (45 \times 2) \times .96}{231}$$

= 112.80 G. P. M.

Q. A duplex double-acting pump 4" X 6" X 6" makes 25 R. P. M.* Slip 4%. What is its actual discharge in G. P. M.?

A. G. P. M.

$$= \frac{\text{Vol. of cylinder X number working strokes X efficiency}}{231}$$

$$= \frac{.7854 \times (6)^2 \times 6 \times (25 \times 4) \times .96}{231} = 70.502 \text{ G. P. M.}$$

Q. A duplex double-acting pump 6" X 4" X 6" makes 50 double strokes per minute. What is its theoretical discharge in gallons per minute?

A.

$$\text{G. P. M.} = \frac{\text{Vol. of cylinder X number working strokes}}{231}$$

$$= \frac{.7854 \times (4)^2 \times 6 \times (50 \times 4)}{231}$$

= 65.28

Q. A simplex double-acting pump 12" X 8" X 8" makes 100 S. P. M. Slip 6%. What is its actual discharge in G. P. M.?

A. G. P. M.

$$= \frac{\text{Vol. of cylinder X number working strokes X efficiency}}{231}$$

$$= \frac{.7854 \times (8)^2 \times 8 \times 100 \times .94}{231}$$

= 163.635 G. P. M.

Q. A simplex single-acting pump 6" X 5" X 5" makes 50 S. P. M. Slip 5%. What is its actual discharge in gallons per minute (G. P. M.)?

A. G. P. M.

$$= \frac{\text{Vol. of cylinder X number working strokes X efficiency}}{231}$$

$$= \frac{.7854 \times (5)^2 \times 5 \times \left(\frac{50}{2}\right) \times .95}{231}$$

= 10.094 G. P. M.

*1 R.P.M. = 1 double stroke.

Q. Find the height of a cylindrical tank which holds 250 gals. and is 24" in diameter.

A. $V = .7854\ D^2H$

$.7854\ D^2H = V$

$H = \frac{V}{.7854\ D^2}$

$= \frac{250\ (231)}{.7854\ (576)}$

$= \frac{57750}{452.39}$

$=$ 127.6" or approximately 10-1/2'

Q. A cylindrical tank, 12' long, holds 2600 gals. when full. What is the diameter of the tank?

A. $V = .7854\ D^2H$

$.7854\ D^2H = V$

$D^2 = \frac{V}{.7854\ H}$

$= \frac{2600\ (231)}{.7854\ (12)^2}$

$= \frac{600\ 600}{113.09}$

$D^2 = 5315$

$D = \sqrt{5315}$

$D = 72.84$" diameter of tank.

Q. A cylindrical tank is 48" diameter and 6' high. It contains 100 gals. of oil. What is the height of the oil in the tank?

A. $V = .7854\ D^2H$

$= .7854\ (48)^2\ 72$

$= 130288.435$ cu. in.

$130288.435 \div 231 = 564$ gals. (when full)

$564 \div 72 = 7.83$ gals/inch

$100 \div 7.83 = 12.7$" of oil in tank

Q. A pressure gage is placed 12' below the main line and registers 128 psi. What is the absolute pressure?

A.

$.434 \times 12 = 5.21$ pressure of liquid due to height

$128 - 5.21 = 122.79$ psig

$122.79 + 14.7 = 137.49$ psia

PREFACE

An applicant sitting for examination on MOTOR will be asked the following groups of questions:

1. For a Motor license over 2000 HP, the applicant will be asked all the usual Steam questions plus the Motor questions.

2. For a Motor license under 2000 HP, the applicant will be asked all the usual Steam questions except the sections on Boilers, Turbines and Reciprocating Engines, plus the Motor questions.

MOTOR

CONSTRUCTION

Q. Sketch a double-acting piston and cylinder.
A. Figure **73.**

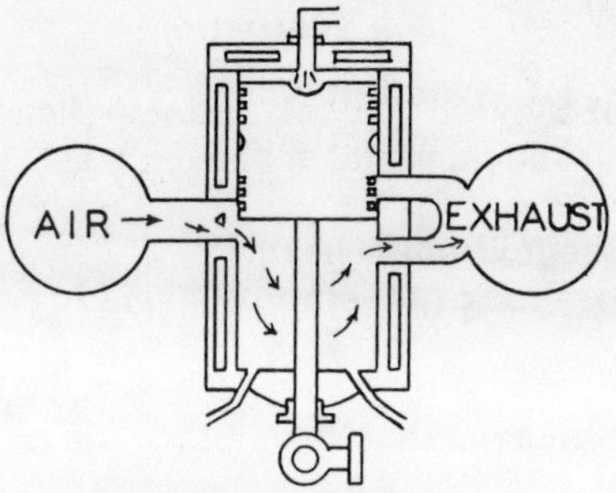

Fig. 73. **Double-acting piston and cylinder.**

Q. In Diesel-engine construction, what metals are used in the following parts: frames, liners, pistons, connecting rods, camshafting, valves, valve springs, bearings, bushings?
A. 1. Frames: Cast iron or steel.
2. Liners: Close-grained cast iron.
3. Pistons: Cast iron or aluminum alloys.
4. Connecting rods: Forged steel.
5. Camshafting: Cast or forged steel.
6. Valves: Nickel chrome steel or alloys.
7. Valve springs: Spring steel.
8. Bearings: (a) Shell—Cast iron, cast steel, bronze; (b) Surface—Babbitt or copper-nickel alloys.
9. Bushings: Bronze and brass.

Q. Describe an open cooling system.
A. The circulating pump takes suction from the sea and discharges salt water through the engine and then overboard, the engine being cooled directly by sea water.

Q. Describe a closed cooling system.
A. The engine is cooled by fresh water from storage tanks, which is recirculated through the engine by a fresh-water pump. This water is cooled by a heat exchanger using sea water (supplied by a salt-water circulating pump) as a cooling medium.

Q. Trace the path of cooling water through a water-cooled piston.
A. When water is used for cooling a piston, it is led through grasshopper joints, telescopic pipes, or rubber hose into the piston. In some cases it is sprayed on the inner head of the piston. Temperature of the water is raised to from 140° to 160° in the piston casting and then it flows back to the supply source where it is cooled and then recirculated.

Q. What is the camshaft speed in relation to the crankshaft speed in a 2-stroke cycle engine; a 4-stroke cycle engine?
A. In a 2-cycle engine, camshaft speed equals crankshaft speed. (Full-time shaft.)

In the 4-cycle engine, the camshaft speed is 1/2 the crankshaft speed. (Half-time shaft.)

Q. Describe a ported liner found in 2-cycle engines. What is the relative position of the exhaust and scavenge ports?
A. The liner used in 2-cycle engine construction usually has two rows of ports which are at slightly different positions. The top row of ports are exhaust ports. At a slightly lower level are the scavenging ports. The position of the exhaust ports is about 75% down the stroke, and the scavenging ports are about 80% down the stroke (on opposite sides).

Q. Name the valves found in a 4-stroke cycle Diesel cylinder head.
A. Fuel, air-starting, relief, indicator-cock, intake and exhaust.

Q. Name the valves found in a 2-stroke cylinder head.
A. Fuel, air-starting, relief, indicator-cock (possibly scavenging valve).

Q. At what part of the cycle does the scavenging take place in a 4-cycle Diesel engine?
A. Scavenging effect takes place during the period from about 10° B. T. D. C. to about 15° A. T. D. C. or when the piston is completing the exhaust stroke, and, during this period, both exhaust and intake valves are open.

Q. Describe the shroud construction of exhaust valves. What are the advantages?
A. Shrouds are fitted to the bottom end of the exhaust-valve guides and are then welded in place. The fitting of shrouds tends to prevent scale, unburned fuel, lubricating oil, or carbon from collecting on the valve stem between the valve stem and valve guide. By this means turbulence of inlet air is increased and sticky valves are eliminated to a certain extent.

Q. Describe various methods of crankshaft construction.
A. Crankshafts for Diesel engines are usually made of open-hearth steel, although in some of the small high-speed engines special alloy steels are used.

Crankshafts are made in three different ways:

1. Forged from one single billet (unit).
2. Made in several sections (semi-unit).
3. Each journal and web made separately, then later shrunk together (built-up).

The forged method is used in case of small engines, while the latter two methods are used more often in construction of crankshafts for large engines. The second method is sometimes called the semi-built-up method, and the third method is sometimes called the built-up method.

Q. Why should cooling-water systems be vented? Where?

A. In order to prevent air pockets in any part of the engine cooling system. The air pockets would cause overheating since there would not be a circulation of water to carry away the heat. Vents should be placed on the circulating pump, the coolers and at sufficient places in the top of the cooling system to bleed off all entrapped air.

Q. What are the advantages of an open cooling system? The disadvantages?
A. Advantages:
1. It is the simplest system.
2. No heat exchanger is needed.
3. Only one water-circulating pump is required.

Disadvantages:
1. There is great danger of scale deposits.
2. Engine must be kept cooler and thus heat efficiency is lost.
3. Foreign materials are more apt to get into the system.
4. There is danger of galvanic action of the sea water.

Q. Describe the operating principles of a double-acting Diesel engine. Why were they developed?
A. In this type of engine there are two combustion chambers and two cylinder heads. The same cycle of events happening at the bottom of the piston occurs on top of the piston, also. The piston is forced downward as in the single-acting engine, but it is then forced upward due to the expansion of the gases of combustion in the lower cylinder. The work done by the piston is due to the action of the gas pressure acting on both faces of the piston. Thus, for the double-acting engine there are twice as many power strokes as in the single-acting engine.

They were developed because it was realized that a very desirable arrangement would be one that would permit operation similar to the steam engine, in which the steam is admitted alternately on top and bottom of the piston so that every stroke of the engine is a power stroke.

It was obvious that as long as the single-acting engine was adhered to, the only means of increase of power in the engine was to increase the number of cylinders, because the cylinder diameter is limited to about 30" due to metal and heat transfer problems. Therefore, the double-acting engine was developed in order to get a greater horsepower for a given weight and size.

Q. State the advantages and disadvantages of a double-acting engine.
A. Advantages: In the double-acting engine, combustion pressure on one side is working against compression on the other, thus giving much smoother running conditions and better working and maneuvering control than in a single-acting cylinder. There are fewer parts in the 2-cycle double-acting engine as compared with the 4-cycle double-acting type so that the cost and weight reduction are outstanding factors on power alone. A reduction in cost and

weight per horsepower is most readily obtained by increasing the horsepower output from an engine of given bore and number of cylinders.

Disadvantages: Scavenging is less perfect, hence the volumetric efficiency is decreased. This, together with the fact that only about 85% of the stroke is effective, due to early uncovering of the exhaust ports, results in the power being increased by only about 75% instead of being doubled. Hence the fuel consumption per H. P. is correspondingly greater. The scavenging pump is large, and more moving parts are required. This increases the weight, the cost, and the percentage of the engine power consumed in nonproductive work, which may be as high as 10%. Greater cylinder heat and correspondingly greater expansion stresses result, making lubrication more difficult.

Q. Describe the construction of a spark arrester and its use.

A. Exhaust gases rising in the stack meet the outside of an inverted cone mounted in or near the muffler. Rising to its top, the gases are redirected downward inside the cone after having been given a whirling rotary motion by curved blading in the cone crown. The gases whirling downward to the apex will deposit any heavy particles of carbon in the bottom of the apex of the cone. The lighter gases will rise again and ascend to the atmosphere through a pipe extended into the top or wide part of the inverted cone.

Q. What is the difference between single- and double-acting engines?

A. In the single-acting engine fuel injection and combustion occur only in one end of the cylinder. The pressure tending to move the piston and cause the shaft to turn is exerted only on one side of the piston, while in the double-acting engine, this action occurs in both ends of the cylinder and the piston exerts power on both upward and downward strokes.

Q. Why are cams, instead of eccentrics, used to operate the valves?

A. Cams can be designed to enable quick opening and closing of the valve, whereas, with an eccentric, this cannot be done. Another desirable feature of the cam is that the length of time that the valve is open can be controlled by the width of the cam.

Q. Why are counterweights applied to the cranks of some Diesel engines?

A. To absorb some of the energy of the power stroke and to give it off again during the compression stroke, which helps to give the engine an even turning motion.

Q. What is the usual crank angle between the cranks in (A) a 2-cylinder, 2-cycle engine; (B) in a 2-cylinder, 4-cycle engine?

A. (A) 180° apart; they fire 1-2.
(B) 360° apart; they fire 1-2.

MOTOR—OPERATION

Q. How can a 4-cycle Diesel engine operate satisfactorily with a higher exhaust back-pressure than a 2-cycle engine?

A. A 4-cycle engine can operate on a higher exhaust back-pressure than a 2-cycle engine because it does not have to depend purely upon the pressure of scavenging air to rid its cylinder of exhaust gases to obtain a good volumetric efficiency. The forcible expelling and suction action of an up-and-down stroke of a piston on a 4-cycle engine will in itself largely accomplish this with no scavenging assistance.

Q. Would you shut off the cooling water immediately after stopping a Diesel engine? Explain.

A. No. The reason for keeping the cooling water circulating after the engine is stopped is to prevent the lubricating oil from being evaporated from the cylinder walls and the piston pins. In case of an engine with a salt-water cooling system, there would also be a formation of scale on account of excess heat if the cooling-water system did not remain in operation after the engine was stopped.

Q. Describe a jacking gear and its use on a Diesel engine.

A. A jacking gear is an arrangement of gears used to rotate the engine crankshaft. It usually consists of a worm gear (driven by steam, electricity, or by hand) which engages a gear-toothed flywheel. There is always some means of disengaging the worm-gear drive when not in use.

Q. Describe a pyrometer—thermocouple and resistant type.

A. A pyrometer is a millivolt meter calibrated in temperature units, which is attached through a selector switch to each individual cylinder exhaust thermocouple. The thermocouple is made of 2 rods of different metals that are welded together at one end. When heated at the welded junction, an electromotive force is produced between the terminals.

Q. Discuss the use of a pyrometer in operating a Diesel engine. What do the high and low temperatures indicate?

A. The temperature reading on a pyrometer is usually a direct interpretation of the amount of power being developed in a cylinder. The greater the pyrometer reading, the higher the cylinder M. E. P., except in cases of faulty injection, incorrect timing, or leaky valves. A very high pyrometer reading means overload, and a very low reading means that the particular cylinder is not carrying its required load.

Q. How is the formation of carbon monoxide in the exhaust gases of a Diesel eliminated?

A. By maintaining proper combustion and scavenging.

Q. At what points in a Diesel cycle is the piston being subjected to the greatest effects of acceleration?
A. At the beginning of each power stroke.

Q. If no starting air is available and air compressor is beyond repair, how may the engine be started?
A. There are two emergency methods of starting the engine. If you have a large supply of CO_2, pipe it up to the starting air line with a reducing valve. (Caution: Never use oxygen.) If no carbon dioxide is available, usually at least one of the auxiliaries can be started without air. Pipe up one cylinder as air compressor, to get enough air to start main engine.

Q. Assuming the above-stated condition and lacking an air compressor, how would you continue your voyage and still keep an air pressure and be able to maneuver your ship?
A. Pipe up one cylinder as an air compressor and cut fuel off. This pipe should run through the air coolers, as it does from the air compressor.

Q. How may a leaking air-starting valve be recognized?
A. The temperature of the air line increases. Another way to check the valve is to break the air line and turn the engine by the jacking gear; if a leak is present, you can hear the air escaping.

Q. Why is the exhaust temperature inherently lower on a 2-cycle Diesel engine than on a 4-cycle Diesel engine?
A. Because the scavenging air of the 2-cycle engine cools off the exhaust gases.

Q. Which is the easier to reverse, a 2-cycle or 4-cycle Diesel engine? Discuss.
A. The 2-cycle Diesel engine is easier to reverse that the 4-cycle Diesel engine. This is because there are fewer moving parts to be changed in direction of rotation, there being fewer valves on the 2-cycle engine.

Q. Which gives the most uniform turning impulse to a crankshaft, 2-cycle or 4-cycle? Discuss fully.
A. The 2-cycle engine gives the most uniform turning effect. The 2-cycle engine would have twice as many power impulses as a 4-cycle engine. The turning is more uniform because the impulses are more frequent.

Q. What precautions should be observed to prevent the formation of scale in the cylinder jackets of Diesel engines equipped with open cooling water systems?
A. The outlet cooling water temperature should not be permitted to exceed 120°F. Large quantities of water should be circulated rapidly and with a small temperature difference, usually 10°F to

20°F., between the inlet and outlet water in order to prevent the formation of local hot spots and scale. The water circulation should be continued after the engines are stopped until the engines are cool.

Q. What effect do leaky piston rings have upon the working of an engine?

A. Leakage of gas past the piston rings reduces the efficiency of the engine and increases the cylinder liner and piston ring wear by blowing the lubricating oil off the rubbing surfaces.

The efficiency of the engine is reduced owing to a low compression pressure causing bad combustion of fuel, and to leakage of gas, resulting in a reduced mean effective pressure.

Q. Describe ignition lag. How is it affected by compression pressure, atomization of fuel, turbulence?

A. Ignition lag is the time interval between the beginning of injection and the beginning of combustion as indicated by the beginning of pressure rise above the compression line on an indicator card. It may be expressed in crank-angle degrees or in seconds of time.

Increase in compression pressure, better atomization, and good turbulence all have a tendency to shorten ignition lag.

Q. How does the combustion differ in gas engines? In Diesel engines?

A. Combustion in the gas engine takes place more nearly according to the Constant-Volume Cycle. In the Constant-Volume Cycle all heat is taken in at a constant volume and all heat is rejected at a constant volume. This means that when combustion or exhaust takes place there is no change in volume. In the gas engine, therefore, combustion occurs practically instantaneously and is, in effect, an explosion. This cycle is also called the Otto Cycle.

Combustion in the Diesel engine takes place more nearly according to the Constant-Pressure Cycle. In the Constant-Pressure Heat Cycle all heat is taken in and rejected at constant pressure. That is, when combustion occurs, there is no appreciable pressure rise, as fuel is being injected and burned constantly. Also, when exhaust takes place, the pressure is constant. In the Diesel engine, therefore, combustion takes place over a period of time as a slow-burning process, not an explosion.

Q. What is meant by the controlled and uncontrolled combustion of a Diesel engine?

A. Uncontrolled combustion is the stage of combustion which takes place just after the delay period. It is called uncontrolled combustion since it is the period in which fuel injected during the delay period burns rapidly with almost no control.

Controlled combustion is the stage of combustion directly after the uncontrolled combustion, during which the remainder of the fuel ignites at the nozzle and burns at a rate controlled by the pump and nozzle.

Q. What is 2-stage or divided combustion?
A. That which is found in precombustion chamber engines. The combustion takes place in the precombustion chamber first and then in the main chamber, thus being divided into two stages.

Q. What is the average pressure or temperature of the following in a Diesel and gas engine?
Compression Pressure — Compression Temperature
Combustion Pressure — Combustion Temperature

A.

	Diesel	Gas Engine
Comp. press. . . .	500 lbs. per sq. in.	120 lbs. per sq. in.
Comp. temp. . . .	1000 degrees F.	240 degrees F.
Comb. press. . . .	600 lbs. per sq. in.	350 lbs. per sq. in.
Comb. temp. . . .	3000 degrees F.	3000 degrees F.

Q. What dopes are used in gasoline and Diesel fuel oils? Why?
A. In gasoline, octane is used to provide slow and uniform burning of the gasoline.

In Diesel oil, cetane is used to speed up the burning.

Q. State the most common troubles encountered with double-acting Diesel engines.
A. Packing troubles, where the piston rod passes through the lower cylinder head; failure of piston rods due to faulty cooling and lubrication; trouble with the piston cooling.

Q. Explain in detail how piston cooling is affected in a double-acting engine.
A. Double-acting pistons must always be directly cooled, because they receive heat on <u>both</u> sides and conduct it only to the cylinder walls. In double-acting engines the cooling fluid is always passed through the piston rod, the swing-joints or telescopic pipes leading to and from the crosshead. The fluid is directed through the piston by baffles to prevent its short-circuiting and finally returns through a tube within the hollow piston rod. The uneven distribution of heat in this type of piston makes it very difficult to cool.

Q. What causes back-pressure on a Diesel-engine exhaust system? How is it prevented and how does it affect the engine efficiency? Discuss fully.
A. A slight back-pressure will always be present in the exhaust system of a Diesel engine (1/2 pound maximum for good efficiency) because of friction within the piping due to restrictions built in its passage or accumulated there as carbon. Incorrect timing of either intake or exhaust valves is another cause. Removal of restrictions from exhaust passages, usually carbon, will correct this. Engine efficiency will drop in direct proportion to the increase in back-pressure. Back-pressure divided by percentage of effective stroke will equal percentage of loss.

Q. What happens if the exhaust valve opens early; Late?
A. Early: There would be a loss of compression, resulting in drop in HP.

Late: There would be too much compression and, therefore, a higher back pressure. This would also cut down the HP.

Q. What happens if the inlet valve opens early; Late?
A. Early: The pressure rise from fuel combustion starts well before maximum compression pressure or TDC is reached. This early firing may be accompanied by a knock.

Late: A drop in HP. Timing will be off.

Q. What is the effect of too much or too little clearance between a cam and its roller?
A. Too much clearance causes late valve opening, early closing, decreased lift and noisy operation. Too little clearance causes early opening, late closing, increased valve lift and may prevent complete closure.

Q. What is the cooling water temperature?
A. The difference in temperature between inlet and outlet should not be greater than 12-1/4° to 15°F. When using sea water, the outgoing water should be carried between 110° and 120°F., but not to exceed this.

Q. Explain how to start up a Diesel installation.
A. Jack the engine to see that it is clear for running; start lubricating oil pump and see that oil reaches all bearings; open bypass valves in fuel oil lines to injection valves and pump oil through by hand to clear out all air and to fill pipes with oil; open stop valves in starting air and injection air lines; give cylinder lubricators a number of turns by hand to get some oil into the cylinders; start jacket cooling water pump; see that the propellers are clear. If the means for heating the cylinders are available and the weather is cold, warming up should be started in time to have all the cylinders properly warmed up before starting.

Q. What observations should be made on watch on a motor ship while under way at sea?
A. If the engine is of the air injection type, watch the air pressure and note any change in the relative readings of the gauges on the air compressor. See that ample supply of fuel is kept in watch tanks. Note pressure and temperature in cooling water and lubricating oil systems; make regular inspections of cylinder head valves; keep a sharp lookout for indications of a hot bearing and watch for indications of misfiring or incomplete combustion in one or more cylinders; take exhaust gas pyrometer readings at frequent intervals. Once every hour open drains on air tanks.

Q. Explain how an internal combustion engine is made reversible.
A. By pushing the camshaft axially with a control lever so that another set of cams on the same shaft is placed under the push-rods or rocker arms. These cams change the timing of the intake and exhaust valves, starting and injection valves, and fuel pump timing, which will reverse the engine's motion when started.

Q. Why is an inlet valve on a 4-cycle Diesel engine open before the exhaust valve closes?
A. The exhaust gases rushing out cause a vacuum in the cylinder; the intake valve opens early to allow clean fresh air to rush in and sweep the cylinder completely clean, and to leave the cylinder filled with fresh air.

Q. How is the air on the intake valve muffled?
A. Some intake pipes lead to the boat deck with a silencer installed, others have a large diameter pipe closed on the end and perforated along its length so that the areas of the slots equals 5 times the area of the pipe.

Q. What would be the effect of a leaky exhaust valve?
A. The cylinder with the leaky valve will lose its compression causing incomplete combustion and loss of power formerly developed by that cylinder. This causes overloading of the other cylinders. Cylinders will carbon up and rings will become gummy and stick. An air blow through the valve may cause a muffled explosion if any unburned gases are present. Replace valve soon.

Q. If the engine was overheated and cold water admitted to jackets, what is outcome?
A. There would be sudden contraction of the hot surfaces such as cylinder heads, lines, causing the head to crack and liner to contract and seize the hot expanded piston. Slow down engine without stalling it, and cool gradually by admitting an increasing supply of water.

Q. If one cylinder were getting more fuel than the others, what would be the effect?
A. The cylinder getting more fuel would be overloaded and would smoke and knock and would soon carbon up, causing the rings to stick and possibly break. The exhaust valve would become overheated and warp due to the carbon. This will cause faulty compression, incomplete combustion, and loss of power. The other cylinders will become overloaded and increase in temperature. Must check the pyrometer continuously and adjust the fuel oil system to distribute the load equally in the cylinders at all times.

Q. What is the cause when the engine slows down?
A. The fuel oil may be a poor grade; water in the fuel; fuel valves

or pump not working properly; fuel pump airbound; engine overloaded; engine may have jumped time due to poor timing chain; and chain jumping teeth on camshaft.

Q. What kind of a Diesel engine requires a scavenging pump?
A. A 2-cycle Diesel and some semi-Diesels need a pump to supply air through scavenging trunk from some outside source to expel the burnt gases and supply fresh air to the cylinder. Scavenging pressure varies 1-1/2—7 psi, 1-1/2 without supercharging and 3—5 psi with supercharging.

Q. What is meant by scavenging on internal combustion engines?
A. Scavenging is the cleaning or blowing of the burnt gases out of the cylinder and supplying the cylinder with fresh air for the following revolutions or cycle of events. Air is admitted through slotted ports in the cylinder liner, and through valves in the head. Types of scavenging: opposed port, uniflow, loop. Uniflow-scavenging valves in cylinder head and exhaust ports around the cylinder. Loop-exhaust port above scavenging port. Air is scavenged by either a piston, pump, blower, or from the crankcase. Pressure ranges from 7 psi for piston to 2 psi for blower.

Q. How are valves for a Diesel engine operated?
A. Valves are operated by rocker arms and pushrods which are actuated by cams on the camshaft. The camshaft is driven by gears or chain or by the crankshaft. An overhead camshaft is driven by gears or chain and actuates valves directly without pushrods.

Q. Should the addition of a supercharger have any adverse effect upon the mechanical efficiency of a 4-cycle Diesel engine?
A. No. It increases the mechanical efficiency because there is more horsepower per pound weight of the engine produced. Thermal efficiency is increased because of more complete combustion.

Q. How would you change the speed of an engine controlled by a shaft governor?
A. By trial and error.

Q. What four common types of governors are used on Diesel engines?
A. Constant-speed; variable-speed; speed-limiting; load-limiting.

Q. What is the advantage of an automatic unloader on a compressor?
A. It is usually used with a constant-speed motor-driven compressor. It permits greater flexibility in operation and capacity. The compressor can start in the unloaded condition thus reducing the starting torque on the motor.

MAINTENANCE

Q. Write out an actual entry you would make in an engineering log after you had refueled the ship.

A. Date Port Ship's Name

Dealer.............. Producer's name

Bill Stating: Flash point—Gallons received—Temperature

Sample Taken: 1/2 pint

Signed: Officer in Charge of Watch

Q. How will wear of the cams of a 4-cycle Diesel engine affect the engine operation?

A. All timing will be off.

Q. How is the compression pressure of a Diesel engine measured?

A. By taking indicator cards and computing.

Q. What effect will excessive wear or wiping of crankshaft journals have on efficient operation of a single-acting Diesel engine?

A. Will affect compression (lower it) due to the fact that the top clearance is increased. Also, the gear teeth mesh will be affected. They will not mesh fully and may cause excessive strain and possible breakage of the teeth.

Q. Why is the pin of a trunk-type piston the most difficult bearing to lubricate and what precautions must be taken with respect to the fit of other bearings in order that the piston pin may be sufficiently lubricated?

A. This pin is at the farthest point in the lube system. The oil must travel through the bore of the piston rod. If the other bearings have too much clearance, the oil will lose its pressure and not reach the piston pin.

Q. How would you run the engine if a cylinder breakdown occurred?

A. The lube oil, air, water, and fuel oil lines to the cylinder would be blanked off. Disconnect the piston rod and hang up the piston by strong backs. Disconnect the valve assembly or take out push-rods. Operate the engine at a reduced speed. It may also be necessary to remove the piston from the cylinder.

FUEL INJECTION

Q. What are the advantages of using the air injection of fuel? State the disadvantages.

A. Advantages:

1. Better control for wide range of loads and speeds.
2. Better turbulence and atomization.
3. Lower grades of fuel may be used.

Disadvantages:

1. More complicated construction.
2. An air compressor must be used.
3. Must carry a higher compression pressure to compensate for the cooling effect of the air injection of fuel.
4. Greater weight per horsepower.
5. Poor maneuverability.

Q. What is meant by a variable-stroke fuel pump? Name three methods of obtaining variable stroke.

A. A variable-stroke pump is a fuel pump in which the effective stroke of the pump plunger may be altered, or in which the effective portion of the stroke may be altered.

Three methods of varying the stroke are:

1. A sliding-cam arrangement.
2. A sliding-wedge arrangement.
3. By rotating the plunger and opening a by-pass port some time in the discharge stroke, thus varying the effective portion of the stroke.

Q. What is meant by injection lag?

A. The lapse (duration) of time (measured in seconds or crank-angle degrees) between the opening of the fuel-pump discharge valve and the beginning of injection at the nozzle. This time lag is due to the stretch of the fuel lines and the compressibility of the fuel. This is found only in independent type of fuel-injection systems.

Q. What are the advantages of having a separate fuel pump and fuel injection nozzle for each cylinder of the Diesel engine?

A. Individual cylinders can be cut out if trouble develops. The power of each cylinder can be adjusted individually.

Q. What is a spray valve?

A. A spray valve controls the injection of fuel oil into the cylinder. Some are mechanically operated by cams, and some actuate against a spring, which is within the valve and holds the needle valve closed on its seat. The valve consists of a body, spring, adjusting nut, stem, valve seat, disk or needle valve, and sprayer plate. The valve allows the fuel to be injected into the cylinder and atomizes the fuel and sprays it into the cylinder in the form of minute particles to promote combustion.

Q. Describe the air starting system.

A. A battery of bottles stores the compressed air for starting. The bottles may be filled by compressing the air with the main compressor when the engine is in operation, or by the auxiliary compressor when the engine is stopped or maneuvering. The filling lines from the compressor to the bottles extend into the bottles about halfway to prevent moisture from getting into the line when filling and discharging at the same time. Each bottle is fitted with a safety valve, gauge, and drain. The starting air leads from the shutoff valves at the bottles through the combined line to a reducing valve, which reduces the air from 1000 psi to approximately 500 psi. It then goes to the main starting valve and from there to the starting valves of each cylinder. There is a gauge behind the reducing valve to show the air pressure going to the engine.

Q. What are the usual causes of a leaky solid injection fuel nozzle on a modern medium-speed Diesel engine? What effects will a leaky nozzle have on the operation of the engine?

A. A leaky valve caused by pitting, grooving or scoring of the valve seat or valve; a leaky gasket; a weak or broken spring. A leaky nozzle will cause a cylinder to fire ahead of time, causing a knock.

Q. The main functions of any fuel injection system are correct timing, metering and atomization of the fuel. Explain how each of these functions is accomplished in a Diesel which is equipped with an individual pump having a helically grooved plunger for each cylinder.

A. The injection pump plunger has a constant stroke but the amount of fuel discharged into the cylinder is determined by the degree of rotation of the pump plunger in its barrel. View 1 of Fig. 74 shows the plunger at the bottom of its stroke when two small ports, in opposite sides of the barrel, are open. A pressure of from 10 to 15 pounds per square inch is maintained outside of the ports, so that, in the position shown, the space above the plunger fills with oil. On the upward stroke, the plunger displaces fuel back through the two small ports until its top edge (A) covers them. The fuel which is then in the barrel is discharged to the cylinder until the scroll edge (B) uncovers the port on the right. Rotation of the plunger in the barrel causes a higher or lower point of the scroll (helical groove) to uncover the port, thereby cutting off the discharge earlier or later.

Views 1 and 2 show the maximum fuel rotation of the plunger. Views 3 and 4 show partial rotation. The effect is quite evident by comparing views 2 and 4. The plunger is almost at the end of its stroke in view 2 before the scroll edge uncovers the right-hand port, but the uncovering of the port is taking place in view 4 long before the end of the stroke. The full by-pass or off position is shown in view 5. When rotated to this position, the straight slot passes up and down over the right-hand port, the fuel passes out of the barrel through this slot after the edge (A) rides above the ports.

When the helical edge (B) of the pump plunger uncovers the port in the pump barrel near the end of the delivery stroke, the pressure of fuel is immediately reduced so that the delivery valve at once drops on its seating, thus cutting off communication between the pump and the nozzle until the next delivery stroke takes place. In coming to its seat to act as a nonreturn valve, the delivery valve is, however, made to perform the other highly important function of Pressure Pipe Release.

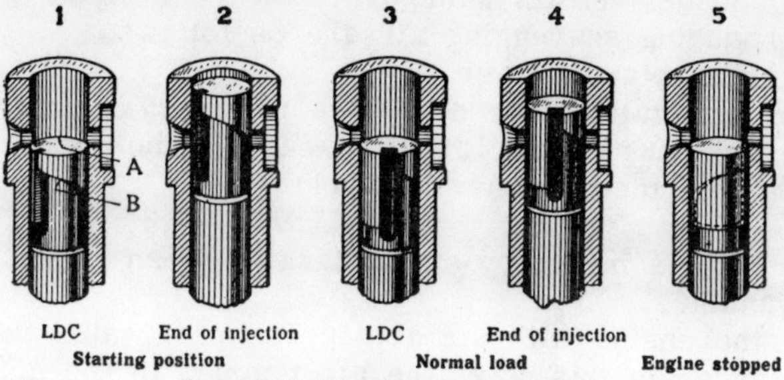

Pump barrel with various plunger positions.

Figure 74.

AIR COMPRESSORS

Q. What is the principal cause of valve troubles in air compressors?

A. Oil works up from crankcases or in from excess lubrication or through faulty parts. This oil forms carbon which mixes with metal dusts from wear of parts, and with dusts drawn in with the intaken air. The resulting mixture is a hard, abrasive product which accumulates around the valves. This has an erosive effect, to which may be added a corrosive effect if the oil is acid or the indrawn dusts or fumes contain acids or acid-forming elements.

Q. Other than the waste of power, is there any probable bad effect of running an air compressor unloaded for long periods of time?

A. Yes. There is an increased tendency to get excess lubricant into the cylinders and other members, leading to possible formation of carbon about the valves, increased need for cleaning, and even to explosion hazards.

Q. If you heard groaning noises from air-compressor cylinders, what troubles would you look for?

A. Lack of lubrication; overheated cylinders; piston rod binding; leaks at discharge valves or at piston packing rings; stuck valves.

Q. How may the amount of air compressed in an air compressor be controlled?

A. 1. Bleeding from one of the stages.
   2. Throttling the compressor suction.
   3. Varying the clearance volume of the stages.

Q. What are the most common troubles encountered with air compressors?
A. With lubrication, cooling, and leaky valves. In case of dirty intake air, there is danger of an explosion.

Q. Describe the use of scavenging air, its origin, and various ways of producing it.
A. It is used to scavenge or rid the cylinders of burned gases of combustion and derives its name from the duty which it performs. Means of producing scavenging air are as follows:

1. By using a rotary blower.
2. By a reciprocating-type compressor (direct or indirect driven).
3. By compressing air in the crankcase of the engine by means of the pistons on their downward stroke.

Q. Describe what is meant by supercharging of an I. C. E. engine and its advantages.
A. Supplying the engine with air at a pressure greater than atmospheric pressure. In this way, the air trapped in the cylinder at the beginning of the compression stroke is at a pressure slightly above atmospheric.

Q. Explain how the temperature of the air drawn into the low-pressure cylinder affects the efficiency of a 3-stage air compressor.
A. More efficient compression due to greater weight per volume if air is cool. Warm moist air may give trouble with moisture in the system.

Q. What causes air-compressor explosions?
A. 1. Excess oil feed resulting in carbon deposits on discharge valves, piping and intercoolers. Under excessive temperatures these deposits become incandescent and give off vapors which combine the excess oil vapors and ignition results.

2. Excessive accumulations of emulsions.
3. Oil vapors in compressor.

Q. Describe in detail how you would remove scale from cylinder jackets.
A. Drain compressor thoroughly. Fill cooling jackets with a 50-50 solution of hydrochloric acid and water (heated). Leave until effervescence stops (approximately 10 or 12 hours). Add more acid. If effervescence starts again, wait until it stops and add more acid. Continue until addition of acid causes no further action. Drain jackets and flush thoroughly with a soda water solution.

Q. What simple tests can be made to check for compressor leaks?
A. Isolate one portion of the system at a time and test all pipe joints with soapsuds or a candle flame.

Q. How would you clean the internal parts of a compressor?
A. Remove all valves and connections and soak in a lye or solvent solution. (Never use kerosene or gasoline.)

Q. What factors would you consider in the selection of compressor oil?
A. High-grade blended oil; high flash point; low viscosity; good stability; non-emulsifying; no carbon residue.

Q. What are the main points in connection with the operation of air compressors which must be given attention if accidents are to be prevented?
A. Prevent oil deposits on coolers. Observe air temperature closely. It is possible for excessive temperature beyond the HP stage to ignite the vaporized oil which is always present. Drain moisture through moisture drains periodically.

Q. What are the effects produced by interstage cooling?
A. Reduces the danger of pre-ignition at fuel valve; reduces volume of air; reduces the deposit of moisture in the air; less work needed to compress air.

Q. What are the advantages, if any, of locating the intake of an air compressor in a cold place?
A. Less chance of moisture accumulation as happens with air of high humidity. Also, cooler air takes up less volume and the machine compresses a greater weight of air on each stroke. The compressor also stays cooler. Roughly there is a 1% increase in the volume of compressed air for each 5° decrease in incoming air.

AIR STARTING

Q. How are air starting and injection air bottles constructed and in what position are they placed in a ship? Why?
A. They are usually of seamless drawn steel, closed at the bottom and having the upper end swaged down and threaded with a forged steel head screwed on. On this head are the inlet and stop valves. The drain pipe for condensation runs down close to the bottom. The average pressure is 1000 lbs.

Air tanks for storing air for starting purposes up to 600 lbs. pressure are usually riveted construction, though in smaller sizes they may be welded. These tanks are cylindrical in shape. Location is a matter of convenience, and may be vertical or horizontal.

Q. Describe the principle of operation of the pilot and master valve in the Diesel air-starting system.
A. Due to the large area of the master starting air valve and high pressure against which it must be worked (usually manually), a small line is connected from the main air line to a ram attached

to the main starting air-valve stem. A small valve called a pilot valve allows the starting air to operate the main starting air valve (by ram effect) with ease, using the main starting pressure.

Q. What are the approximate opening and closing angles of an air-starter valve?
A. 50 ATC to 100° ATC.

Q. If the air-starting valve were turned on and the engine failed to turn over at all, what troubles would you look for?
A. 1. Check to see if there is sufficient air pressure.
2. Check to see if jacking gear is disengaged.
3. Check starting valve — it may be stuck.
4. Check timing.

Q. Why is a separator required in an air line?
A. To remove the grit, dirt, water, or other impurities from the air. A filter at the suction side of the compressor separates the dirt or grit from the air before it enters the LP stage. Intercoolers between the different stages remove the heat of compressions, moisture, oil vapor and carbon particles, all of which are drained off intermittently. Other intercoolers are used for the same purpose.

LUBRICATION

Q. How does combustion interfere with lubrication?
A. Long ignition lag, or early injection, with the resultant excessive pressure and temperature, causes the oil film to break down, due to temperature, and overloads the oil film in the bearings. Incomplete combustion leaving carbon and carbon residue causes the piston rings to stick, allowing blow-by, which also breaks down the oil film.

Q. What is the effect of over-lubrication in a Diesel cylinder?
A. Carbonization of oil on valve seats. Possible explosive mixture.

Q. Explain how Diesel cylinder lubricating oil with a flash point of 400°F. at atmospheric pressure can withstand the high pressures and temperatures which exist in a Diesel cylinder.
A. The cylinder liners are cooled.

Q. How is lube oil cleaned so that it can be used again?
A. By batch or continuous cleaning system. In the batch system, all the oil in the engine is drained out and cleaned in an oil reclaimer while a clean batch is run into the engine. In the continuous system, the oil drains by gravity from the crankcase to the sump tank. A pump with a low suction picks up the oil and discharges it through a heater into a centrifuge separator which removes the dirt, sludge and water from the oil. It then goes back to the sump where another pump takes it and sends it through strainers and coolers to the engine. The centrifuge is operated

so that it will pass all the oil through it in a given period of time.

Q. Why is oil used instead of water for cooling pistons of a Diesel engine?

A. If water were used and a leak occurred in the piston cooling system through a faulty connection or crack, it would leak into the crankcase, emulsifying the oil and spoiling its lubricating properties, thus burning out bearings and causing great damage to the engine.

MATHEMATICS

Q. The exhaust valve of a Diesel engine is of the poppet valve type and 4" in diameter. How far must it lift vertically off its seat to give full port opening?

A. 1". (See page 136 in 3rd Assistant Section.)

Q. A 4-cycle motor has 4 cylinders, 4" diameter, 5" stroke; the MEP averages 65 pounds psi when the revolutions are 500 per minute. What is IHP?

A.

$$\frac{PLAN}{33000} = \frac{(65)(\frac{5}{12})(.7854)(4)^2(250 \text{ X } 4)}{33000} = 10.3$$

Q. A gravity tank is 24" diameter. How many pounds of fuel are consumed by an engine if the level is reduced 1"? The fuel weighs 7.3 lbs./gal.

A. Volume = height x area

Volume = 1" x $(24)^2$ x .7854 = 452.3904 cu. ins.

452.3904 ÷ 231 = 1.9584 gals.

1.9584 x 7.3 = 14.2963 lbs.

Q. A flywheel with a 4-ft. radius turns at 80 RPM. What is the peripheral speed in feet per second?

A.

$$\text{Peripheral speed} = \frac{\text{Circumference} \times \text{RPM}}{60}$$

$$= \frac{\pi D \times 80}{60}$$

$$= \frac{(3.1416)(8)(80)}{60}$$

$$= 33.5 \text{ ft./sec.}$$

Q. A diesel engine develops 120 H. P. It drives a generator that is 90% efficient. What is the amperage when the voltage is 220?

A. Output = Input

= (120)(746)(.90)

= 80568 watts

W = EI

$$I = \frac{W}{E} = \frac{80568}{220} = 366 \text{ amps}$$

Q. A Diesel engine consumes .45 pounds of fuel per B. H. P. Hour. What is the efficiency if the heating value of the fuel is 18,000 B. T. U. /Lb?

A. $\frac{33,000 \times 60}{778} = 2545$ B. T. U. /Hp. hr.

$18,000 \times .45 = 8100$ B. T. U.

$\text{Efficiency} = \frac{\text{output}}{\text{input}} = \frac{2545}{8100} = 31.41\%$

Q. Two main engines, 2800 Hp. each: fuel oil consumption for 24 hours is 8750 gals. Each gallon weighs 7.3 Lbs. What is the fuel consumption per Horsepower Hour?

A. $8750 \div 24 = 364.5$ gals. /hr.

$364.5 \times 7.3 = 2660.85$ Lbs. /Hr.

Total Hp. $2800 \times 2 = 5600$

$2660.85 \div 5600 = .475$ Lbs. /Hp. hr.

Q. A shaft turns 1400 R. P. M. driving a centrifugal pump. It is desired to turn the pump at 1225 R. P. M. There are 28 teeth on the driving shaft. How many teeth are required on the driven shaft?

A. Let X = Number of teeth required

∴ The product of the driving end must equal the product of the driven end.

$1225 \times X = 1400 \times 28$

$1225X = 39,200$

$X = 32$ teeth

Q. The fuel consumption of an engine is .44 lbs/H. P. hr. The calorific value of the fuel is 17,500 B. T. U. What is the thermal efficiency?

A. $\frac{33,000 \times 60}{778} = 2545$ **B. T. U. /H. P. hr.**

$17,500 \times .44 = 7700$

$\text{Efficiency} = \frac{\text{Output}}{\text{Input}} = \frac{2545}{7700} = 33.05\%$

Q. What is the volume of a cylinder if the clearance volume is 6.25 cu. in. with the piston at T. D. C.? The piston is 3" diameter and the stroke is 6.5".

A. $V = PD + CV$

$= .7854\,(3)^2\,(6.5) + 6.25$

$= 45.95 + 6.25$

$= 52.20$ **cu. in.**

Material Required for All Grades

CONTENTS

Answers to Questions Appear at End of Each Section

MULTIPLE-CHOICE QUESTIONS & ANSWERS

ELECTRICITY

1. The unit of electrical pressure is the:
   A. amp B. volt C. watt D. ohm

2. The unit of electrical resistance is the:
   A. amp B. volt C. watt D. ohm

3. The unit of electrical current flow is the:
   A. amp B. volt C. watt D. ohm

4. A galvanometer is used to measure:
   A. small voltages C. large currents
   B. small currents D. large voltages

5. Volts × amps equals:
   A. kilowatts B. watts C. ohms D. watt-hours

6. The chemical symbol for carbon dioxide is:
   A. CO_2 B. CO C. C_2O D. C_2O_2

7. If three resistances of 2, 4, 6 ohms are connected in parallel, the total resistance is:
   A. 2. 05 ohms B. 1. 09 ohms C. 0. 05 ohms D. 12 ohms

8. Which of the following formulas would solve for amperage?
   A. R/E B. RE C. E/R D. R + E

9. A circuit that has one wire in contact with the hull of the ship is a:
   A. series circuit C. short circuit
   B. grounded circuit D. closed circuit

10. A circuit that has two or more wires in contact with the hull of the ship is a:
    A. closed circuit C. open circuit
    B. short circuit D. any of the above

11. In a three-wire system, the ground wire is:
    A. black B. green C. white D. red

12. The capacity of #14 rubber-insulated wire is:
    A. 30 amps B. 20 amps C. 15 amps D. 60 amps

13. The horsepower of an 1800 KW motor is:
    A. 1800 horsepower C. 2142 horsepower
    B. 2412 horsepower D. 2421 horsepower

14. In a three-wire 115-230 volt distribution system, the voltage between the neutral and negative is:
    A. 230 volts B. no voltage C. 345 volts D. 115 volts

15. When using Ohm's Law, $\frac{E}{R}$ would solve for:
    A. amperage
    B. voltage
    C. watts
    D. resistance

16. The counter EMF of a motor is zero when the:
    A. motor is at rated speed
    B. motor is almost up to rated speed
    C. armature is not turning
    D. armature just begins to turn

17. The first thing to do if a generator overspeeds is to:
    A. trip the circuit breaker
    B. secure the steam
    C. adjust the rheostat
    D. trip the overspeed trip

18. In a wiring diagram where two wires come together, it is indicated by a:
    A. dot B. cross C. loop D. circle

19. The ratio of real power to apparent power is the:
    A. safety factor
    B. power factor
    C. design factor
    D. A or C

20. An open coil can be detected by:
    A. high resistance
    B. low resistance
    C. overheated coil
    D. A and C

21. A shorted coil can be detected by:
    A. high resistance
    B. low resistance
    C. overheated coil
    D. B and C

22. Emergency generators can be stopped with the:
    A. overvoltage trip
    B. overspeed trip
    C. manual trip
    D. undervoltage trip

23. The horsepower of a motor with an effective resistance of 2 ohms on a 110-volt line is:
    A. 16 horsepower
    B. 12 horsepower
    C. 10 horsepower
    D. 8 horsepower

24. An ammeter reads 230 amps and the voltmeter 115. The power in KW at the time of the reading is:
    A. 26,450 B. 26.45 C. 264.5 D. 2645

25. #14 wire is not:
    A. larger than #16
    B. smaller than #12
    C. larger than #18
    D. smaller than #16

26. A mil is:
 A. 1/10" B. 1/100" C. 1/1000" D. 1/1,000,000"

27. In a series circuit, the total current is:
 A. the same as that of the largest branch circuit
 B. the same throughout all parts of the circuit
 C. the same as that of the smallest branch circuit
 D. none of the above

28. When using Ohm's Law, $\frac{E}{I}$ would solve for:
 A. amperage
 B. voltage
 C. resistance
 D. watts

29. The size of electrical wires is measured in:
 A. centimeters
 B. B. W. G.
 C. circular mils
 D. cu. in.

30. The resistance of a 100-watt bulb in a 120-volt circuit is:
 A. 142 ohms B. 144. 5 ohms C. 142. 5 ohms D. 143 ohms

31. In the flow of 1 cycle of AC current past any given point in a circuit, the maximum current flow occurs:
 A. once B. twice C. 4 times D. 3 times

32. Which of the following are fitted on or near a switchboard?
 A. wooden handrail
 B. rubber mat or wood platform
 C. drip pan over the switchboard
 D. all of the above

33. A generator interpole always has the same polarity as the:
 A. pole following it
 B. pole preceding it
 C. opposite main pole
 D. none of the above

34. When resistance is increased in a shunt motor field, the speed:
 A. decreases
 B. remains the same
 C. increases
 D. none of the above

35. Which of the following is the distinguishing feature of a shunt motor?
 A. it has a high starting torque
 B. a load will not affect it if running at high speed
 C. it has a stable speed through a wide load range
 D. it will not drop in speed if overloaded

36. In a series circuit, the total resistance is:
 A. the total of all resistances
 B. the reciprocal of all the resistances
 C. the average of all the resistances
 D. the same as the largest resistance

37. Which of the following methods can be used to remove moisture from a generator?
    - A. steam radiators
    - B. electric heaters
    - C. hot air
    - D. any of the above

38. Why is AC used on modern ships?
    - A. less maintenance
    - B. less weight
    - C. simpler operation
    - D. all of the above

39. When a load is applied to a shunt-wound motor, which of the following will happen?
    - A. speed increases - CEMF increases
    - B. speed decreases - CEMF decreases
    - C. speed decreases - CEMF increases
    - D. speed increases - CEMF decreases

40. The effects of armature reaction in a generator are reduced by the use of:
    - A. interpoles
    - B. commutator windings and interpoles
    - C. commutating windings
    - D. compensating windings and commutating poles

41. One Joule is equal to:
    - A. 10 watt seconds
    - B. 1 watt hour
    - C. 1 watt second
    - D. 1 watt minute

42. Which of the following instruments is not needed in the operation of 2 DC generators operating in parallel?
    - A. voltmeter
    - B. ammeter
    - C. wattmeter
    - D. overcurrent relay

43. AC generator voltage is increased by:
    - A. increasing the KVA
    - B. increasing the speed of the prime mover
    - C. increasing the PF
    - D. A or C

44. Branch lighting circuits are allowed to carry:
    - A. 880 watts
    - B. 1200 watts
    - C. 440 watts
    - D. no maximum wattage but circuit must be limited to 30 amps

45. A synchroscope is used to:
    - A. parallel AC generators
    - B. parallel DC generators
    - C. synchronize propulsion motor and generator
    - D. none of the above

46. A power factor meter shows the:
    A. true power
    B. apparent power
    C. relationship between volts and amps
    D. relationship between true power and apparent power

47. Which of the following statements is true?
    A. an increase in current decreases temperature
    B. an increase in current increases temperature
    C. an increase in current has no effect on temperature
    D. any increase in current will double the temperature

48. The reverse power relay:
    A. keeps amperage at a safe level
    B. keeps voltage at a safe level
    C. protects generator from motorizing
    D. none of the above

49. Horizontal cable supports should be placed every:
    A. 14" B. 24" C. 18" D. 12"

50. The purpose of an unloader on a motor-driven air compressor is to:
    A. reduce starting torque
    B. give smoother operation
    C. increase torque
    D. A and B

51. If the length of a wire is doubled and the cross sectional area is reduced to one-half, the change in resistance will be:
    A. halved B. quadrupled C. doubled D. quartered

52. If a DC generator was rotated in the wrong direction, it would fail to come up to voltage because the:
    A. brushes would burn out
    B. generator would motorize
    C. armature field would oppose the field current
    D. circuit breaker would not energize

53. Which of the following is not a standard circuit?
    A. 3-phase, AC C. 3-wire, DC
    B. 4-phase, 4-wire, AC D. 3-phase, 4-wire, AC

54. The navigation lighting branch circuits are equipped with:
    A. 3-amp fuses C. 10-amp fuses
    B. 5-amp fuses D. 15-amp fuses

55. One could determine if a synchronous propulsion motor was about to go out of step by checking the:
    A. ammeter C. stability meter
    B. synchroscope D. voltmeter

56. Varnished cambric insulated power and light cables are not allowed in sizes smaller than:
    A. #10 B. #14 C. #12 D. #18

57. Rubber or asbestos varnished cambric insulated power and light cables are not allowed in sizes smaller than:
    A. #14 B. #12 C. #18 D. #16

58. A synchronous motor differs from an induction motor in that it:
    A. is not self-starting
    B. requires an AC and DC power supply
    C. may be used for power factor correction
    D. all of the above

59. The direction of rotation of a split-phase induction motor can be reversed by:
    A. reversing the starting winding leads
    B. reversing the main winding leads
    C. reversing the leads of one phase
    D. reversing any two leads

60. Circuit breakers are required on generators of over:
    A. 50 KW B. 100 KW C. 75 KW D. 25 KW

61. Synchronous converters are used to:
    A. change AC to DC
    B. change DC to AC
    C. change synchronous to induction
    D. change induction to synchronous

62. This symbol represents a:
    A. thermostat
    B. rheostat
    C. resistor
    D. capacitor

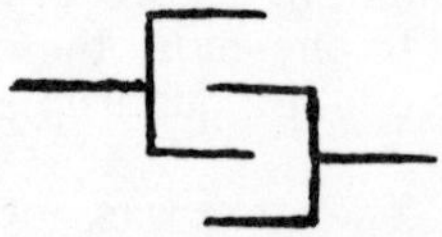

63. If it becomes necessary to operate a motor at a slight overload for a short period of time, you should:
    A. install larger fuses
    B. check bearing and motor temperatures frequently
    C. call the chief engineer
    D. jump the overload relay

64. Before taking an AC generator off the board that is operating in parallel with others, you should:
    A. reduce the voltage
    B. reduce field excitation
    C. reduce steam input with governor mechanism
    D. close in on the throttle

65. DC motors are provided with starters and controllers to:
    A. provide local control for the motor
    B. improve power factor
    C. provide good speed control at light loads
    D. start motor at less than line voltage

66. Before synchronizing alternators, they must have:
    A. the load equalized
    B. the same frequency
    C. the same voltage
    D. the same amperage

67. The advantage of DC motors over AC motors is that they:
    A. have higher allowable speeds
    B. can lift heavier loads
    C. have better speed control over a wide range
    D. have a longer life

68. If two alternators are operating in parallel with a load greater than either one is capable of carrying and one of them has to be cut out, you should:
    A. increase excitation
    B. increase voltage of the machine that is to stay on the line
    C. speed up the machine that is to stay on the line
    D. eliminate some branch circuits to reduce the load

69. A generator has been exposed to water. Which of the following instruments would be used to determine if it is safe to operate?
    A. a megger
    B. a voltmeter
    C. a hydrometer
    D. a wattmeter

70. Motor leads are insulated by:
    A. using heat-resistant enamel
    B. using heat-resistant varnish
    C. using heat-resistant white lead
    D. using heat-resistant red lead

71. A synchronous motor is a:
    A. machine to turn AC into power
    B. machine to turn DC into power
    C. compound-wound motor
    D. synchronous machine that uses AC for excitation

72. Small hand tools are grounded to:
    A. prevent short circuits
    B. equalize voltage with hull of vessel
    C. guard against personnel shock
    D. all of the above

73. The advantages of turboelectric drive are:
    A. better bridge control
    B. simpler operation
    C. short shaft
    D. all of the above

74. Which of the following will not cause a generator to vibrate?
    A. misalignment
    B. loose bolts
    C. loose pigtails
    D. faulty speed governor

75. The safety factor in using a double-pole switch is the fact that:

A. both line wires are dead when switch is turned off
B. it can be replaced easily
C. it will stand greater loads
D. it can be used on any voltage

76. Which of the following devices prevents overload on a generator?
A. rheostat
B. circuit breaker
C. fusestat
D. any of the above

77. Which of the following will cause generator brushes to burn?
A. brushes not in neutral plane
B. generator overloaded
C. weak brush springs
D. all of the above

78. Static electricity is most often produced by:
A. pressure B. heat C. magnetism D. friction

79. A "Universal" motor must be a:
A. series-wound motor
B. shunt-wound motor
C. compound-wound motor
D. A or C

80. A circuit that does not provide a complete path for the flow of current is:
A. an open circuit
B. a closed circuit
C. a series circuit
D. a grounded circuit

81. A "dead board" is:
A. a board with no power
B. a board with no switches
C. a board with no circuit breaker
D. none of the above

82. Which of the following can be used to convert AC to DC?
A. motor generator
B. synchronizer
C. rectifier
D. A and C

83. What protective device is used on a "dead front" switchboard to take the place of the fused knife switch on a "live front" board?
A. open-type circuit breaker
B. motor-operated circuit breaker
C. enclosed-type circuit breaker
D. a fusestat

84. When would you ground an electric hand tool?
A. when it is a DC machine
B. when it is an AC machine
C. any machine that operates on over 100 volts
D. only when working on wet steel plates

85. Which of the following would cause sparking at the brushes?
A. high mica
B. dirty commutator
C. grooved commutator
D. any of the above

86. The safest solvent to use for cleaning electric motors is:
    A. carbon tetrachloride
    B. diluted muriatic acid
    C. carbon dioxide
    D. trichloride ethylene

87. Which of the following controls the voltage output?
    A. the number of field coils
    B. the size of the field coils
    C. the amount of current flowing through the field coils
    D. the speed of the prime mover

88. Which of the following would cause a hot bearing?
    A. poor alignment
    B. overload
    C. bearings too tight
    D. all of the above

89. Which of the following will not cause a hot motor bearing?
    A. insufficient lubrication
    B. loose brushes
    C. overload
    D. misalignment

90. The correct amount of grease to put in motor bearings is to fill the bearing:
    A. completely
    B. race and half-fill the housing
    C. race with heavy grease and the housing with light grease
    D. race with a light grease and the housing with a heavy grease

91. An open armature coil can be detected by:
    A. an increase in motor speed
    B. a decrease in motor speed
    C. sparking at the brushes
    D. a ring of fire running around the commutator

92. Most generators will withstand an overload of:
    A. 30% B. 25% C. 50% D. 35%

93. The USCG R&R require that engine-room switchboards be equipped with:
    A. overhead drip pans
    B. a wooden handrail
    C. a rubber mat or wood platform
    D. all of the above

94. Which of the following would cause a generator to fail to build up?
    A. field connections reversed
    B. speed too low
    C. machine short-circuited
    D. any of the above

95. Interpoles are connected in:
    A. series with the armature
    B. parallel with the armature
    C. parallel with the series field
    D. series with the shunt field

96. A "dead front" board is a board with:
    A. insulated switches and no open terminals
    B. no switches on it
    C. no circuit breaker
    D. no safety handrail or rubber mat

97. Which of the following DC loads will be found on vessels with AC ship's service systems?
    A. general alarm system
    B. emergency lighting
    C. rectifying for battery charging
    D. all of the above

98. Which of the following is required to induce a voltage and current in a wire?
    A. a magnetic field
    B. a conductor in a closed circuit
    C. motion between A and B
    D. all of the above

99. Ambient temperature is the:
    A. normal surrounding compartment temperature
    B. maximum temperature at which motors can be operated
    C. minimum temperature at which motors can be operated
    D. normal operating temperature of a motor

100. Any motor can be constructed to be:
    A. dripproof
    B. short-proof
    C. explosion-proof
    D. A or C

101. All electric cables run through watertight bulkheads must be:
    A. welded on both sides of the bulkhead
    B. installed with watertight packing glands
    C. grounded on both sides of the bulkhead
    D. fitted with unions on each side of the bulkhead

102. Emergency lights supplied by the emergency lighting system shall be lighted continuously at all times except in the:
    A. machinery spaces
    B. passageways
    C. embarkation decks
    D. crew's messroom

103. A reverse power relay is used to:
    A. reverse the rotation of a motor
    B. protect the generator from a power reversal
    C. change AC to DC or DC to AC
    D. reverse the rotation of a generator

104. T. P. S. T. stands for:
    A. triple-pole standard terminal
    B. triple-pole single-throw switch
    C. temporary-pole standard terminal
    D. tandem-pole single-throw switch

105. D. P. D. T. stands for:
A. duplex pole dead-throw switch
B. double-pole duplex terminal
C. double-pole double-throw switch
D. direct power to direct terminal

106. Small switches are classified as:
A. rotary B. pushbutton C. tumbler D. all of the above

107. Another name for an overcurrent protection device is a:
A. magnetic breaker C. fuse
B. thermal breaker D. all of the above

108. If a motor caught on fire, the first thing to do is:
A. use a CO_2 extinguisher on it
B. use a dry chemical extinguisher on it
C. secure the motor
D. slow down the motor until it cools off

109. Armature coils are laminated to:
A. increase the lines of force
B. increase the magnetism
C. cut down on losses by eddy currents
D. decrease heat losses

110. One ground lamp burning bright and the other burning dim would indicate a:
A. ground on both legs of the circuit
B. short circuit
C. ground on the side of the dim light
D. ground on the side of the bright light

111. All electric splices shall be electrically and mechanically secure without solder and unless an approved splicing device is used, they shall be:
A. welded B. soldered C. brazed D. any of the above

112. Which of the following splices must be used on board ship?
A. pigtail B. "T" tap C. Western Union D. Underwriters

113. When selecting the size of wire to be used in a circuit, the most important item to consider is the:
A. resistance of the circuit C. amperage of the circuit
B. voltage of the circuit D. amount of wire to be used

114. Which of the following valves could a rheostat be compared with?
A. globe valve C. throttle valve
B. check valve D. safety valve

115. Before working on an electrical circuit:
   A. open the switch
   B. tag the switch "do not close"
   C. have required tools and meters ready
   D. all of the above

116. Which of the following is not a good conductor of electricity?
   A. copper B. mica C. silver D. aluminum

117. An insulator is a:
   A. substance that offers a resistance to current flow
   B. substance that offers a low resistance to current flow
   C. substance that absorbs electricity
   D. form of a condenser

118. A synchronous motor is excited with:
   A. a synchronizer C. DC current
   B. AC current D. 3-phase current

119. A universal series motor can be operated on:
   A. DC current only C. AC or DC
   B. AC 3-phase current only D. AC current only

120. The advantage of DC motors over AC motors is the fact that they:
   A. require less maintenance
   B. offer a better means of controlling speed
   C. are less expensive
   D. can be started across the line

121. The magnetic brake on an electric winch is operated by:
   A. a power failure C. the stoppage of the motor
   B. heavy weights D. A or C

122. DC electric winches are usually:
   A. shunt-wound C. compound-wound
   B. series-wound D. series-parallel wound

123. Which of the following switches will be found on an electric lifeboat winch?
   A. limit switch C. emergency disconnect switch
   B. master switch D. all of the above

124. A rheostat is a device that regulates the strength of an electric current by:
   A. varying the resistance in the circuit
   B. varying the voltage in the circuit
   C. increasing magnetic field in the circuit
   D. varying the current in the circuit

125. An "exciting current" is one that is required to:
  A. build up a dead circuit C. excite a synchronous motor
  B. create a magnetic field D. build up the voltage in a battery

126. Damp armature windings will cause:
  A. reduced voltage C. overheating
  B. reduced current D. increased resistance

127. A short circuit may be measured by:
  A. ammeter or megger C. dynamometer or megger
  B. voltmeter or megger D. ammeter and ohmmeter

128. When a bar-to-bar test is made with a volt-ohmmeter an open coil will be indicated by:
  A. a high reading on open bars and lower on the others
  B. the same reading as across bars and slightly lower on others
  C. a lower reading on open bars and slightly higher on others
  D. a zero reading across open bars and higher on others

129. When removing roller bearings the force of the pulley should be placed on the:
  A. outer race B. raceway C. inner race D. retainer plate

130. Which of the following would cause one bar of a commutator to blacken?
  A. a grounded coil C. an open coil
  B. a shorted coil D. any of the above

131. The word EXIT in exit lights shall be in block letters not less than:
  A. 4" B. 3" C. 2" D. 1"

132. If the steering motor indicating light goes on at the propulsion control station it means that the:
  A. steering motor has started C. steering motor has tripped out
  B. steering motor has stalled D. steering motor is overloaded

133. The opening of a steering gear motor circuit breaker is indicated automatically by:
  A. a visual alarm on the bridge
  B. an audible alarm at the propulsion control station
  C. an audible alarm on the bridge
  D. a visual alarm at the propulsion control station

134. In lieu of the overcurrent protection on the steering gear motor there shall be:
  A. a visual alarm in the engine room
  B. a visual alarm in the wheelhouse
  C. an audible alarm on the bridge deck
  D. an audible alarm at the propulsion control station

135. Which of the following types of motors is most commonly used for crane hoists?
   - A. DC series-wound
   - B. universal
   - C. split-phase
   - D. synchronous-induction

136. The total resistance in a series circuit is:
   - A. the same as is found in its largest branch
   - B. equal to the sum of the resistance in its branches
   - C. equal to the difference between that in its largest branch and that in its smallest branch
   - D. none of the above

137. The horsepower of a 50 KW motor is:
   A. 67 B. 50 C. 75 D. 100

138. How many KW will a generator put out if it is driven by a 75 horsepower engine?
   A. 60. 75 B. 50. 25 C. 55. 95 D. 54. 50

139. The resistance of a 60-watt bulb in a 120-volt line is:
   A. 200 ohms B. 220 ohms C. 240 ohms D. 110 ohms

140. One kilowatt is equal to:
   - A. 1. 25 horsepower
   - B. 1. 33 horsepower
   - C. 1. 50 horsepower
   - D. 2. 00 horsepower

141. What is the smallest sized wire allowed aboard ship?
   A. 12 B. 16 C. 10 D. 14

142. The electrical power in watts used by a toaster using 5 amps with a resistance of 22 ohms is:
   A. 5. 50 B. 550 C. 110 D. 220

143. Twelve (12) 100-watt lamps in parallel in 120-volt circuit will draw:
   A. 100 amps B. 1 amp C. 10 amps D. 15 amps

144. The total resistance of a parallel circuit is always:
   - A. larger than that of the branch with the greatest resistance
   - B. equal to the sum of the individual branch resistances
   - C. equal to the reciprocal of the sum of the individual branch resistances
   - D. smaller than that of the branch with the lowest resistance

145. If a parallel circuit has three branches of unequal resistance:
   - A. the current through each branch is the same
   - B. the total resistance is equal to the sum of the individual branches
   - C. the voltage drop across each branch is the same
   - D. the voltage drop across each branch is directly proportional to the resistance of that branch

146. The total resistance in this circuit is:
A. 1 ohm
B. 1-1/3 ohms
C. 1-1/2 ohms
D. 1-1/6 ohms

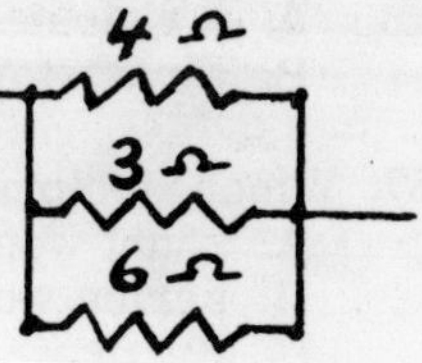

147. The total resistance in this circuit is:
A. 20 ohms
B. 6-1/3 ohms
C. 12-2/9 ohms
D. 14-1/2 ohms

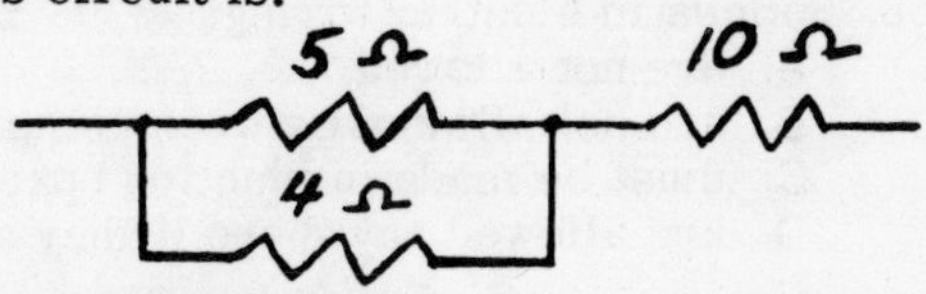

148. The total resistance in this circuit is:
A. 10-1/2 ohms
B. 4-1/3 ohms
C. 4 ohms
D. 4-2/3 ohms

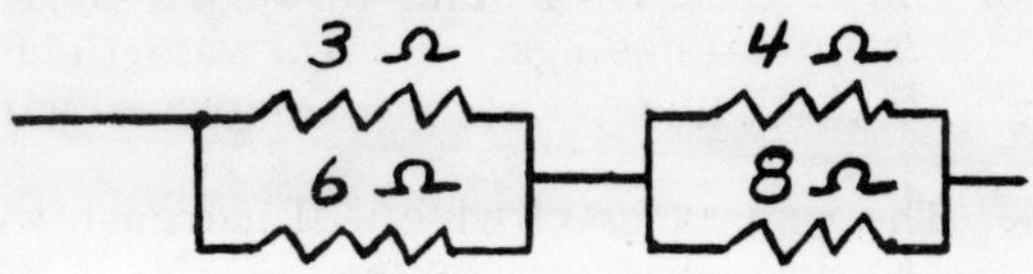

149. An ammeter reads 230 and the voltmeter 115. The power output in KW is:
A. 2. 645 B. 26. 54 C. 26. 45 D. 264. 5

150. One horsepower equals:
A. 1000 watts B. 746 watts C. 100 watts D. 940 watts

151. The electrical power in kilowatts used by a 220-volt motor drawing 15 amps is:
A. 3. 8 B. 3. 6 C. 4. 0 D. 3. 3

152. The resistance of a 110-volt circuit using 5 amps of current is:
A. 2 ohms B. 22 ohms C. 1 ohm D. 2. 2 ohms

153. If the resistance of a circuit is doubled and the applied voltage kept constant, the current will:
A. be quadrupled
B. be doubled
C. be cut in half
D. remain the same

154. A generator driven by a 75-horsepower engine will theoretically produce:
A. 55. 95 KW B. 5. 559 KW C. 75 KW D. 750 KW

155. Electron tubes can:
A. convert currents and voltages
B. amplify weak signals
C. generate high frequencies
D. all of the above

156. Which of the following is used to protect motors on across-the-line starting?

A. a rheostat C. a balance coil
B. a resistor D. an auto-transformer

157. Winch motors are usually:
A. shunt wound C. compound wound
B. series wound D. synchronous

158. Splices in electric wiring:
A. are not allowed
B. are not allowed between compartments
C. must be made in junction boxes
D. are allowed anywhere if they are soldered

159. Electric cables leading through a deck are protected by:
A. metal casings C. watertight unions
B. kick pipes D. ground wires

160. The resistance of wire will increase with:
A. reduced temperature C. increased length of wire
B. increased cross-sectional area D. none of the above

161. DC generator voltage is increased by:
A. cutting-in field resistance
B. cutting-out field resistance
C. cutting-in armature resistance
D. cutting-out armature resistance

162. DC shunt motor speed is increased by:
A. shifting brushes
B. cutting-out rheostat resistance in shunt field
C. cutting-in rheostat resistance in shunt field
D. none of the above

163. You can be sure a circuit is safe to work on if:
A. you pull and tag the switch yourself
B. the chief engineer says so
C. the electrician says so
D. any of the above

164. If an AC generator is run at 10% overload:
A. the commutator will overheat
B. the brushes will spark
C. the circuit breaker will kick out
D. no danger will occur

165. This symbol represents:
A. potentiometer
B. rheostat
C. capacitor
D. relay

166. If a wire is increased in circular mils:
  A. its size is smaller in diameter
  B. its size is larger in diameter
  C. its resistance is lower per foot
  D. B and C.

167. The insulation resistance of a field wire is determined by:
  A. the speed of rotation C. the current flow in the field
  B. using solid wire D. using braided wire

168. To indicate a ground on one phase of a 3-phase panel board:
  A. 1 light will brighten C. 3 lights will brighten
  B. 1 light will dim D. 3 lights will dim

169. The maximum current allowed on lighting circuits is:
  A. 30 amps with #10 wire C. 30 amps with #8 wire
  B. 30 amps with #14 wire D. 15 amps with #14 wire

170. In the sketch if the resistance is increased, the speed of the motor will:
  A. increase
  B. decrease
  C. remain the same
  D. double

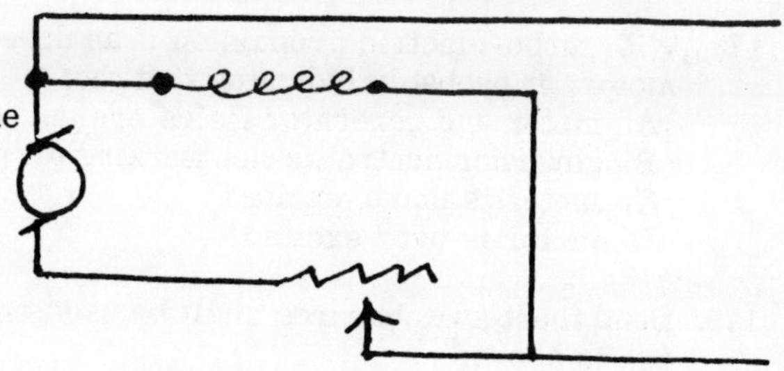

171. The maximum wattage on a 15-amp circuit is:
  A. 440 B. 220 C. 110 D. 55

172. If the field control rheostat on a DC generator is considerably off its normal setting, the most likely cause would be:
  A. prime mover not at right speed C. overload
  B. voltage too high D. open field coil

173. In the sketch if the resistance is increased, the speed of the motor will:
  A. increase
  B. decrease
  C. remain the same
  D. be cut in half

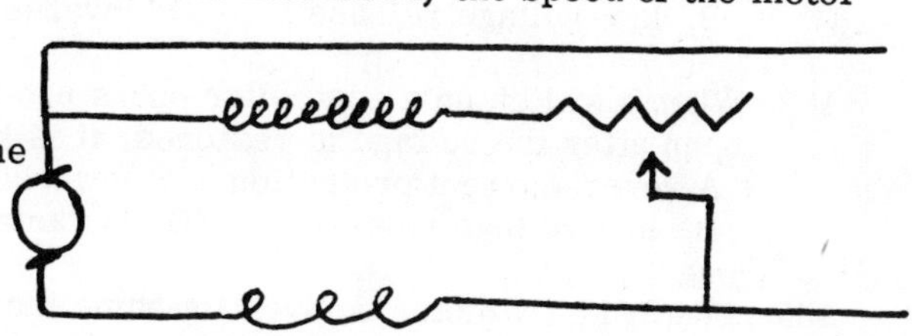

174. If the polarity of a DC line is reversed, the connecting motors will:
  A. stop C. be reversed
  B. increase in speed D. rotate in the same direction

175. When operating alternators in parallel and the load is unequal, the correct procedure is to:
   A. change the frequency
   B. change speed of prime mover
   C. decrease field excitation
   D. increase field excitation

176. A generator is protected from overspeed by:
   A. a phase balance relay
   B. a reverse current relay
   C. an overspeed trip on the prime mover
   D. a circuit breaker

177. Before a new generator is started, the air gap between the rotor and stator (or armature and field coils) should measure:
   A. more on top than bottom
   B. more on bottom than top
   C. even all around
   D. any of the above as long as it does not touch

178. With turbo-electric propulsion if an unusual noise is heard in the motor, it probably indicates that the:
   A. motor and generator fields are out of step
   B. governor control is not working properly
   C. motor is under excited
   D. motor is over excited

179. Dead front switchboards shall be used on DC boards when the voltage is above:
   A. 110   B. 250   C. 220   D. 115

180. Dead front switchboards shall be used on AC boards when the voltage is above:
   A. 110   B. 440   C. 220   D. 55

181. When a switch on a controller opens upon voltage failure and then closes again after the voltage is restored, it is known as:
   A. low-voltage protection   C. over-current protection
   B. low-voltage release   D. inverse time relay

182. When a switch on a controller opens upon voltage failure and stays open after the voltage is restored, it is known as:
   A. over-current protection   C. low-voltage protection
   B. low-voltage release   D. instantaneous release

183. The SELSYN motor derives its name from:
   A. the inventor's name   C. the abbreviation of self-synchronous
   B. a trade brand name   D. none of the above

184. An AC armature is always the:
   A. conductors into which voltage is induced
   B. rotating part of the generator
   B. stationary part of the generator
   D. conductors through which DC exciter current flows

185. Which of the following parts of an AC motor corresponds to the armature of a DC motor?
   A. the field coils C. the rotor
   B. the stator D. the armature

186. Which of the following parts of an AC motor corresponds to the field of a DC motor?
   A. the stator C. the rotor
   B. the field coils D. the armature

187. A wound rotor motor is best suited for:
   A. fans and blowers C. capstans
   B. winches D. davits

188. The interpole of a generator:
   A. is located between the main field poles
   B. reduces magnetic distortion
   C. aids in commutation
   D. all of the above

189. What type windings are placed on reciprocating-driven generators to dampen speed fluctuations?
   A. compound C. shunt
   B. amortisseur D. not necessary due to flywheel

190. Series motors drive their load through:
   A. chain drives C. pulley drives
   B. belt drives D. direct couplings

191. The principal parts of a transformer are the:
   A. core, insulation and windings
   B. primary windings and magnetic flux
   C. core, primary and secondary windings
   D. primary core and secondary windings

192. A lighted pilot light at the propulsion station indicates that the steering motor is:
   A. running C. overloaded
   B. stopped D. on standby

193. Which of the following is used to protect the moving coil of a voltmeter?
   A. a shunt C. a high resistance in parallel
   B. a low resistance in parallel D. a high resistance in series

194. When service cables are attached to a box or casing by means of a stuffing box or bushing, the purpose is to:
   A. give added strength
   B. make a ground connection
   C. prevent moisture from entering
   D. any of the above

195. Interpoles (commutating) are connected in compound DC generators in:
   A. series with the series field
   B. series with the armature
   C. series with the shunt field
   D. series with the series and shunt fields

196. To test a generator's overspeed trip you would:
   A. put under load and let it trip itself out
   B. do not touch it except in emergency
   C. bring up to full load, allow to trip out, then reset
   D. bring up speed slowly, trip out, reset while running

197. Which of the following statements is true?
   A. any increase in temperature will double the resistance
   B. an increase in temperature has no effect on resistance
   C. an increase in temperature lowers resistance
   D. an increase in temperature increases resistance

198. A magnet that is heated will:
   A. increase in magnetism
   B. decrease in magnetism
   C. become demagnetized
   D. not change its magnetism

199. Heating a magnet will:
   A. have no effect on it
   B. destroy its magnetic properties
   C. increase its magnetic properties
   D. decrease its magnetic properties

200. A bar magnet has:
   A. one pole
   B. two poles
   C. four poles
   D. all of the above

201. A horseshoe magnet has:
   A. one pole B. three poles C. four poles D. two poles

202. Which of the following statements is true?
   A. like poles attract each other
   B. like poles repel each other
   C. unlike poles repel each other
   D. none of the above

203. Soft iron is most suitable for use in a:
   A. permanent magnet
   B. temporary magnet
   C. natural magnet
   D. solid magnet

204. Residual magnetism is the:
   A. magnetism in a natural magnet
   B. magnetism in a solenoid magnet
   C. magnetism remaining in a substance after it has been removed from the influence of a magnetic field
   D. magnetism in a transformer

205. Permeability is the:
   A. power a piece of metal has on a magnet
   B. power found in a transformer
   C. power found in an electromagnet
   D. ease with which metals may become magnetized or demagnetized

206. Retentivity is the power:
   A. a metal has for retaining magnetic lines of force
   B. an electromagnet can maintain
   C. a natural magnet can maintain
   D. relationship between two motors

207. Residual magnetism is the magnetism:
   A. in a field coil
   B. in the rotor
   C. remaining in a substance after it has been removed from a magnetic field
   D. gained in converting DC to AC

208. High mica between commutator bars is removed by:
   A. removing   C. undercutting
   B. burning   D. chiseling

209. If the manufacturer's specifications for brush pressure are not known, a safe pressure would be:
   A. 2 to 3 psi   C. 1 to 1-1/2 psi
   B. 1 to 3 psi   D. none of the above

210. New brushes are fitted by:
   A. drawing sandpaper between the brush and commutator
   B. drawing crocus cloth between brush and commutator
   C. drawing emery cloth between brush and commutator
   D. using a half-round file to shape them

211. A commutator is made of:
   A. copper bars   C. nickel bars
   B. bronze bars   D. steel bars

212. Brushes are made of:
   A. lead   B. castile   C. carbon   D. lignite

213. The proper color for a commutator is:

A. dark brown
B. shiny brown
C. glazed chocolate
D. reddish brown

214. The purpose of staggering brushes is to:
A. even out the load
B. prevent the uneven wear on the commutator
C. reduce chattering
D. cover more area on the commutator for more power

215. Grooves around the circumference of a commutator indicate:
A. improper brushes
B. high mica
C. improper brush staggering
D. open armature circuit

216. Sparking and grooving of commutator may be caused by:
A. overload
B. incorrect spring tension
C. wrong type of brushes
D. any of the above

217. Commutator segments are made of:
A. zinc B. brass C. copper D. nickel

218. Which of the following materials would you use to clean a commutator?
A. sandpaper B. oilcloth C. emery cloth D. monk's cloth

219. The purpose of a commutator is to:
A. keep a constant voltage
B. rectify armature current
C. convert magnetic lines of force to flux
D. keep a constant amperage

220. An auto transformer is used in the motor-starting circuit to:
A. limit the current
B. reduce the voltage
C. control the speed
D. increase the voltage

221. A step-up transformer:
A. raises voltage and decreases amperage
B. raises voltage and amperage
C. decreases voltage and amperage
D. decreases voltage and increases amperage

222. If a transformer was plugged into a DC circuit:
A. it would blow a fuse
B. the primary would overload and the secondary would be dead
C. the secondary would burn out
D. any of the above

223. Transformers are used with:
A. DC current
B. AC or DC current
C. AC current
D. synchronous current

224. A step-down transformer lowers:
    A. voltage and amperage
    B. voltage and increases amperage
    C. amperage and increases voltage
    D. none of the above

225. If the secondary of a transformer is stepped down, the primary will have:
    A. fewer turns C. twice as many turns
    B. more turns D. half as many turns

226. Power transformers are usually rated in:
    A. ampere turns B. KVA C. KWH D. KW

227. The heat in a transformer core is dissipated by:
    A. insulating the jacket
    B. varnishing the windings
    C. covering with insulating coating
    D. steel punchings

228. Which of the following is not important with transformers?
    A. casing C. contacts
    B. primary windings D. secondary windings

229. Laminations in a transformer:
    A. reduce eddy currents C. increase output
    B. increase PF D. all of the above

230. If a lead-acid battery overheats, it will:
    A. evaporate water C. discharge itself
    B. decrease in specific gravity D. all of the above

231. When mixing sulphuric acid and water for an electrolyte for a battery:
    A. add water to the acid C. add acid and water together
    B. add acid to the water D. alternate acid and water

232. A fully charged lead-acid battery reads from:
    A. 1. 280 - 1. 300 C. 1. 050 - 1. 350
    B. 1. 025 - 1. 075 D. 1. 200 - 1. 500

233. Dirty lead-acid type batteries should be cleaned off with:
    A. soap and water C. baking soda (sodium bicarbonate)
    B. sodium chloride D. potassium hydroxide

234. The normal cell voltage of a fully charged lead-acid battery is:
    A. 2 to 2. 2 volts C. 1. 2 to 1. 4 volts
    B. 1. 4 to 1. 8 volts D. none of the above

235. A 6-cell lead-acid battery produces:
A. 6 volts B. 12 volts C. 9 volts D. none of the above

236. If the electrolyte in an alkaline battery were low, you would add:
A. alkali solution
B. acid
C. distilled water
D. nothing; it will come back itself on charging

237. When the electrolyte contracts in a battery, the specific gravity:
A. rises
B. lowers
C. stays the same
D. doubles itself

238. Which of the following gases is given off by lead-acid storage batteries?
A. oxygen B. hydrogen C. nitrogen D. carbon monoxide

239. What determines the voltage of a lead-acid cell?
A. the type of electrodes
B. the strength of the electrolyte
C. the size of the plates
D. none of the above

240. The ordinary 6-volt lead-acid storage battery consists of:
A. 6 cells B. 3 cells C. 4 cells D. 12 cells

241. The electrolyte in a lead-acid storage battery is:
A. HC B. H_2SO_4 C. CaC D. MgC

242. The lead plates in storage batteries are separated by:
A. rubber B. wood C. glass D. any of the above

243. A 24-volt lead-acid storage battery consists of:
A. 8 cells B. 6 cells C. 12 cells D. none of the above

244. The normal cell voltage of a fully charged alkaline battery is:
A. 1.0 volt B. 2.0 volts C. 1.5 volts D. 2.5 volts

245. In cold weather the specific gravity of a battery:
A. rises
B. lowers
C. remains the same
D. none of the above

246. Which of the following items is necessary to keep a storage battery in good operation?
A. maintain proper specific gravity
B. keep cool and well ventilated
C. maintain proper level of electrolyte
D. all of the above

247. If the charging rate to a battery was too high, it would:
A. increase the terminal voltage
B. increase the specific gravity
C. increase the rate of hydrogen liberation
D. decrease the terminal voltage

248. H_2SO_4 is:
A. sulphuric acid
B. hydrochloric acid
C. hydraulic acid
D. muriatic acid

249. Electricity in a battery is produced by:
A. chemical action
B. chemical reaction
C. a chemical acting on metallic plates
D. all of the above

250. Salt water in contact with storage batteries will develop:
A. carbon dioxide
B. nitrogen gas
C. carbon monoxide
D. chlorine gas

251. To protect battery terminals from corroding, they should be covered with:
A. grease B. tape C. 3-in-1 oil D. white lead

252. Storage batteries are not allowed to be stored in the:
A. engine room
B. radio shack
C. pump room
D. any of the above

253. A hydrometer is used to measure the:
A. internal temperature of a battery
B. acid content of a battery
C. specific gravity of the battery electrolyte
D. water content of the battery

254. To test an alkali battery for cell charge:
A. use a hydrometer
B. measure the specific gravity
C. measure voltage drop between terminals
D. use a hygrometer

255. In a lead-acid battery if the SG decreases on discharge, then:
A. water is made on discharge
B. water is made on charge
C. acid is made on discharge
D. no change takes place

256. Ventilation of battery rooms is required to:
A. prevent accumulation of explosive gases
B. prevent accumulation of toxic gases
C. supply necessary oxygen for battery action
D. none of the above

257. The resistance of two 60-amp 6-volt batteries connected in series is:
A. .15 ohms B. .20 ohms C. .10 ohms D. .25 ohms

258. If a number of storage batteries are connected in series, the terminal voltage is:
A. the same C. decreased
B. increased D. none of the above

259. How many 1.5 volt batteries are required to supply a load of 12 volts if the batteries are connected in parallel?
A. 12 B. 6 C. 3 D. none of the above

260. How many 1.5-volt batteries are required to supply a load of 12 volts if the batteries are connected in series?
A. 12 B. 6 C. 8 D. 10

261. The resistance of two 60-amp 6-volt batteries connected in parallel is
A. .20 ohms B. .10 ohms C. .05 ohms D. .15 ohms

262. The total voltage and amperage of two 50-amp 6-volt batteries connected in series is:
A. 12 volts, 100 amps C. 6 volts, 100 amps
B. 12 volts, 50 amps D. 6 volts, 50 amps

263. The total voltage and amperage of two 50-amp 6-volt batteries connected in parallel:
A. 6 volts, 100 amps C. 12 volts, 100 amps
B. 6 volts, 50 amps D. 12 volts, 50 amps

264. The rating of a storage battery that delivers 15 amps for 12 hours is
A. 180 ampere hours C. 27 ampere hours
B. 150 ampere hours D. 360 ampere hours

265. Which of the following types of fixtures are allowed in the battery room?
A. explosion-proof C. totally enclosed
B. drip-proof D. all of the above

266. Battery rooms require lead-lined trays for which of the following types of batteries?
A. alkaline B. nickel-iron C. lead-acid cell D. Edison

267. The correct level of the electrolyte to be maintained in a battery is:
A. automatically controlled C. 1/4" below the fill plug
B. just above the separators D. not too important

268. The negative plate in a nickel-iron storage battery is:
A. iron B. nickel C. lead D. steel

269. The electrolyte in a nickel-iron storage battery is:

A. sulphuric acid
B. carbonic acid
C. potassium hydroxide
D. calcium hydroxide

270. The symbol represents a:
A. normally open contactor
B. normally closed switch
C. normally open switch
D. normally closed contactor

271. The symbol represents a:
A. circuit breaker
B. fuse
C. ammeter
D. rheostat

272. The symbol represents a:
A. normally open switch
B. normally open contactor
C. normally closed switch
D. normally closed contactor

273. This diagram represents a:
A. compound-wound motor
B. series-wound motor
C. shunt-wound motor
D. universal motor

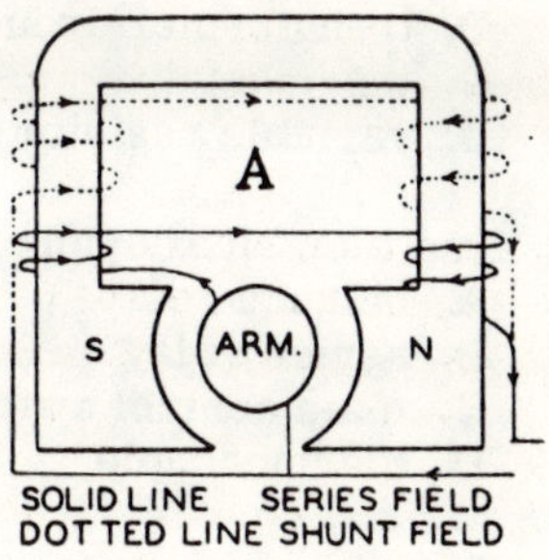

274. This diagram represents a:
A. shunt-wound motor
B. flat-compounded motor
C. series-wound motor
D. compound-wound motor

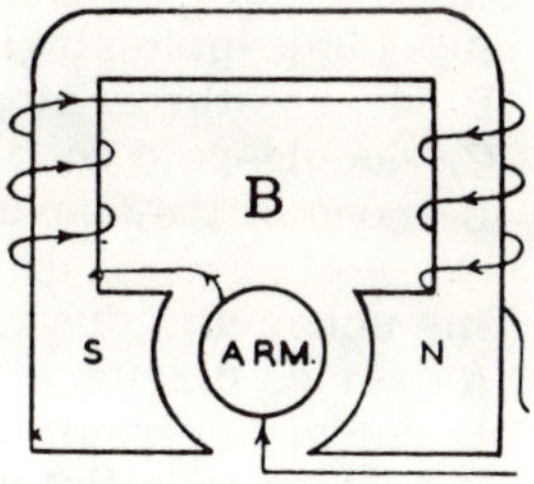

275. This diagram represents a:
A. flat-compounded motor
B. shunt-wound motor
C. series-wound motor
D. compound-wound motor

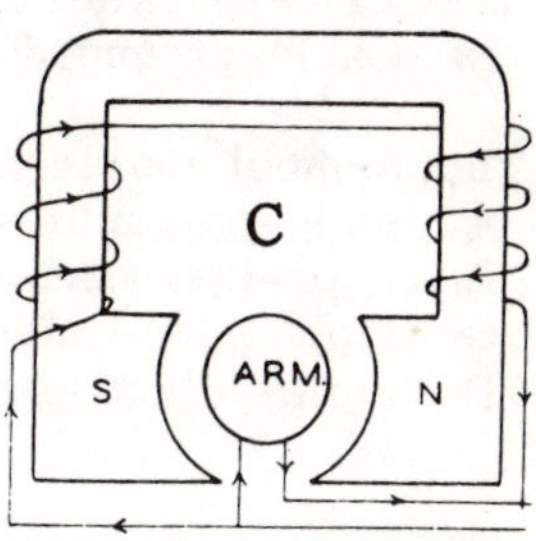

276. This symbol represents a:
   A. 4-way switch
   B. triple-pole switch
   C. 3-way rotary switch
   D. double-pole switch

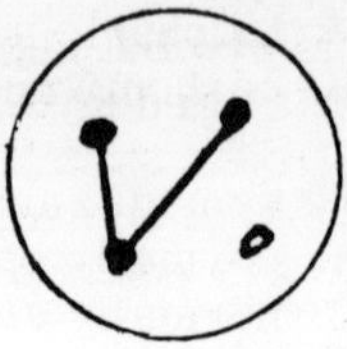

277. This electrical symbol represents a:
   A. rheostat
   B. thermal element
   C. variable resistor
   D. circuit breaker

278. This diagram represents a:
   A. double splice
   B. "T" tap
   C. pigtail splice
   D. Western Union splice

279. This electrical symbol represents a:
   A. circuit breaker
   B. transformer-iron core
   C. thermostat
   D. variable resistor

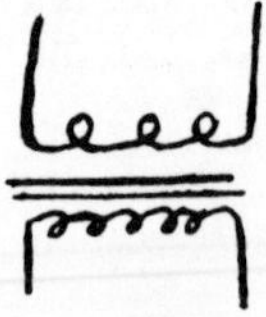

280. This electrical symbol represents a:
   A. balance relay
   B. series relay
   C. main control switch
   D. thermocouple

281. This symbol represents a:
   A. double-pole single-throw switch
   B. double-throw single-pole switch
   C. double-pole double-throw switch
   D. none of the above

282. This schematic diagram represents a:
   A. series circuit
   B. parallel circuit
   C. series-parallel circuit
   D. double-grounded circuit

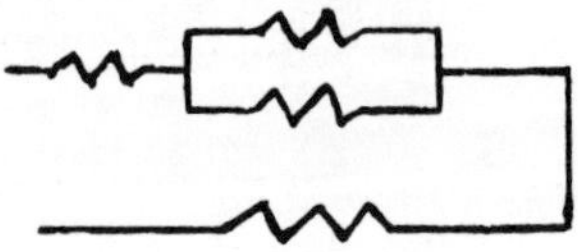

283. This symbol represents a:
   A. triple-pole multi-throw switch
   B. triple-throw triple-pole switch
   C. triple-pole double-throw switch
   D. triple-pole single-throw switch

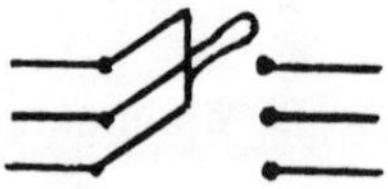

284. This symbol represents a normally:
  A. closed switch
  B. open switch
  C. open contactor
  D. closed contactor

285. This symbol represents a:
  A. normally open switch
  B. normally open contactor
  C. normally closed switch
  D. automatic relay

286. This symbol represents a:
  A. series-wound motor
  B. shunt-wound motor
  C. compound-wound motor
  D. flat-compound motor

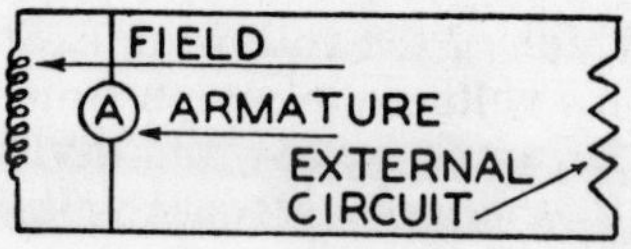

287. This symbol represents a:
  A. compound-wound motor
  B. series-wound motor
  C. shunt-wound motor
  D. over-compounded motor

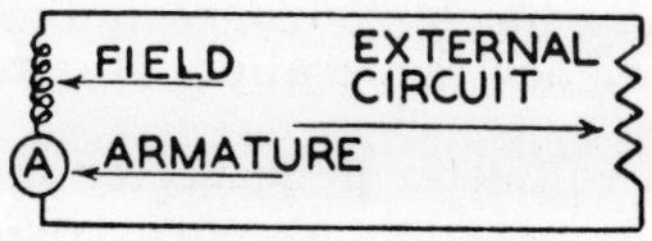

288. This symbol represents a:
  A. shunt-wound motor
  B. series-wound motor
  C. universal motor
  D. compound-wound motor

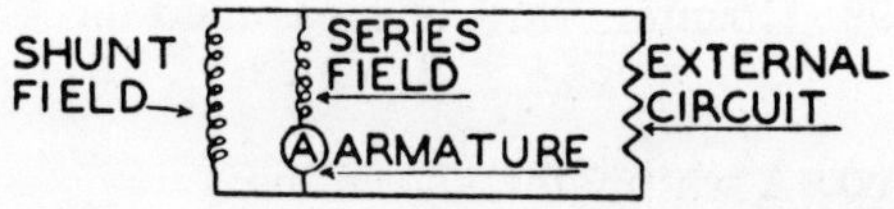

289. Before using a megger it should be tested. This is done by placing the test leads together and turning the crank. A perfect reading, indicating that the leads and megger are in good condition, reads:
  A. 100 B. 500 C. zero D. 220

290. A megger is always used on a:
  A. grounded circuit C. live circuit
  B. short circuit D. deenergized circuit

291. A megger measures:
  A. insulation resistance C. voltage
  B. grounded voltage D. synchronism

292. The power supply for a megger is obtained from:
  A. any 115-volt DC circuit C. flashlight batteries
  B. any 115-volt AC circuit D. a magneto

293. A megger is used to measure:
A. kilohms B. megohms C. milliohms D. microhms

294. A megger is:
A. an instrument for measuring current
B. a hand-cranked AC generator
C. a hand-cranked DC generator
D. a meter for measuring the thickness of insulation

295. When megger-testing a new armature coil to ground, the reading under normal conditions should be at least:
A. 1 meg B. 10 megs C. 100 megs D. 1000 megs

296. Which of the following expressions correctly states Ohm's Law?
A. volts equal amps times resistance
B. amps equal volts divided by resistance
C. resistance equal volts divided by amps
D. all the above are correct

297. An ammeter is connected in a circuit in:
A. series-parallel B. parallel C. series D. contact

298. A voltmeter is connected in a circuit in:
A. parallel B. series-parallel C. series D. relay

299. Counter EMF is measured in:
A. amps B. ohms C. volts D. coulombs

300. A wattmeter shows the:
A. voltage C. combined volts and amps or load delivery
B. amperage D. resistance

301. An ohmmeter measures:
A. voltage C. resistance
B. insulation resistance D. amperage

302. Current measuring instruments must always be connected in:
A. series with the circuit C. series-parallel connection
B. parallel with the circuit D. shunt

303. If a voltmeter is connected across a line, it must be protected with a
A. shunt
B. high-resistance coil in the armature circuit
C. low-resistance coil in the armature circuit
D. shunt and a coil of high resistance

304. Which of the following connections would be most likely to injure the instrument attached?

A. an ammeter in series in the circuit
B. a voltmeter connected across the line
C. an ammeter connected across the line
D. a voltmeter in series with the line

305. Which of the following instruments can be used to measure wire?
A. anemometer B. megger C. micrometer D. dividers

306. The voltage of a DC generator depends on which of the following?
A. speed of armature C. number of armature conductors
B. strength of field D. all of the above

307. A DC compound-wound generator that has a voltage drop from no-load to full-load is said to be:
A. under-compounded C. flat-compounded
B. over-compounded D. none of the above

308. A DC compound-wound generator that does not change speed with variations in load is said to be:
A. over-compounded C. under-compounded
B. flat-compounded D. over-flat-compounded

309. Which of the following is not found on a DC generator?
A. pigtails C. stationary armature
B. brushes D. brush holders

310. DC generators are rated in:
A. KVA B. KWA C. KW D. HP

311. A DC generator that has a voltage rise from no-load to full-load is said to be:
A. under-compounded C. flat-compounded
B. over-compounded D. under-flat-compounded

312. The procedure for starting a DC generator and cutting it in on the line is:
A. start the prime mover, adjust the voltage, close the main switch, close the circuit breaker
B. close the circuit breaker, start the prime mover, adjust the voltage, close the main switch
C. start the prime mover, adjust the voltage, close the circuit breaker, close the main switch
D. adjust the voltage, start the prime mover, close the circuit breaker, close the main switch

313. The purpose of the commutator and brushes on a DC generator is to:
A. change AC to DC current C. neutralize armature reaction
B. change DC to AC current D. carry current to outside circuit

314. Which of the following devices is not found on a DC switchboard?
A. frequency meter C. current transformer
B. synchroscope D. all of the above

315. Which of the following devices is not found on a DC board?
A. ammeter B. voltmeter C. rheostat D. synchroscope

316. In DC circuits, the power is expressed as the product of:
A. coulombs and amperes C. amperes and ohms
B. amperes and volts D. coulombs and volts

317. A series-wound motor is used to run a pump driven with a belt. If the belt breaks, the motor will:
A. overspeed and run out of control
B. stop
C. slow down
D. keep running at the same speed

318. If the load is removed from a series motor, the speed will:
A. decrease C. increase
B. remain the same D. none of the above

319. In a compound-wound motor, part of the line current flows through
A. stator B. shunt field coils C. interpoles D. frame

320. A short in a DC motor would be detected by:
A. brush sparking C. blackened bars
B. ground lights D. A or C

321. The type of DC motor used for high-starting torque and low-starting current is the:
A. squirrel-cage motor C. wound-rotor motor
B. compound-wound motor D. synchronous motor

322. If the load was removed from a series-wound motor, it would:
A. slow down C. speed up
B. stop D. remain at the same speed

323. If the load was removed from a shunt-wound motor, it would:
A. slow down C. speed up
B. stop D. remain at the same speed

324. To reverse the rotation of a DC motor, reverse:
A. the field leads C. the field or armature leads
B. the armature leads D. any of the above

325. As the load of a series motor is increased, the speed will:
A. increase C. remain the same
B. decrease D. none of the above

326. If the load was suddenly released from a series motor, it would:
A. speed up C. stop
B. slow down D. continue to operate at the same speed

327. If the load was suddenly released from a shunt motor, it would:
A. speed up C. continue to operate at same speed
B. slow down D. stop

328. Which of the following groups of motors are DC motors?
A. compound and synchronous
B. series and induction
C. series, shunt and compound
D. induction and synchronous

329. A shunt motor would be best suited for:
A. constant speed results C. a cargo winch
B. an anchor windlass D. any of the above

330. In a three-phase circuit, the phases are:
A. 360° apart B. 180° apart C. 160° apart D. 120° apart

331. An AC series-wound motor will operate on:
A. AC only C. AC or DC
B. DC only D. any of the above

332. On a three-phase three-lamp ground detector, what would be the indicatinn if one phase was grounded?
A. all three lamps would be bright
B. one lamp dark and the others bright
C. one lamp bright and the others dark
D. none of the above

333. Synchronization means:
A. in synchrony C. equal speeds
B. at the same time as D. cycle-for-cycle

334. A synchronous motor is one that:
A. is synchronized
B. turns in synchronization with the alternator
C. converts DC to AC
D. none of the above

335. Frequency depends on the:
A. amount of excitation current applied
B. number of poles
C. speed of the motor
D. B and C

336. If two AC generators are to be operated in parallel, the load is distributed evenly by:

A. means of rheostat C. adjusting the governor settings
B. a balance coil D. changing excitation

337. Frequency depends on the:
A. speed of the machine C. number of poles
B. size of the machine D. A and C

338. If a step-down transformer has 120 volts in the primary and the turns ratio is 6 to 1, the secondary will put out:
A. 60 volts B. 16 volts C. 20 volts D. 720 volts

339. The frequency of an AC current is:
A. cycles per minute C. cycles per second
B. cycles per minute ÷ 60 D. B or C

340. The output frequency of an AC generator which operates at 60 RPM and has 12 field poles is:
A. 12 cps B. 6 cps C. 18 cps D. 24 cps

341. 60-cycle means that the cycle is repeated:
A. 60 times a minute C. 60 times an hour
B. 60 times a second D. none of the above

342. A synchronous motor is one that turns:
A. at twice the speed of the generator
B. at half the speed of the generator
C. in a given ratio with the number of poles involved
D. at the same speed as the generator

343. In AC generators:
A. the armature coils are stationary and the field coils revolve
B. the armature and the field coils both revolve
C. the armature coils revolve and the field coils are stationary
D. A or C

344. Ship's service and emergency alternators used aboard ship are:
A. three-phase C. four-phase
B. two-phase D. single-phase

345. To produce a frequency of 60 cps, an 8-pole alternator must operate at:
A. 960 rpm B. 480 rpm C. 1920 rpm D. 900 rpm

346. The type of AC motor used for high-starting torque and low-starting current is the:
A. squirrel-cage motor C. wound-rotor motor
B. compound-wound motor D. synchronous motor

347. To reverse the rotation of an AC motor, reverse:

A. any two of the three leads
B. the field leads
C. the armature leads
D. the field and armature leads

348. Which of the following AC motors has its rotor energized by DC?
A. synchronous C. repulsion
B. squirrel-cage D. wound-rotor

349. The synchronous speed of a 4-pole 60-cycle AC motor is:
A. 1800 rpm B. 2400 rpm C. 1200 rpm D. 3600 rpm

350. To reverse the rotation of a 3-phase AC propulsion motor, change:
A. the first-phase and third-phase leads
B. any 2 of the 3 leads
C. the positive and ground leads
D. the second-phase and third-phase leads

351. Two AC generators are to be connected in parallel. The synchronizing switch should be closed when:
A. one light is bright and one dim
B. both lights are dim
C. both lights are bright
D. one is bright and the other out

352. The frequency of a 2-pole turboalternator turning at 3000 RPM is:
A. 50 cycles B. 5. 5 cycles C. 6. 0 cycles D. 60 cycles

353. Which of the following is not found on AC generators?
A. slip rings B. commutator C. brushes D. field coils

354. Excitation for an AC generator can be taken from a:
A. battery C. converter
B. separate exciter D. any of the above

355. Another name for an AC generator is:
A. a dynamometer C. an alternator
B. a dynamotor D. any of the above

356. The approximate brush pressure on generator collector rings is:
A. 2 psi B. 4 psi C. 6 psi D. 8 psi

357. The symbol below is an indication for a/ an: (a) D. C. voltmeter; (b) Rheostat; (c) Ohmmeter; (d) Plug fuse; (e) Revolution counter.

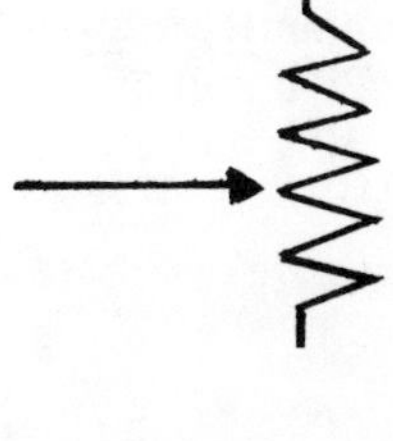

358. The symbol below indicates a/an: (a) Two-pole, single-throw knife switch; (b) D. C. oil type circuit breaker; (c) Double-pole automatic relay; (d) Normally open main contact for D. C.; (e) A. C. single-phase motor connector.

359. A synchronous converter will: (a) Reverse the power factor by 180°; (b) Convert A. C. current into D. C. current; (c) Change a synchronous motor to a wound-rotor type; (d) Not be found on board ships.

360. Bad commutation can be caused by:
A. improper brush setting
B. wrong type of brushes
C. overload
D. any of the above

361. A service factor of 1.25 on the nameplate of a generator means that in an emergency the generator can operate for a short period of time at:
A. 125% overload
B. 25% overload
C. 25% above normal voltage
D. none of the above

362. Failure of a generator to build up could be caused by:
A. speed of prime mover too slow
B. field connections reversed
C. brushes set improperly
D. any of the above

363. The circular mil area of 1/8" wire is:
A. 12,500 B. 15,625 C. 25,000 D. 125

364. A commutator will become badly worn or grooved if:
A. the brushes are too hard
B. the brushes are staggered improperly
C. A and B
D. none of the above

365. The slip in an induction motor usually varies between:
A. 3% and 8% C. 20% and 30%
B. 1% and 4% D. 10% and 20%

366. Doubling the diameter of a wire:
   A. increases the resistance 4 times
   B. decreases the resistance 4 times
   C. decreases the resistance 2 times
   D. increases the resistance 2 times

367. Doubling the length of a wire:
   A. triples the resistance
   B. quadruples the resistance
   C. doubles the resistance
   D. has no effect on resistance

368. In a rotary converter the input to the AC end is 110 volts. The output at the DC end will be approximately:
   A. 110 volts
   B. 75 volts
   C. 220 volts
   D. 100 volts

369. A synchronous converter will:
   A. reverse the power factor
   B. change a synchronous motor to a wound-rotor type
   C. not be found on board ship
   D. convert AC current into DC current

370. Rectifiers are used for:
   A. changing DC current to AC current
   B. changing AC current to DC current
   C. a power supply for switchboard indicator lamps
   D. reducing the resistance in a circuit

371. Adding resistances in series:
   A. has no effect on the flow in the combined circuit
   B. increases the flow of current in the combined circuit
   C. decreases the flow of current in the combined circuit
   D. none of the above

372. Field excitation where required is always supplied from:
   A. a source of direct current
   B. an alternator
   C. a magnet
   D. selineum rectifier

373. Which of the following is not ordinarily used for determining power factor?
   A. voltmeter   B. tachometer   C. ammeter   D. wattmeter

ANSWERS TO ELECTRICITY QUESTIONS

1. B
2. D
3. A
4. B
5. B
6. A
7. B
8. C
9. B
10. B
11. B
12. C
13. B
14. D
15. A
16. C
17. B
18. A
19. B
20. D
21. D
22. C
23. D
24. B
25. D
26. C
27. B
28. C
29. C
30. B
31. B
32. D
33. B
34. C
35. C
36. A
37. D
38. D
39. B
40. D
41. C
42. C
43. B
44. D
45. A
46. D
47. B
48. C
49. A
50. A
51. B
52. C
53. B
54. A
55. A
56. C
57. A
58. D
59. A
60. D
61. A
62. D
63. B
64. C
65. D
66. B
67. C
68. D
69. A
70. B
71. A
72. C
73. D
74. C
75. A
76. B
77. D
78. D
79. A
80. A
81. A
82. D
83. C
84. C
85. D
86. D
87. C
88. D
89. B
90. B
91. D
92. B
93. D
94. D
95. A
96. A
97. D
98. D
99. A
100. D
101. B
102. C
103. B
104. B
105. C
106. D
107. D
108. C
109. C
110. C
111. D
112. C
113. C
114. C
115. D
116. B
117. A
118. C
119. C
120. B
121. D
122. C
123. D
124. A
125. B
126. C
127. D
128. A
129. C
130. A
131. C
132. D
133. B
134. A
135. A
136. B
137. A
138. C
139. C
140. B
141. D
142. B
143. C
144. D
145. C
146. B
147. C
148. D
149. C
150. B
151. D
152. B
153. C
154. A
155. D
156. D
157. C
158. C
159. B
160. C
161. B
162. C
163. A
164. D
165. B
166. D
167. C
168. B
169. A
170. B
171. A
172. A
173. A
174. D
175. B
176. C
177. C
178. A
179. B
180. D
181. B
182. C
183. C
184. A
185. B
186. C
187. A
188. D
189. B
190. D
191. C
192. A
193. D
194. C
195. B
196. D
197. D
198. C
199. B
200. B

201. D
202. B
203. B
204. C
205. D
206. A
207. C
208. C
209. C
210. A
211. A
212. C
213. C
214. B
215. C
216. D
217. C
218. A
219. B
220. B
221. A
222. B
223. C
224. B
225. B
226. B
227. D
228. C
229. A
230. D
231. B
232. A
233. C
234. B
235. B
236. C
237. A
238. B
239. B
240. B
241. B
242. D
243. C
244. C
245. A
246. D
247. C
248. A
249. D
250. D
251. A
252. D
253. C
254. C
255. A
256. A
257. B
258. B
259. D
260. C
261. C
262. B
263. A
264. A
265. A
266. C
267. B
268. A
269. C
270. D
271. B
272. C
273. A
274. C
275. B
276. C
277. C
278. D
279. B
280. B
281. C
282. C
283. D
284. C
285. A
286. B
287. B
288. D
289. C
290. D
291. A
292. D
293. B
294. C
295. A
296. D
297. C
298. A
299. C
300. C
301. C
302. A
303. B
304. C
305. C
306. D
307. A
308. B
309. C
310. C
311. B
312. C
313. A
314. D
315. D
316. B
317. A
318. C
319. B
320. D
321. B
322. C
323. D
324. D
325. B
326. A
327. C
328. C
329. A
330. D
331. C
332. B
333. D
334. B
335. D
336. C
337. D
338. C
339. D
340. B
341. B
342. C
343. D
344. A
345. D
346. C
347. A
348. A
349. A
350. B
351. B
352. A
353. B
354. D
355. C
356. A
357. B
358. A
359. B
360. D
361. B
362. D
363. B
364. C
365. A
366. B
367. C
368. B
369. D
370. B
371. C
372. A
373. B

GENERAL – EMERGENCY – FIRE FIGHTING

1. It is all right to use a solid stream of water on a:
   A. Class C fire C. Class A fire
   B. Class B fire D. all of the above

2. Which of the following fire extinguishers would be effective on an oil fire but dangerous to use on an electrical fire?
   A. foam B. CO_2 C. soda acid D. dry chemical

3. Fire can be prevented from spreading by:
   A. sealing off the affected area
   B. removing combustibles from the immediate area
   C. cooling the surrounding area
   D. all or any of the above

4. A smoke-detecting system indicates the presence of fire by which of the following means?
   A. smell C. alarm signal
   B. visual D. all of the above

5. If a fire in the engine room of an oil-burning vessel got out of control, one should use:
   A. the 40-gallon foam extinguisher
   B. all the portable extinguishers in the engine room
   C. the fixed CO_2 system
   D. all the fire extinguishers available

6. Portable fire extinguishers must be recharged every:
   A. 12 months B. 6 months C. 18 months D. 24 months

7. A foam fire extinguisher can be used on a:
   A. Class B fire C. Class C fire
   B. Class A fire D. A and B

8. A dry chemical fire extinguisher can be used on a:
   A. Class B fire B. Class C fire C. Class A fire D. A and B

9. According to G. R. &R. life jackets must be dyed:
   A. any bright color C. sun orange
   B. India orange D. bright orange

10. On hearing the fire-alarm signal, a crewman should:
    A. man the nearest fire hose
    B. report to his fire station
    C. report to the bridge
    D. report to his fire station wearing life jacket

11. How many spare charges must be carried for each type of portable fire extinguishers?

A. at least 100% B. at least 50% C. at least 20% D. at least 25%

12. Each person's duties at fire drills and emergency drills are:
A. in the engine-room log C. in the bridge log
B. on the station bill D. on the certificate of inspection

13. Before a fire can start and continue to burn, there must be present:
A. fuel, heat and oxygen C. fuel, heat and hydrogen
B. paper, oily rags and heat D. paper, wood and oxygen

14. Which of the following designates the proper class, type of fire, and type of fire extinguisher?
A. Class C, electrical, foam
B. Class B, trash, dry chemical
C. Class B, paints, carbon dioxide
D. Class A, wood, carbon dioxide

15. A Class C fire is one which consists of burning:
A. rags B. wood C. oil D. none of the above

16. The best portable extinguisher to use on a Class B fire would be the:
A. dry chemical C. soda acid
B. CO_2 D. A or B

17. When fighting an oil or gasoline fire, never use a:
A. high-velocity fog C. foam
B. low-velocity fog D. solid stream of water

18. Which of the following types of extinguishers is equipped with a safety valve?
A. foam type B. CO_2 C. dry chemical D. all of the above

19. A fire in wood, paper, rags, etc. is considered a:
A. Class D fire B. Class C fire C. Class A fire D. Class B fire

20. A soda-acid fire extinguisher can be used on a:
A. Class A or C fire C. Class B fire
B. Class A fire D. Class B or C fire

21. A CO_2 extinguisher can be used on a:
A. Class A fire C. Class C fire
B. Class B fire D. any of the above

22. The fire signal is:
A. continuous ringing of the general alarm for not more than 10 seconds
B. six short and two long blasts on the ship's whistle
C. continuous ringing of the general alarm for not less than 10 seconds and same on the ship's bell or whistle
D. none of the above

23. On reporting to a vessel for the first time, your fire station can be found by looking at the:
    A. ship's articles C. certificate of inspection
    B. station bill D. crew list in engine room

24. A cofferdam is:
    A. an empty space between tank tops and bilges
    B. a tank for storing chemicals
    C. a cement dam in a fresh-water tank to prevent "sloshing"
    D. an empty space separating compartments to prevent the contents of one compartment from entering another in case of leakage

25. An explosion meter is used to detect:
    A. oxygen content of air
    B. concentrations of flammable gases and vapors
    C. hydrogen content of air
    D. combustibility of hydrogen and oxygen

26. Lifeboat electric gear must be checked every:
    A. 6 months B. 3 months C. 12 months D. month

27. Emergency generators must be located:
    A. above bulkhead or freeboard deck
    B. in any dry place
    C. close to the engine room
    D. near the bridge

28. Diesel-emergency generators must be operated:
    A. once a week for an hour C. once a week for two hours
    B. once a month for two hours D. once a month for one hour

29. If a fireman pulled a "hot" burner, the first thing he must do is:
    A. notify the engineer on watch C. shut off the oil supply
    B. get out of the fireroom D. ring the fire alarm

30. What is used on passenger vessels to prevent fire from spreading through the ventilating system?
    A. fireproof materials
    B. water sprinklers in the vent system
    C. CO_2 extinguishers in the vent system
    D. automatic closing fire dampers

31. The purpose of wire screen mesh on the gooseneck vents from fuel-oil tanks is to prevent:
    A. rags and other foreign matter from getting into the tank
    B. water from entering the tank in rough seas
    C. flame entry into the tank in case of fire on deck
    D. A and C

32. Class A fire-resistant bulkheads must resist fire and smoke penetration for at least:
    A. 15 minutes B. 30 minutes C. 60 minutes D. 45 minutes

33. The abandon-ship signal is:
    A. six or more blasts on the ship's whistle
    B. seven or more blasts followed by one long blast
    C. seven or more blasts, followed by one long blast on ship's whistle
    D. seven long rings and one long ring on general-alarm system

34. Watertight bulkheads are used to:
    A. make water tanks airtight
    B. store water
    C. divide the ship into watertight sections for safety purposes
    D. protect passenger quarters

35. The collision bulkhead is:
    A. at the bow of the ship
    B. the first watertight bulkhead in the ship
    C. on the bridge deck
    D. between the passenger and cargo areas

36. When entering an empty oil tank, even after authorized inspection, one should always:
    A. have someone stand by entrance or manhole
    B. secure a lifeline to oneself
    C. take along a flame safety lamp
    D. all of the above

37. Battery rooms should be well ventilated because batteries liberate:
    A. oxygen gas C. hydrogen fumes
    B. acid fumes D. nitrogen gas

38. If you were on watch and you heard a continuous ringing on the general alarm for a period of more than ten seconds, you should:
    A. shut down the plant C. go to your fire station
    B. start up the fire pump D. get out of the engine room

39. The minimum temperature for boiler feedwater entering a boiler is:
    A. 212° F. B. 150° F. C. 100° F. D. any temperature

40. All steam valves over 6" in diameter must be fitted with:
    A. a rising-stem type valve C. flange unions
    B. a by-pass line and valve D. steam lagging

41. Normal air contains:
    A. 16% oxygen C. 40% oxygen
    B. 21% oxygen D. 25% oxygen

42. Unconsciousness will occur in human beings when the percentage of oxygen gets down to:
A. 16% B. 14% C. 10% D. 12%

43. The valve on the bilge suction of the main circulator must be a:
A. globe valve C. non-return valve
B. gate valve D. angle valve

44. What type of power is required for emergency lighting?
A. diesel engine C. semi-diesel engine
B. batteries D. any of the above

45. How often must mechanical emergency lighting equipment be checked?
A. operated and checked once a week
B. operated under load for 2 hours, once a month
C. once a week for one hour
D. A and B

46. A fire in electrical equipment is considered a:
A. Class C fire C. Class A fire
B. Class B fire D. Class D fire

47. A fire in oil, grease, etc., is considered a:
A. Class C fire B. Class A fire C. Class B fire D. Class D fire

48. Class B fire-resistant bulkheads must resist fire and smoke penetration for at least:
A. 15 minutes B. 30 minutes C. 60 minutes D. 45 minutes

49. Cargo ships over 1000 gross tons are required to have a:
A. fresh-air breathing apparatus
B. self-contained gas mask
C. flame safety lamp and self-contained mask
D. all-purpose breathing apparatus

50. All electric cables run through decks must be protected with:
A. welded joints C. kick pipes
B. packing boxes D. a ground connection

51. Emergency lights must be marked with a:
A. letter "E", 1/2" high C. red arrow pointing to exit
B. letter "E" in red D. letter "E" in red, 1/2" high

52. Spontaneous combustion can occur in:
A. oily rags B. coal C. grain D. any of the above

53. The first thing to do with an electrical fire is to:
A. sound the fire alarm C. start fighting the fire
B. call the chief engineer D. stop the flow of electricity

54. The first thing to do for a person suffering from asphyxiation is to:
    A. apply artificial respiration
    B. check the victim's temperature
    C. check the victim's pulse
    D. loosen victim's clothing

55. As an engineer on watch, you hear a continuous ringing of the ship's general alarm system for 10 seconds. You would:
    A. go to your fire station
    B. secure the plant
    C. put pressure on the fire main
    D. send wiper to locate the trouble

56. The primary consideration when fighting a Class C fire is:
    A. a blanketing effect
    B. a non-conducting effect
    C. a quenching effect
    D. a cooling effect

57. The vessel's fire equipment is listed on the:
    A. station bill
    B. official ship's papers
    C. hull certificate
    D. certificate of inspection

58. Which of the following vessels is not required to have cargo compartments fitted with a fixed fire-fighting system?
    A. refrigerated vessels
    B. vessels carrying coal or grain in bulk
    C. container carrying vessels
    D. A and B

59. Batteries used for emergency lighting must be checked:
    A. once a month
    B. every other month
    C. every 60 days
    D. once every 6 months

60. According to the G. R. &R., engine-room spaces require at least:
    A. two means of escape
    B. one means of escape
    C. three means of escape
    D. four means of escape

61. All fire-fighting equipment is required to be inspected:
    A. at each annual inspection
    B. every 6 months
    C. every 2 years
    D. as often as necessary

62. Fire-detecting systems are required on:
    A. any vessel on an international voyage
    B. any vessel over 150 feet in length having sleeping accommodations for passengers
    C. any vessel over 150 feet in length not on an international voyage but having sleeping accommodations for 50 or more passengers
    D. all of the above

63. The power for power-operated watertight doors is obtained from the:
    A. main generator
    B. emergency light and power system
    C. auxiliary generator
    D. radio room

64. Instantaneous water supply to a sprinkling system is assured by:
    A. a pressure tank
    B. high pressure in the pipe
    C. an automatically controlled pump
    D. any of the above

65. If the lettering or tagging on a fire extinguisher is illegible:
    A. hold until the next inspection
    B. replace with an approved tag type
    C. send back to the manufacturer
    D. restamp or retag

66. If a person removed from a vapor-filled tank was not breathing, you should:
    A. notify the captain
    B. notify the chief engineer
    C. loosen his clothing and keep him warm
    D. apply artificial respiration

67. Fire-screen doors are equipped with:
    A. a closing magnet
    B. a control from the radio room
    C. a control from the bridge
    D. an electric eye so they will not close accidentally

68. A fire is under control when:
    A. it stops spreading
    B. no smoke can be seen
    C. no flame can be seen
    D. you have all the available fire-fighting equipment in use

69. The minimum flash point for fuels used in internal combustion emergency generators is:
    A. 110° F. B. 100° F. C. 140° F. D. 90° F.

70. On which of the following vessels is a fire-detecting system required?
    A. all cargo vessels
    B. all vessels over 400 feet long
    C. all passenger vessels on international voyages
    D. all vessels over 150 feet long

71. The master switch for the lifeboats is located:
    A. on the opposite bulkhead from the lifeboat

B. on the ship's rail
C. in the emergency generator room
D. in the nearest inside passageway

72. To produce water in a steady stream from an all-purpose nozzle:
A. push handle forward
B. pull handle back
C. put handle in mid-position
D. A or B

73. Steam, foam and CO_2 smothering systems shall be marked in red letters at least:
A. 2" high
B. 1" high
C. 1-1/2" high
D. 1/2" high

74. With the all-purpose nozzle, the stream of water is changed from high velocity to fog by:
A. pushing the handle forward
B. pulling the handle back
C. putting the handle in a vertical position
D. none of the above

75. The special fire-fighting nozzle that can deliver either a solid stream of water or a fog effect is called a:
A. combination nozzle
B. fire nozzle
C. fog nozzle
D. pitot nozzle

76. To change from a high-velocity fog to a low-velocity fog, one would attach:
A. a different nozzle
B. an applicator
C. a different combination nozzle
D. any of the above

77. The all-purpose fog nozzle is secured to the hose by a:
A. quick closing clamp
B. patented snap clamp
C. press-on clip
D. screw thread

78. The function of the fog produced by a fog nozzle is to:
A. blanket the fire
B. exclude air from the fire
C. cause steam to absorb the heat of the fire
D. none of the above

79. The advantage of low-velocity fog is that it:
A. cools the area quickly
B. turns to steam and smothers the fire
C. protects personnel
D. all of the above

80. Fire hoses can be used only for:
A. fire drills
B. fire extinguishing
C. washing down decks
D. A and B

81. A spanner wrench is a special wrench for:
    A. spanning flanges
    B. use on the OBA
    C. use on hose couplings
    D. use on the flame safety lamp

82. There is always less pressure at the fire-hose outlet than is found at the discharge of the pump. This loss in pressure is caused by:
    A. leaky-pump discharge valves
    B. leaky-pump suction valves
    C. friction in piping and valves
    D. leaky pitot valves

83. Fire hoses can be disconnected when:
    A. taking seas in heavy weather
    B. fueling ship
    C. there is danger of damage when loading cargo
    D. A and C

84. Fire-hydrant valves cannot be installed pointing:
    A. vertically up
    B. vertically down
    C. horizontal
    D. facing inboard

85. Fire hoses can be disconnected:
    A. when using them for washing down
    B. at no time
    C. when taking seas in heavy weather
    D. when fueling ship

86. The pitot tube is for:
    A. measuring water pressure in fire lines
    B. measuring discharge pressure of CO_2 extinguishers
    C. measuring pressure in CO_2 bottles
    D. none of the above

87. The type of hose thread required on water fire-hose equipment is the:
    A. U. S. standard
    B. National standard
    C. NC standard
    D. American standard

88. On the discharge side of a fire pump you will find a:
    A. pressure gauge and safety valve
    B. air chamber and relief valve
    C. pressure gauge and relief valve
    D. air chamber and safety valve

89. For the safety of fire hoses, fire pumps are fitted with:
    A. safety valves
    B. pressure gauge and relief valve on discharge side
    C. relief valve on discharge side
    D. automatic suction valve shutoff

90. A steam-smothering system shall be tested:
   A. with full boiler pressure
   B. with a 50-pound air test
   C. by a hammer test
   D. B and C

91. Steam-smothering for cargo tanks on tank vessels must be connected to:
   A. shore steam connection
   B. main boilers
   C. donkey boilers
   D. all of the above

92. The minimum size of piping for steam smothering lines to paint lockers and similar small spaces is:
   A. 1" B. 3/4" C. 1/2" D. 1-1/2"

93. No steam-smothering shall be built into ships constructed after:
   A. Jan. 1965 B. Jan. 1967 C. Jan. 1962 D. Jan. 1966

94. The minimum size of piping for steam-smothering lines to cargo holds is:
   A. 2" B. 1-1/2" C. 1" D. 3/4"

95. The minimum pressure allowed on steam-smothering systems is:
   A. 100 psi B. 125 psi C. 75 psi D. 150 psi

96. On the steam-smothering system found on tank vessels, which of the following valves are left open?
   A. all the valves
   B. the master valve
   C. the manifold valves
   D. B and C

97. A dry chemical extinguisher is recommended for use on:
   A. oil fires
   B. small electrical fires
   C. fires in mattresses or clothing
   D. A and B

98. In a dry chemical extinguisher, the propellant is:
   A. CO_2 B. a hand pump C. CCL_4 D. HCL

99. When using a foam-type fire extinguisher, direct the stream of foam:
   A. directly on the fire
   B. back and forth across the flames
   C. directly in the path of the fire
   D. against a bulkhead or other nearly vertical area so that foam runs down and spreads or fans out over the fire

100. In a portable foam extinguisher, the propellant is:
   A. CO_2 B. CCL_4 C. SO_2 D. HC

101. There must be enough extra safety valves for foam-type fire extinguishers for at least:
   A. 25% of the extinguishers
   B. 35% of the extinguishers
   C. 50% of the extinguishers
   D. 10% of the extinguishers

102. Before releasing the fixed CO_2 extinguishing system into the engine room:
    A. secure all fans and ventilation
    B. warn all personnel
    C. secure fires
    D. all of the above

103. The control valve for the fixed CO_2 system in the engine room is located:
    A. on the bridge
    B. near the main throttle
    C. adjacent to the engine-room escape
    D. in the chief engineer's room

104. In the fixed CO_2 system, the CO_2 is discharged through a:
    A. perforated pipe
    B. horn
    C. nozzle
    D. any of the above

105. A CO_2 fire extinguisher must be recharged when found to be:
    A. 15% below required weight
    B. 10% below required weight
    C. 25% below required weight
    D. empty

106. To operate a CO_2 extinguisher:
    A. turn extinguisher upside down and strike on deck to break seal
    B. remove locking pin on valve at top of cylinder and open valve
    C. direct horn at the fire and squeeze down on handle
    D. turn extinguisher upside down and open valve

107. Before releasing fixed CO_2 system to engine room:
    A. warn all personnel on watch
    B. secure fires
    C. shut down fans and blowers
    D. all of the above

108. Between the time of the fixed CO_2 alarm signal and actual delivery of the CO_2 to the system, there is a:
    A. 10-second delay
    B. 20-second delay
    C. 30-second delay
    D. no delay

109. After the engine room has been subjected to the discharge of the fixed CO_2 fire-extinguishing system, no one should be allowed into the engine or fire room until:
    A. the engine room has been ventilated for 24 hours
    B. a flame safety lamp burns in all parts of the engine and fire room
    C. the entire area has been washed down with fresh water
    D. a sufficient amount of oxygen has been released into the space to make it safe

110. If you were in a compartment where a fixed CO_2 extinguisher was being released, you should:
    A. notify the bridge
    B. ring the general alarm
    C. evacuate immediately
    D. put on an OBA

111. A CO_2 extinguisher is dangerous to use in:
    A. extremely hot areas
    B. extremely cold areas
    C. small enclosed areas
    D. passenger quarters

112. The control for the fixed CO_2 system for the engine room is located:
A. near the main throttle C. in the chief engineer's room
B. on the bridge D. outside the engine-room escape

113. To release fire extinguishing CO_2 gas to the engine-room areas, one would open the valve control to the engine spaces and then:
A. discharge all gas to the engine area
B. operate the separate manual control to release the gas
C. release sufficient gas to the area to put out the fire
D. stand by for the automatic control to release the gas

114. The fixed CO_2 fire-extinguishing system to the engine room is released:
A. electrically
B. by a remote-controlled cable outside the CO_2 room
C. by a manually-controlled valve on the CO_2 cylinders
D. by a remote-controlled cable at the throttle platform

115. The number of control stations on a fixed CO_2 system to cargo spaces is:
A. one B. two C. three D. four

116. A fixed CO_2 system is required to have:
A. a manual alarm
B. a connection to the general alarm system
C. an automatic alarm sounded before the gas is released
D. all of the above

117. The maximum temperature allowed in the CO_2 bottle storage room is:
A. 130° F. B. 100° F. C. 100° C. D. 150° F.

118. The maximum discharge time of the CO_2 extinguishing system to the main propulsion unit is:
A. 2 minutes B. 3 minutes C. 1 minute D. 4 minutes

119. CO_2 fire extinguishers are periodically checked for proper charge by:
A. checking with a gauge
B. cracking the relief valve
C. checking the internal temperature
D. weighing

120. The maximum percentage of poisonous gases a canister gas mask can handle is:
A. 2% B. 4% C. 5% D. 10%

121. An all-purpose gas mask canister is painted:
A. white B. green C. red D. blue

122. A gas mask supplies its own oxygen, independent of the atmosphere, for:

A. 60 minutes
B. 45 minutes
C. 30 minutes
D. none of the above

123. A gas mask can only be used when there is:
   A. at least 16% oxygen present
   B. less than 16% oxygen present
   C. less than 10% oxygen present
   D. none of the above

124. A gas mask:
   A. purifies air
   B. generates oxygen
   C. generates pure air
   D. none of the above

125. A canister-type gas mask works on the principle of:
   A. filtering
   B. chemical combination
   C. adsorption
   D. all of the above

126. To test a canister-type gas mask:
   A. squeeze the breathing tube and try to inhale
   B. blow into the mouthpiece
   C. hold hand over hole in bottom of canister and try to inhale
   D. hold hand over hole in bottom of canister and try to exhale

127. The canister on an all-purpose gas mask is installed by:
   A. removing the top seal and inserting
   B. removing the bottom seal and inserting
   C. removing the top seal, inverting and inserting
   D. removing the top and bottom seals, inverting and inserting

128. Which of the following safety devices should be used when using canister type gas masks?
   A. a flame safety lamp
   B. a safety line
   C. a watch timer
   D. all of the above

129. If a flame safety lamp burns normally with a dim flame, it shows the presence of:
   A. less than 10% oxygen
   B. at least 16% oxygen
   C. no oxygen
   D. no hydrogen

130. Which of the following is the safety feature of the flame safety lamp?
   A. the safety cap
   B. the key
   C. the low combustion fuel
   D. the low wick control

131. Which of the following should be used when it is necessary to enter a compartment which has not been declared gas-free?
   A. a canister-type gas mask
   B. an all-purpose gas mask
   C. a fresh-air hose mask and harness and safety line
   D. any of the above along with a flame safety lamp

132. Before entering a compartment with a flame safety lamp:
   A. lock it and leave the key outside the space to be inspected
   B. be sure to empty the strainer
   C. lock it
   D. replace the safety glass

133. A flame safety lamp will go out when the percentage of oxygen goes below:
   A. 16%   B. 10%   C. 20%   D. 25%

134. If there was insufficient air for breathing, the flame safety lamp would:
   A. flare up and go out
   B. go out
   C. keep burning with a blue flame
   D. keep burning with a yellow flame

135. If a flame safety lamp suddenly flared up, it would indicate:
   A. excessive oxygen
   B. lack of oxygen
   C. explosive gases present
   D. wick needs adjusting

136. Flame safety lamps use:
   A. white gasoline
   B. leaded gasoline
   C. light diesel oil
   D. naphtha-type fuels

137. A flame safety lamp:
   A. determines hydrogen deficiency
   B. detects flames in a fire
   C. determines oxygen deficiency
   D. will go out if no flames are present

138. The fresh-air supply for a fresh-air breathing apparatus is supplied by:
   A. an air compressor
   B. a hand-operated pump
   C. an oxygen cylinder
   D. A or B

139. The safety valve in the OBA (oxygen-breathing apparatus) warns of:
   A. excessive pressure in cylinder
   B. time to leave compartment because oxygen supply is getting low
   C. too much bad air being breathed in
   D. time to purge the bottle

140. The fresh-air supply for the OBA is:
   A. supplied by an air compressor
   B. supplied by an oxygen cylinder
   C. generated by a chemical in a canister
   D. supplied by a hand-operated pump

141. If you were in a compartment wearing an OBA, you would evacuate the area if:
   A. you had trouble breathing

B. the safety whistle sounded
C. the pressure gauge sank to the red mark
D. any of the above

142. The metal mouthpiece of an OBA consists of:
A. inhalation, exhalation, compression and relief valves
B. inhalation, exhalation, saliva and relief valves
C. inhalation, exhalation, relief and reducing valves
D. intake, exhalation, saliva and admission valves

143. Which of the following valves is not contained in the OBA mouthpiece?
A. inhalation C. exhalation
B. saliva D. reducing

144. The reducing valve on the OBA reduces the pressure from:
A. 500 psi to 5 psi
B. 130 atmospheres to 3 psi
C. 100 psi to 1 atmosphere
D. 100 atmospheres to 1 atmosphere

145. A full oxygen cylinder for the self-contained breathing apparatus registers a pressure of:
A. 1000 psi C. 50 - 100 atmospheres
B. 125 - 135 atmospheres D. 500 psi

146. The fresh air supply for the conventional type OBA is supplied by:
A. an air compressor C. a hand-operated pump
B. an oxygen cylinder D. any of the above

147. The OBA can be used safely by:
A. two persons
B. as many persons as there are hoses attached
C. one person
D. a person holding a safety certificate

148. When using an OBA, the nitrogen accumulation should be removed every:
A. time you think of it C. 5 minutes
B. 30 minutes D. 15 to 20 minutes

149. The admission valve on an OBA will close when:
A. oxygen is almost gone C. you exhale
B. by-pass is opened D. safety valve is jammed

150. Crew training in use of the oxygen-breathing apparatus (OBA) is the responsibility of the:
A. master B. chief mate C. chief engineer D. A and C

151. An OBA supplies its own oxygen, under normal breathing conditions,

independent of the atmosphere, for approximately:
A. 60 minutes B. 30 minutes C. 45 minutes D. 90 minutes

152. The by-pass on the OBA should be used:
A. when testing the apparatus
B. in normal operation
C. only in an emergency in case some part ceases to function properly
D. never

153. The admission valve on an OBA will close when the:
A. bag is deflated C. oxygen bottle is empty
B. bumper plate is raised D. by-pass is opened

154. The first operation in testing an OBA for tightness is to:
A. open the main valve C. open the by-pass
B. deflate the bag D. close the by-pass

155. An OBA tests airtight if:
A. the bag deflates C. the bag inflates
B. the safety valve releases D. none of the above

156. Which one of the following is not a part of the oxygen cylinder-type breathing apparatus?
A. an oxygen-generating canister
B. a carbon-dioxide-removing chemical
C. a reducing valve
D. a high-pressure gauge

157. When entering a fuel tank which has not been certified gas-free, you would use:
A. a canister type gas mask
B. a fresh air hose mask
C. a flame safety lamp
D. none of the above

158. Flame safety lamps use:
A. naphtha type fuel C. diesel type fuel
B. gasoline type fuel D. alcohol type fuel

159. If a person removed from a vapor-filled tank was not breathing, you would:
A. start artificial respiration immediately
B. loosen his collar and keep him warm
C. notify the chief engineer
D. send for a doctor

160. Cargo vessels over 1,000 G. T. on an international voyage are required to carry on board a minimum of:
   A. one fresh air breathing apparatus
   B. one all-purpose gas mask
   C. one self-contained breathing apparatus
   D. two gas masks

161. Fresh air breathing apparatus can be used only in a compartment:
   A. having at least 16% oxygen
   B. having at least 21% oxygen
   C. that has been certified gas-free
   D. none of the above

162. You would release nitrogen accumulation from an oxygen breathing apparatus every:
   A. 15 to 20 minutes
   B. 25 to 30 seconds
   C. time you feel it is necessary
   D. time oxygen cylinders are changed

163. An oxygen breathing apparatus tests airtight if:
   A. the bag deflates
   B. the safety valve relieves
   C. the bag inflates
   D. none of the above

164. Fresh air breathing apparatus is required on board all:
   A. tank ships over 1,000 G. T.
   B. passenger vessels over 1,000 G. T.
   C. cargo vessels over 1,000 G. T.
   D. of the above

165. The canister type gas mask works on the principle of:
   A. filtering
   B. absorption
   C. chemical combination
   D. all of the above

166. In fighting fire with foam equipment the stream of foam should be directed:
   A. in a sweeping motion across the fire
   B. on the deck and ahead of the fire
   C. directly at yellow portion of the flame
   D. at a level as high as the flame

ANSWERS TO EMERGENCY, FIRE FIGHTING & GENERAL QUESTIONS

1. C
2. A
3. D
4. D
5. C
6. A
7. D
8. D
9. B
10. D
11. B
12. B
13. A
14. D
15. D
16. D
17. D
18. A
19. C
20. B
21. D
22. C
23. B
24. D
25. B
26. B
27. A
28. B
29. C
30. D
31. D
32. C
33. B
34. C
35. B
36. D
37. C
38. B
39. C
40. B
41. B
42. C
43. C
44. D
45. D
46. A
47. C
48. B
49. A
50. C
51. D
52. D
53. D
54. A
55. C
56. B
57. D
58. B
59. D
60. A
61. A
62. D
63. B
64. A
65. A
66. D
67. C
68. A
69. A
70. C
71. A
72. B
73. A
74. C
75. A
76. B
77. D
78. B
79. D
80. D
81. C
82. C
83. D
84. A
85. C
86. A
87. B
88. C
89. B
90. D
91. D
92. B
93. C
94. B
95. A
96. C
97. D
98. A
99. D
100. A
101. C
102. D
103. C
104. D
105. B
106. B
107. D
108. B
109. B
110. C
111. C
112. D
113. B
114. B
115. B
116. C
117. A
118. A
119. D
120. A
121. C
122. D
123. A
124. A
125. D
126. C
127. D
128. D
129. B
130. B
131. C
132. A
133. A
134. B
135. C
136. D
137. C
138. B
139. B
140. C
141. D
142. B
143. D
144. B
145. B
146. B
147. C
148. D
149. C
150. D
151. C
152. C
153. B
154. A
155. D
156. A
157. B
158. A
159. A
160. C
161. D
162. A
163. D
164. A
165. D
166. B

REFRIGERATION

1. Which of the following could be used to check a leak in an ammonia system?
   A. litmus paper B. halide torch C. sulphur stick D. A and C

2. Ammonia leaks in the condenser can be detected by:
   A. smelling the discharge water
   B. applying litmus paper to the circulating water discharge
   C. adding oil of peppermint to the system and tracing the smell
   D. applying a soapy solution to the condenser heads and looking for bubbles

3. A sulphur stick burning in the presence of ammonia will give off a:
   A. dense yellow smoke C. dense red smoke
   B. dense white smoke D. dense green smoke

4. An ammonia leak will turn litmus paper:
   A. blue B. red C. green D. yellow

5. In an ammonia system the oil gauge must be kept:
   A. closed except when checking oil level
   B. open at all times
   C. closed when machine is shut down
   D. open when machine is shut down

6. Valves and piping in an ammonia system are made of:
   A. brass B. iron C. bronze D. copper

7. The relief valve on an ammonia machine is located:
   A. on the discharge pipe from the condenser
   B. on the discharge pipe from the compressor
   C. in the compressor head
   D. A or B

8. When purging an ammonia condenser into a bucket of water, one can tell when the air is out and ammonia starts to come through by the:
   A. smell of the ammonia being liberated from the water
   B. color of the water turning green
   C. color of the water turning bluish
   D. change of bubbling sound of air to the crackling sound of ammonia

9. The crossover connection in an ammonia system can be used to:
   A. hot-gas defrost
   B. pump air out of the system
   C. add refrigerant to the system
   D. reduce the pressure on the discharge side of the condenser

10. How is an ammonia system purged so that the operator will not be overcome by the fumes?

A. into a bucket of lube oil
B. into a bucket of water
C. back into the compressor
D. into the atmospheric line

11. Thermal expansion valves are usually of the:
A. diaphragm type
B. bellows type
C. magnetic type
D. A or B

12. When checking zinc plates in a condenser, one should:
A. install all new plates
B. clean the plates and renew worn out ones
C. ground each plate to the shell
D. paint the plates with red lead

13. The scale trap is located between the:
A. compressor and oil separator
B. expansion valve and evaporator coils
C. king (liquid) valve and expansion valve
D. evaporator coils and compressor

14. When the evaporator coils are located in the icebox, the system is known as a:
A. indirect system
B. high-pressure system
C. direct system
D. low-pressure system

15. The purge valve is located:
A. on the receiver discharge
B. in the highest part of the system
C. in the lowest part of the system
D. on the evaporator coils

16. The solenoid valve is:
A. a pressure-controlled stop valve
B. a temperature-controlled stop valve
C. a Freon-controlled check valve
D. none of the above

17. The thermal expansion valve is located between the:
A. solenoid valve and the evaporator coils
B. charging valve and the solenoid valve
C. receiver and the king valve
D. king valve and the solenoid valve

18. The purpose of the scale trap is to:
A. dissolve scale and dirt in the system
B. remove insoluble gases from the refrigerant
C. remove dirt, scale and metal chips from the refrigerant
D. control the amount of scale going to the compressor

19. The oil separator (trap) is located between the:
A. compressor discharge valve and the condenser

B. receiver and the expansion valve
C. receiver and the king valve
D. condenser and the receiver

20. The charging connection in a refrigerating system is located:
A. before the receiver
B. between the condenser and the receiver
C. between the receiver and the king valve
D. between the king valve and the solenoid valve

21. The dehydrator is located between the:
A. condenser and receiver
B. condenser and the king valve
C. receiver and the expansion valve
D. receiver and the king valve

22. The charging valve is located between the:
A. king valve and the expansion valve
B. evaporator coils and the compressor
C. compressor and the receiver
D. receiver and the condenser

23. The solenoid valve is located between the:
A. thermal expansion valve and the evaporator
B. scale trap and the thermal expansion valve
C. king valve and the scale trap
D. automatic and manual expansion valves

24. A device used to keep moisture from passing through the system is called a:
A. humidifier B. aerator C. dehydrator D. trap

25. The relief valve is located on the:
A. discharge side on the condenser
B. outlet of the evaporator coils
C. receiver tank
D. discharge side of the compressor

26. The dehydrator is used:
A. to remove moisture from the system
B. when adding refrigerant to the system
C. to remove air from the system
D. A and B

27. Another name for the liquid valve is the:
A. Freon valve B. shutoff valve C. king valve D. master valve

28. If no gaskets are used in the piping joints of a Freon system, the joints must be:
A. welded joints
B. ground joints
C. finished joints
D. soldered joints

29. A device for holding open the suction valve and drawing gas from the suction manifold and returning it to the suction line without compressing it is called a:
   A. relief valve C. suction line by-pass
   B. cylinder unloader D. discharge line by-pass

30. The solenoid valve can by typed as a:
   A. bellows valve C. bimetallic valve
   B. thermal valve D. magnetic stop valve

31. The bulb for the thermal expansion valve is located:
   A. near the evaporator coil outlet
   B. in the middle of evaporator coils
   C. near the evaporator coil inlet
   D. on the bottom row of evaporator coils

32. The elements of a thermostat switch are usually of the:
   A. pilot-valve type C. bimetal type
   B. diaphragm type D. valve type

33. The expansion valve is located between the:
   A. compressor and condenser C. evaporator and compressor
   B. receiver and evaporator D. condenser and receiver

34. The oil separator is located between the:
   A. condenser and dehydrator
   B. compressor and condenser
   C. evaporator and compressor
   D. solenoid valve and thermal expansion valve

35. Zinc rods are found in the:
   A. salt-water side of condenser C. compressor crankcase
   B. evaporator coils D. refrigerant side of condenser

36. A double-trunk piston is used to:
   A. prevent oil from mixing with the refrigerant
   B. absorb some of the side thrust
   C. prevent gas from getting to crankcase
   D. all of the above

37. Which of the following gasket materials should be used on a Freon system:
   A. asbestos B. rubber C. metallic D. A and C

38. Many pressure gauges on a Freon system have two dials or graduations on one gauge. The two dials represent:
   A. suction and discharge pressure
   B. pressure and temperature
   C. liquid and gas pressure
   D. cooling water inlet and outlet temperatures

39. A double-pipe condenser has:
    A. two pipes for cooling water and one for refrigerant
    B. a small pipe inside a larger pipe, the cooling water passing through the small pipe and the refrigerant through the large pipe
    C. two piping systems side by side, one with cooling water and one with refrigerant
    D. none of the above

40. When there is a Freon leak, the halide torch will burn:
    A. orange B. blue C. green D. white

41. Large leaks in a Freon system cannot always be detected with a halide torch because it changes color with the slightest amount of Freon present. A large leak can be detected easier by applying:
    A. sheets of litmus paper to all joints and watch for color change
    B. a soapsuds solution, mixed with a little glycerin to hold the solution together, and watch for bubbles
    C. a thin layer of mineral oil to all joints and watch for bubbles
    D. a lighted candle at the joints and watch for leaky spots blowing candle flame

42. The relief valve on a CO_2 machine is located:
    A. next to the king valve
    B. in the compressor head
    C. on the discharge pipe between the compressor and the discharge valve
    D. on the discharge pipe from the condenser

43. The discharge pressure of the compressor should be:
    A. the pressure which corresponds to a temperature from 5° to 15° below that of the condenser discharge
    B. the pressure which corresponds to a temperature equal to that of the condenser discharge
    C. the pressure which corresponds to a temperature from 5° to 15° higher than the condenser discharge
    D. none of the above

44. The purpose of relief valves on refrigeration machines is to:
    A. by-pass the compressor when dehydrating
    B. prevent overloading in the iceboxes
    C. prevent excessive pressure in case of stoppage on the discharge side of the system
    D. A and B

45. A leaky suction valve can usually be detected by:
    A. a higher suction pressure
    B. a fluctuating suction pressure gauge
    C. closing in on the suction valve having no effect on the suction pressure
    D. any of the above

46. Too low suction pressure could be caused by:
    A. too much oil in the system
    B. dirty scale traps
    C. shortage of refrigerant gas
    D. any of the above

47. When the evaporator coils are located in a brine solution and the brine is pumped through the icebox, the system is known as:
    A. an indirect system
    B. a direct system
    C. a double-evaporator system
    D. a low-pressure system

48. Before securing a compressor to do maintenance work on it, be sure to:
    A. purge the system
    B. pump down the system
    C. have spare parts ready
    D. B and C

49. Obstruction of the expansion valve is usually caused by:
    A. scale
    B. water in the system
    C. congealed oil in the system
    D. any of the above

50. Excess frost on the evaporator coils:
    A. does not affect the system
    B. reduces efficiency of the plant
    C. takes load off compressor
    D. keeps the icebox cooler

51. Air circulation in the icebox is accomplished by the use of:
    A. hollow sidewalls
    B. louver doors
    C. diffuser fans
    D. air vents to deck

52. The cooling-water side of the condenser should be opened for inspection every:
    A. six months B. year C. two years D. three months

53. Some causes of a noisy compressor are:
    A. worn bearings, pins, etc.
    B. slugging due to flooding back of refrigerant
    C. too much oil in crankcase
    D. any of the above

54. If the thermal expansion valve becomes inoperative, the iceboxes will have to be controlled by the:
    A. king valve
    B. manual solenoid valve
    C. manual expansion valve
    D. solenoid valve

55. Sweating of the crankcase is caused by:
    A. too much oil in the system
    B. too much superheat
    C. insufficient superheat
    D. expansion valve hung up

56. Slugging is usually caused by:
    A. too much refrigerant in the system
    B. too much oil in the system
    C. expansion valve not operating properly
    D. too much cooling water to condenser

57. The low-pressure cutout switch is operated by:
    A. a bellows C. spring tension
    B. a magnet D. water pressure

58. If an automatic Freon system will not start up, check the:
    A. high-pressure cutout C. reset mechanism
    B. low-pressure cutout D. all of the above

59. A hot suction line might be caused by:
    A. insufficient lubrication C. too much refrigerant
    B. insufficient refrigerant D. expansion valve closed too much

60. The device used for low-pressure control and high-pressure cutout on a compressor is called a:
    A. pressure controller C. controller switch
    B. cutout D. cutout switch

61. If the solenoid valve closed by accident, the compressor would be stopped by the:
    A. automatic trip C. low-pressure cutout switch
    B. low-water cutout switch D. high-pressure cutout switch

62. The purpose of the low-pressure cutout switch is to:
    A. cut out the compressor at a set pressure
    B. maintain a preset suction pressure to the compressor
    C. maintain liquid refrigerant at the suction of the compressor
    D. cut compressor in and out at a preset pressure

63. If the cooling water to the condenser should suddenly fail:
    A. an alarm will ring to notify the engineer
    B. the compressor will shut down
    C. the expansion valve will close
    D. the solenoid valve will close

64. The most likely cause of high superheat would be:
    A. too much refrigerant
    B. expansion valve open too wide
    C. expansion valve closed too much
    D. back-pressure valve set too high

65. Liquid reaching the compressor through the suction line is called:
    A. superheating B. flooding back C. overflowing D. recycling

66. The suction pressure switch is operated by:
    A. electric current C. pressure on a bellows
    B. a relay cutout D. thermocouple

67. If the compressor short-cycles on the high-pressure cutout, which of the following would you check?

A. check for too much refrigerant in the system
B. if plenty of cooling water is running through but it is not picking up heat, the condenser tubes need cleaning
C. be sure system is getting cooling water
D. all of the above

68. A Freon unit will tend to short-cycle when operating under:
A. normal conditions C. heavy loads
B. light loads D. all of the above

69. Air is removed from the system by:
A. opening the purge valve
B. increasing the amount of cooling water
C. running the refrigerant through an aerator
D. running the refrigerant through a deaerator

70. Short-cycling means that the machine:
A. runs too slow C. stops and starts frequently
B. runs too fast D. grounds out frequently

71. The suction pressure in a Freon system should be:
A. the pressure which corresponds with a temperature about 20° F. above the temperature of the icebox
B. the pressure which corresponds with a temperature about 20° F. below the temperature of the icebox
C. the pressure which corresponds with a temperature equal to the temperature of the icebox
D. none of the above

72. If any of the electrically controlled devices in a Freon system malfunction, which of the following valves will also automatically shut off?
A. king valve C. expansion valve
B. condenser cooling-water inlet valve D. solenoid valve

73. A leaky discharge valve can usually be detected by:
A. a drop in icebox temperature
B. a discharge pressure lower than normal
C. a fluctuating high-pressure gauge
D. any of the above

74. The dehydrating agent in a Freon system is usually:
A. sodium chloride C. calcium chloride
B. activated alumina D. slaked lime

75. If the compressor short-cycles on the low-pressure cutout, the trouble might be:
A. too much frost on the evaporator coils C. lack of refrigerant
B. dirty traps and strainers D. any of the above

76. If an electrically operated compressor failed to start, the cause might be:
   A. a blown fuse
   B. an open switch
   C. burned out holding coils in solenoid valve
   D. any of the above

77. The high-pressure side of the system is sometimes referred to as the:
   A. hot side D. cold side C. suction side D. cooling side

78. If the compressor were to run continuously without lowering the temperature, the trouble would probably be:
   A. leaky discharge valves
   B. leaks in the system
   C. insufficient refrigerant in the system
   D. any of the above

79. Which of the following would cause a high head pressure?
   A. suction valve not open enough C. insufficient cooling water
   B. too much cooling water D. icebox door left open

80. If frost forms on the cylinders, the cause would be:
   A. charging valve left open
   B. expansion valve not open wide enough
   C. expansion valve open too wide
   D. dehydrator not working properly

81. Which of the following would cause low head pressure?
   A. insufficient cooling water C. too much cooling water
   B. insufficient refrigerant gas D. B and C

82. Too high suction pressure could be caused by:
   A. leaky suction valves
   B. expansion valve bulb not working properly
   C. expansion valve open too wide
   D. any of the above

83. Which of the following would cause high head pressure?
   A. air in the condenser C. insufficient cooling water
   B. dirty condenser D. any of the above

84. An excessively high head pressure could be caused by:
   A. insufficient cooling water to the condenser
   B. insufficient cooling water to the evaporator coils
   C. solenoid valve shutoff
   D. too much cooling water to the condenser

85. Which of the following would cause a high suction pressure?
   A. expansion valve open too wide B. dirty dehydrator

C. king valve not open wide enough
D. expansion valve not open wide enough

86. If a compressor runs continuously, the cause might be a:
A. clogged scale trap
B. stuck high-pressure switch
C.. defective thermal bulb
D. stuck low-pressure switch

87. Low suction pressure is caused by:
A. expansion valve causing flooding back
B. solenoid valve not functioning properly
C. leaky compressor suction valves
D. air in the system

88. How is a Freon system purged?
A. with a reefer pump
B. back to the extra supply tank
C. the same as an ammonia system
D. the same as a CO_2 system

89. If the compressor discharge temperature is higher than the receiver temperature:
A. add more refrigerant to the system
B. decrease the amount of cooling water to the condenser
C. increase the amount of cooling water to the condenser
D. remove some of the refrigerant from the system

90. How is a CO_2 system purged?
A. when CO_2 comes out of the purge valve, frost will form on a piece of metal held near the outlet
B. through the king valve
C. pumped out with a suction pump
D. the CO_2 will come out of the purge valve in liquid form

91. If the head pressure is too high:
A. the relief valve should open and let excess refrigerant pass to receiver
B. close in on the suction valve
C. the relief should open before the high-pressure cutout
D. the high-pressure cutout switch should operate before the relief valve opens

92. The system should be purged:
A. while starting up
B. while system is operating
C. after the system has been shut down for a few hours
D. once a week

93. If the compressor had been running satisfactorily for a long period of time but suddenly the compartment temperature started to rise, the trouble might be:
A. a refrigerant leak has developed

B. the expansion valve may contain frozen water
C. the solenoid valve has jammed shut
D. any of the above

94. If the compressor had been running satisfactorily for a long period of time but the oil level was rising slowly, one should:
A. check the dehydrator cartridge
B. check to see if there is too much refrigerant in the system
C. shut down the compressor and check the oil level with the machine stopped
D. drain out sufficient oil to bring it down to the proper running lev

95. The purpose of the oil trap is:
A. to add oil to the compressor
B. to remove oil from the refrigerating gas
C. to remove oil from the charging tank
D. none of the above

96. Too much oil in the compressor would:
A. absorb too much refrigerant from the system
B. deposit oil on the condenser tubes
C. damage the expansion valve
D. cause leakage through the shaft seals

97. The oil level in the compressor should be checked:
A. just before starting the compressor
B. after a long period of operation
C. after an extended lay-up period
D. while the compressor is in operation

98. The agent used in an indirect reefer system is:
A. calcium chloride
B. sodium chloride
C. potassium chloride
D. A or B

99. Air can be prevented from getting into the system by:
A. running the refrigerant through an aerator
B. keeping the dehydrator clean at all times
C. keeping all glands and stuffing boxes on the high-pressure side tight
D. keeping all glands and stuffing boxes on the low-pressure side tight

100. A method of reducing capacity without reducing compressor speed is called:
A. short-cycling
B. hot-gas by-passing
C. high-pressure by-passing
D. low-pressure by-passing

101. If the discharge side of the thermal expansion valve is warmer than the inlet side, the:

A. expansion valve is not working properly
B. expansion valve is working normally
C. solenoid valve is not working normally
D. scale trap is dirty

102. Abnormal discharge temperatures would be caused by:
A. faulty piston rings
B. leaky discharge valves
C. leaky suction valves
D. any of the above

103. If the compressor runs continually, the cause might be:
A. a clogged scale trap
B. thermal bulb not operating properly
C. low-pressure switch jammed shut
D. high-pressure cutout switch is jammed open

104. The capacity of a refrigerating unit can be regulated by:
A. changing the amount of refrigerant in the system
B. changing the amount of cooling water to the condenser
C. changing the speed of the compressor
D. any of the above

105. If there is too much Freon in the system:
A. the safety valve will discharge
B. the compressor will short-cycle
C. the compressor will run continuously
D. it will mix with the oil in the crankcase

106. If the temperature in the icebox is too high, the trouble could be:
A. automatic controls not functioning properly
B. a clogged scale trap
C. air in the system
D. insufficient cooling water to the condenser

107. Frost on the outside of the thermal expansion valve may be caused by:
A. too much superheat
B. air in the system
C. moisture in the system
D. any of the above

108. Which of the following would not cause high suction pressure?
A. suction valve not adjusted properly
B. expansion valve stuck open
C. leaky suction valves
D. insufficient refrigeration

109. Water in the refrigerant is liable to:
A. freeze on the expansion valve seat and cut the flow of refrigerant
B. clog the oil trap
C. freeze in the king (liquid) valve
D. emulsify the oil in the compressor

110. The function of the expansion valve is to:
    A. regulate the amount of liquid refrigerant to the expansion coils
    B. change the gas refrigerant to a liquid
    C. shut off the flow of refrigerant to the condenser
    D. change the high-pressure liquid to a low-pressure liquid

111. When starting a refrigerating system, always:
    A. vent the evaporator
    B. by-pass the condenser
    C. vent the condenser
    D. none of the above

112. When heavy electrical currents are involved, the thermostat will be operated by a:
    A. pressure pipe
    B. relay
    C. fusetron
    D. small circuit breaker

113. Before securing a compressor to do maintenance work on it, be sure to:
    A. have gas mask handy
    B. make arrangements to have perishables taken care of
    C. notify the Captain
    D. A and B

114. When securing a Freon system for repairs:
    A. pump down to 1 or 2 pounds pressure
    B. pump down to a slight vacuum
    C. pump down to 10 to 15 pounds pressure
    D. remove all refrigerant from the system

115. When charging a Freon system, all the valves should be in their normal position except the:
    A. solenoid valve
    B. king (liquid) valve
    C. purge valve
    D. expansion valve

116. The purpose of the receiver is to:
    A. cool the refrigerant gas
    B. separate the oil from the refrigerant
    C. store the refrigerant
    D. condense the refrigerant

117. The solenoid valve controls the:
    A. amount of refrigerant entering the evaporator coils
    B. amount of refrigerant going to the expansion valve
    C. amount of refrigerant going to the compressor
    D. pressure of the refrigerant going to the evaporator coils

118. The purpose of the evaporator is to:
    A. absorb latent heat of vaporization
    B. absorb latent heat of fusion
    C. transfer latent heat of vaporization
    D. transfer latent heat of fusion

119. The purpose of the dehydrator is to:
   A. add more refrigerant to the system
   B. remove oil from the refrigerant
   C. remove moisture from the crankcase oil
   D. remove moisture from the refrigerant

120. The principle of mechanical refrigeration is:
   A. the conversion of a liquid to a gas
   B. the absorption of temperature under heat, pressure, compression and expansion
   C. the compression of a liquid under temperature and expansion
   D. the absorption of heat under temperature, compression, pressure and expansion

121. A thermostat is a:
   A. temperature-operated switch
   B. pressure-operated switch
   C. superheat-operated switch
   D. back-pressure-operated switch

122. The thermal expansion valve responds to the:
   A. amount of superheat in the vapor
   B. amount of superheat in the liquid
   C. temperature in the evaporator coils
   D. pressure in the evaporator coils

123. The expansion valve on a Freon system controls the:
   A. superheat of the gas leaving the compressor
   B. back pressure in the evaporator
   C. temperature of the icebox
   D. superheat of the gas leaving the evaporator

124. The purpose of the expansion valve by-pass is to:
   A. increase the efficiency of the plant
   B. increase the capacity of the evaporator
   C. control the refrigerant to the evaporator in case the automatic valves fail
   D. bypass the compressor

125. The thermal expansion valve:
   A. controls the amount of gas coming from the dehydrator
   B. controls the quantity of liquid refrigerant going to the evaporator coils
   C. controls the amount of gas going to the receiver
   D. removes trapped oil from the refrigerant

126. The function of the compressor is to:
   A. pull the refrigerant gas through the system
   B. increase the pressure of the refrigerant
   C. discharge the refrigerant to the condenser
   D. all of the above

127. The solenoid valve is controlled by:
    A. the amount of liquid in the system
    B. the amount of gas in the system
    C. the temperature in the condenser
    D. the temperature in the icebox

128. Oil is added to a Freon compressor by:
    A. shutting down the machine and pouring in through the crankcase inspection plate opening
    B. pumping in with an electric-driven pump
    C. pumping in with a hand pump
    D. pouring through oil hole in base

129. When adding oil to a Freon system, one must be sure that:
    A. the condenser is shut down
    B. all air is removed from the pump and fittings
    C. there is not too high a suction pressure
    D. the discharge pressure is not too high

130. To help a person who had been overexposed to ammonia gas, one would:
    A. apply cold compresses C. apply artificial respiration
    B. douse with cold water D. wrap in warm blankets

131. Ammonia will corrode:
    A. brass B. bronze C. copper D. all of the above

132. A refrigerating unit of one (1) ton capacity can remove:
    A. 100 BTU's per min. C. 288 BTU's per min.
    B. 200 BTU's per min. D. 500 BTU's per min.

133. The refrigerant with the lowest boiling point is:
    A. NH_3 B. CO_2 C. F_{12} D. F_{22}

134. A ton of refrigeration is equal to the removal of:
    A. 288,000 BTU per 24 hrs. C. 28,800 BTU per 24 hrs.
    B. 28,000 BTU per 24 hrs. D. 280,000 BTU per 24 hrs.

135. The boiling point of Freon-12 at atmospheric pressure is:
    A. + 22° F. B. + 22° C. C. − 22° F. D. − 22° C.

136. The boiling point of CO_2 at atmospheric pressure is:
    A. − 110° F. B. + 110° C. C. + 110° F. D. − 110° C.

137. The boiling point of NH_3 at atmospheric pressure is:
    A. + 28° C. B. − 28° C. C. + 28° F. D. − 28° F.

138. Which of the following would you apply if a person got Freon in his eyes?

A. clean water C. soapy water
B. sodium bicarbonate D. sterile mineral oil

139. A double-seated valve allows the valve to be:
A. packed only in the closed position
B. packed in the wide open or closed position
C. operated as a suction or discharge valve
D. removed for replacement without shutting down

140. The amount of CO_2 or Freon in a cylinder is measured by:
A. pressure B. volume C. weight D. psi

141. The latent heat of fusion of ice is:
A. 500 BTU B. 144 BTU C. 188 BTU D. 970 BTU

142. Latent heat:
A. can be measured with a pyrometer
B. cannot be measured with a thermometer
C. changes as the refrigerant cools
D. can be measured with a thermometer

143. Absolute zero is:
A. 144° below zero on the fahrenheit scale
B. the same as zero on the fahrenheit scale
C. 970° below zero on the fahrenheit scale
D. 460° below zero on the fahrenheit scale

144. A ton of refrigeration is equal to the cooling effect of:
A. 2,000 lbs. of ice melting in 24 hrs.
B. 2,000 lbs. of water being converted to ice
C. 2,000 lbs. of ice melting in 12 hrs.
D. 2,240 lbs. of ice melting in 24 hrs.

145. Freon is:
A. odorless B. colorless C. non-poisonous D. all of the above

146. Another name for discharge pressure is:
A. absolute pressure C. head pressure
B. suction pressure D. condenser pressure

147. Oil used in a refrigeration system is:
A. vegetable oil C. straight mineral oil
B. lube oil SAE 20 D. lube oil SAE 10

148. One disadvantage of a CO_2 system is the fact that:
A. it is difficult to condense the refrigerant if the circulating water temperature is too high
B. it takes more refrigerant to keep the iceboxes cold
C. due to high pressure it is difficult to keep oil from mixing with the refrigerant

D. it is difficult to condense the refrigerant if the circulating water temperature is too low

149. The disadvantage of a CO_2 system over an ammonia system is the fact that:
   A. the pipes and fittings of a CO_2 system must be of the high-pressure type
   B. the CO_2 system operates at a much higher pressure
   C. the CO_2 system requires a larger prime mover
   D. all of the above

150. A good refrigerant should be:
   A. non-inflammable C. non-poisonous
   B. non-explosive D. all of the above

151. If a refrigerating system extracted 48,000 BTU per hour, the tonnage of the machine would be:
   A. 2 tons B. 4 tons C. 5 tons D. 3 tons

152. A combination of a wet and dry bulb thermometer is called a:
   A. hygrometer B. hydrometer C. A or D D. psychrometer

153. A pressure controller is usually operated by the movement of a:
   A. bellows B. diaphragm C. syphon D. A or B

154. The critical temperature of a refrigerant is:
   A. the temperature at which it will freeze
   B. the temperature above which it cannot be liquified
   C. the temperature below which it cannot be liquified
   D. none of the above

155. If the critical temperature of a refrigerant is too close to the desired condensing temperature, the equipment must necessarily be of:
   A. extra heavy construction C. the direct expansion type
   B. light construction D. the indirect expansion type

156. A good refrigerant should have a:
   A. high sensible heat C. low sensible heat
   B. high latent heat D. low latent heat

157. The most common method of heat flow in refrigeration work is by:
   A. conduction B. radiation C. expulsion D. convection

158. The largest quantities of heat are available during the process of:
   A. vaporization B. fusion C. melting D. cooling

159. The most common Freon gas used in centrifugal compressors is:
   A. F 12 B. F 22 C. F 11 D. F 21

160. The most common Freon gas used in reciprocating compressors is:
A. F 12 B. F 22 C. F 11 D. F 21

161. A hygrometer is also known as a:
A. hydrometer B. manometer C. psychrometer D. A or C

162. The temperature in the meat and fish box should be approximately:
A. 10° to 20° F. C. 0° to -10° F.
B. -10° to 0° F. D. -10° to 10° F.

163. The temperature in the vegetable box should be approximately:
A. 35° to 45° F. C. 40° to 50° F.
B. 10° to 20° F. D. 15° to 30° F.

164. The temperature in the dairy box should be approximately:
A. 20° to 30° F. C. 10° to 20° F.
B. 0° to 10° F. D. 30° to 40° F.

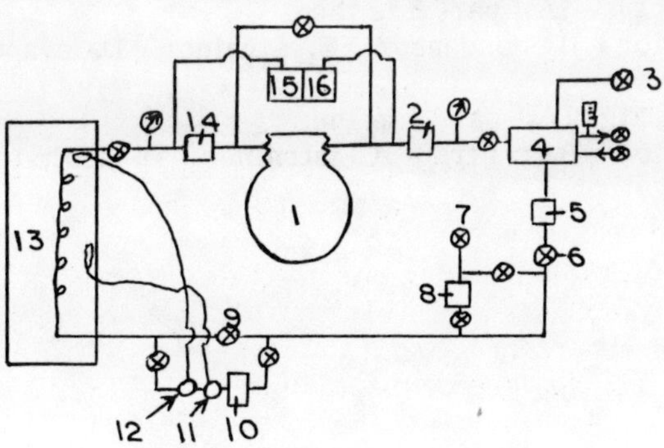

Figure 1A.

165. Referring to Fig. 1A, part 3 is the:
A. charging valve B. liquid valve C. purge valve D. scale trap

166. Referring to Fig. 1A, part 5 is the:
A. condenser B. receiver C. evaporator D. dehydrator

167. Referring to Fig. 1A, part 6 is the:
A. solenoid valve B. charging valve C. purge valve D. liquid valve

168. Referring to Fig. 1A, part 9 is the:
A. thermal expansion valve C. solenoid valve
B. manual expansion valve D. liquid valve

169. Referring to Fig. 1A, part 11 is the:
A. solenoid valve C. liquid valve
B. thermal expansion valve D. none of the above

170. Referring to Fig. 1A, part 15 is the:
A. high-pressure cutout C. relief valve
B. low-pressure cutout D. scale trap

171. Referring to Fig. 1A, part 7 is the:
A. purge valve B. liquid valve C. charging valve D. solenoid valve

172. Referring to Fig. 1A, part 12 is the:
A. thermal expansion valve C. liquid valve
B. solenoid valve D. manual expansion valve

173. Referring to Fig. 1A, part 10 is the:
A. solenoid valve B. strainer C. liquid valve D. purge valve

174. Referring to Fig. 1A, part 4 is the:
A. evaporator B. dehydrator C. condenser D. receiver

175. Referring to Fig. 1A, part 8 is the:
A. dehydrator B. condenser C. strainer D. evaporator

176. Referring to Fig. 1A, part 2 is the:
A. dehydrator B. oil trap C. strainer D. scale trap

ANSWERS TO REFRIGERATION QUESTIONS

1. D
2. B
3. B
4. A
5. A
6. B
7. C
8. D
9. A
10. B
11. D
12. B
13. C
14. C
15. B
16. B
17. A
18. C
19. A
20. D
21. C
22. A
23. B
24. C
25. D
26. D
27. C
28. B
29. B
30. D
31. A
32. C
33. B
34. B
35. A
36. D
37. D
38. B
39. B
40. C
41. B
42. C
43. C
44. C
45. D
46. D
47. A
48. D
49. D
50. B
51. C
52. D
53. D
54. C
55. A
56. C
57. A
58. D
59. B
60. A
61. C
62. D
63. B
64. C
65. B
66. C
67. D
68. B
69. A
70. C
71. B
72. D
73. D
74. B
75. D
76. D
77. A
78. D
79. C
80. C
81. D
82. D
83. D
84. A
85. A
86. D
87. B
88. D
89. B
90. A
91. D
92. C
93. D
94. C
95. B
96. A
97. B
98. D
99. D
100. B
101. A
102. D
103. C
104. D
105. B
106. A
107. C
108. D
109. A
110. A
111. C
112. B
113. D
114. A
115. B
116. C
117. B
118. A
119. D
120. D
121. A
122. A
123. D
124. C
125. B
126. D
127. D
128. C
129. B
130. C
131. D
132. B
133. B
134. A
135. C
136. A
137. D
138. D
139. B
140. C
141. B
142. B
143. D
144. A
145. D
146. C
147. C
148. A
149. D
150. D
151. B
152. C
153. D
154. B
155. A
156. B
157. D
158. A
159. C
160. A
161. C
162. A
163. A
164. D
165. C
166. B
167. D
168. B
169. A
170. B
171. C
172. A
173. B
174. C
175. A
176. B

DRAWING

The applicant for either engineer's license must make several drawings. One must draw three views — front, top and side — from a given isometric drawing. To give a complete course in drawing in a book like this would be impossible. A few samples are given to help in this project.

Probably the easiest way to make a three-view drawing is to number each corner on the original isometric drawing and proceed with the 45° mitre line method, as follows:

1. Working drawings are always set down in a standard pattern — top, front and side views, as shown in Fig. 2A.
2. Number each corner in the given isometric drawing.
3. Mark off the general shape (if not required to make exact scale drawing) for front and top views.
4. Remember that lines from the front view must project vertically to lines on the top view (or vice versa) and lines from the front view must project horizontally to the side view.
5. Extend the lower line in the top view horizontally one to two inches (depending on size of paper) and mark a point X.
6. From X draw a 45° line XY.
7. Now start with point 1 in the top view. Extend a horizontal line to the 45° line (this happens to be point X). Drop a perpendicular line.
8. Extend a horizontal line from point 1 in the front view until it hits the vertical line from X. This locates point 1 in the side view.
9. Take point 2 in the top view. Extend horizontal line to the 45° line, then drop a perpendicular. Extend a horizontal line from point 2 in front view until it hits this vertical line. We now have point 2 in the side view.
10. Continue these operations of extending horizontal lines from a point in the top view to the 45° line, then dropping vertical lines until they coincide with horizontal lines drawn from the same points in the front view.
11. After completing a few simple drawings (check Fig. 3A) in this fashion, it will not be too difficult to go on to more complicated ones.

Note. In Figs. 4A **and** 5A, a hole through a piece of metal in the front view appears as a circle. In the side and top views, it appears as two horizontal lines represented by the diameter of the hole.

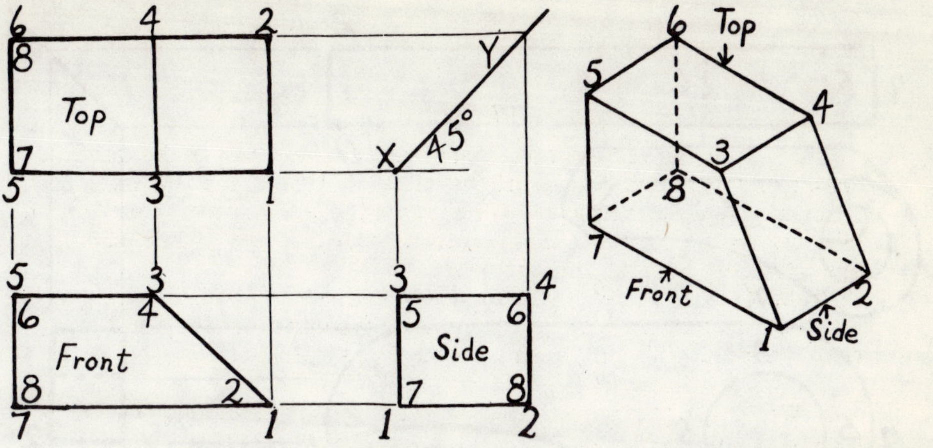

Figure 2A.

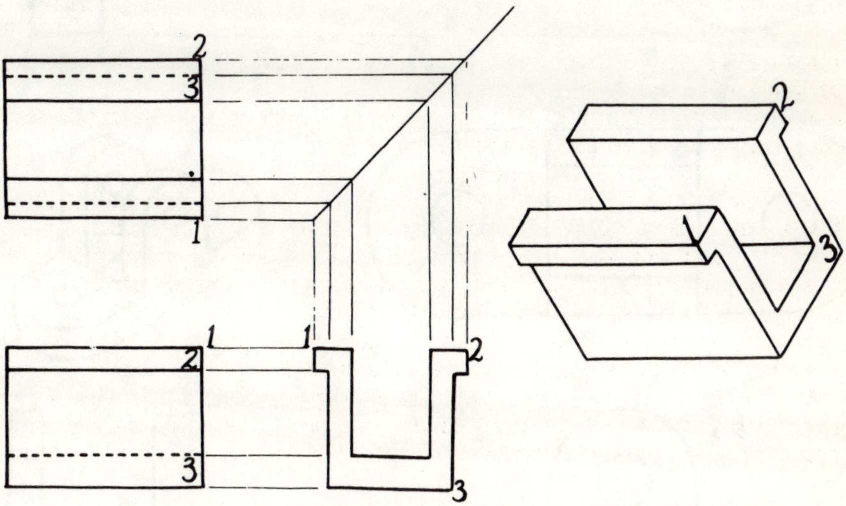

Figure 3A.

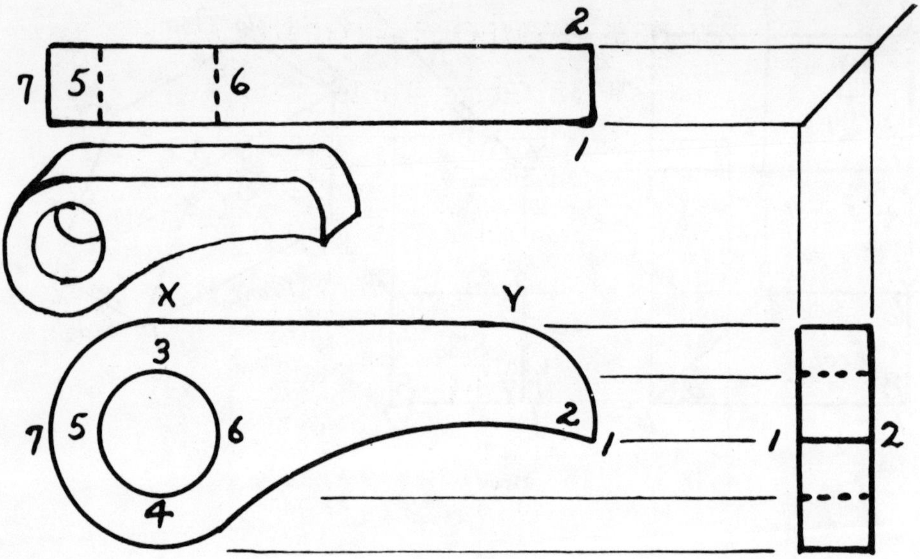

Figure 4A.

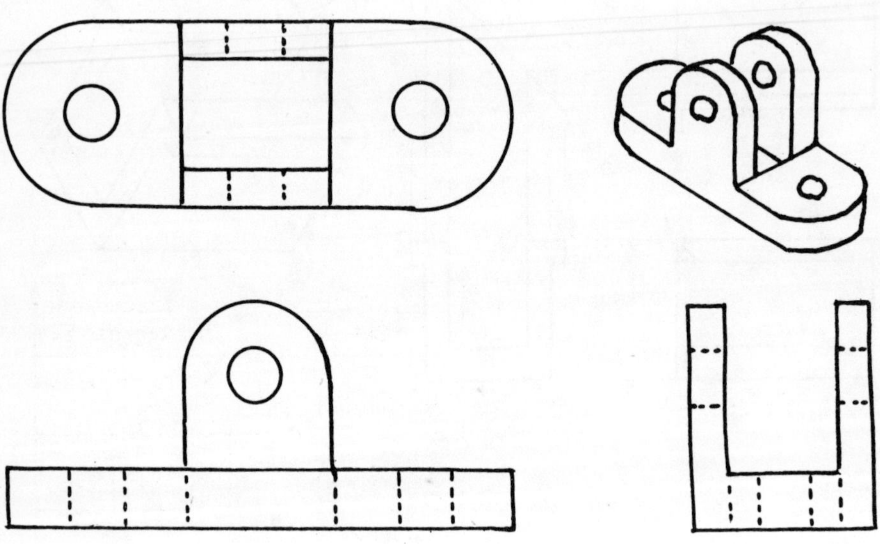

Figure 5A.

TABLES—SYMBOLS—FORMULAS

Wire Gauge Table

Size AWG	Diameter of Solid Wires in Mils	Area in Circular Mils (dia.)^2
18	40.3	1,624
16	50.8	2,583
14	64.1	4,107
12	80.8	6,530
10	101.9	10,380
8	128.5	16,510
6	162.0	26,250
5	181.9	33,100
4	204.3	41,740
3	229.4	52,630
2	257.6	66,370
1	289.3	83,690
0	325.0	105,500
00	364.8	133,100
000	409.6	167,800
0000	460.0	211,600

1 Mil 0.001 inch.

Battery Data

Specific Gravity (Hydrometer Reading)	Approximate Cell Voltage (on open circuit)	Condition of Battery	Freezing Point of Electrolyte (in degrees F.)
1.280 to 1.300	2.15 to 2.2	Fully charged	90° to 95° below zero
1.260	2.1	Three-quarters charged	75° below zero
1.215	2.0	One-half charged	27° below zero
1.180	1.95	One-quarter charged	6° below zero
1.150(or below)	1.9 (or less)	Completely discharged	5° above zero or higher

AMERICAN N.C. AND N.F. THREADS

Size of tap	*Threads per inch*	*Size of tap drill*
1/4 NC	20	7
1/4 NF	28	3
5/16 NC	18	F
5/16 NF	24	I
3/8 NC	16	5/16
3/8 NF	24	Q
7/16 NC	14	U
7/16 NF	20	25/64
1/2 NC	13	27/64
1/2 NF	20	29/64
9/16 NC	12	31/64
9/16 NF	18	33/64
5/8 NC	11	17/32
5/8 NF	18	37/64
11/16 NC	11	19/32
11/16 NF	16	5/8
3/4 NC	10	21/32
3/4 NF	16	11/16
13/16 NC	10	23/32
7/8 NC	9	49/64
7/8 NF	14	13/16
15/16 NC	9	53/64
1 NC	8	7/8
1 NF	14	15/16
1 1/8 NC	7	63/64
1 1/8 NF	12	1 3/64
1 1/4 NC	7	1 7/64
1 1/4 NF	12	1 11/64
1 3/8 NC	6	1 7/32
1 3/8 NF	12	1 19/64
1 1/2 NC	6	1 11/32
1 1/2 NF	12	1 27/64
1 5/8 NC	5 1/2	1 29/64
1 3/4 NC	5	1 9/16
1 7/8 NC	5	1 11/16
2 NC	4 1/2	1 25/32

Threads and Thread Cutting. Threads cut on the outside of a piece of round stock are called external threads. Threads cut on the inside of a cylindrical hole are called inside threads. Dies are used for cutting the external or outside threads and taps are used for cutting inside threads.

The two types of general threads used are the National Coarse (N. C.) and National Fine (N. F.). Pitch is the distance from the top of one thread to the top of an adjacent thread. The pitch is usually stated as the number of threads per inch, such as 8, 10, 12, etc. A tap or die marked 1/4-20 indicates 1/4" diameter and 20 threads per inch.

Die threads are tapered from one face only, so be sure to start the cut with that face.

To start the thread, press down firmly on the stock, while turning to the right (clockwise). When the die catches, start using a cutting oil. Continue turning, following the same procedure as with tapping, half a turn clockwise and a quarter turn back, etc.

Sheet Metal (and wire) comes in various thicknesses measured by number according to the American Wire Gage system. Sheet metal usually comes in sheets 4' × 6'. Some of the more common sizes are listed with their corresponding thicknesses in thousandths of an inch. See following table.

Gage Number	American Wire Gage or Brown and Sharpe-Diameter or Thickness in Decimal Parts of Inch
0	.325
2	.258
4	.204
6	.162
8	.128
10	.102
12	.081
14	.064
16	.051
18	.040
20	.032
22	.025
24	.020
26	.0159
28	.0126
30	.010
32	.008
34	.0063
36	.005
38	.004
40	.003

PIPING SYMBOLS

Valves

 Globe, angle or gate

 Globe

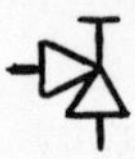 Angle

 Gate

 Cross

 Swing check

 Lift check

 Stop check

 Hose globe

 Hose gate

 Hose angle

 Relief

Valves

 Solenoid

 Needle

 Motor operated

 Manifold stop valves

 Manifold stop valves with operating rods above bulkhead deck

Fittings

 Flanged joint

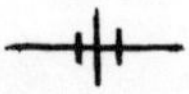 Union

 Elbow

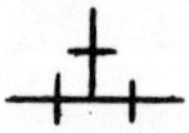 Tee

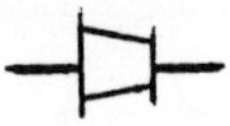 Reducer

 Safety

ELECTRICAL SYMBOLS

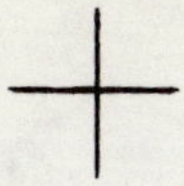
Cross wires

Connecting wires

Connector—male

Connector—female

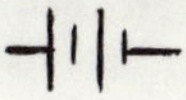
Battery

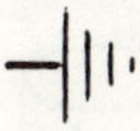
Ground

Fuse

Circuit breaker

Rheostat

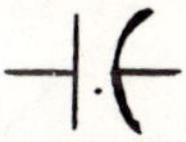
Capacitor—fixed

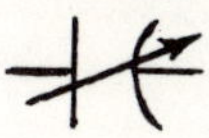
Capacitor—variable

Resistor—fixed

Resistor—tapped

Resistor—adjustable

Transformer—air core

Transformer—iron core

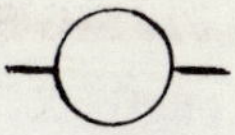
Motor

Squirrel-cage motor

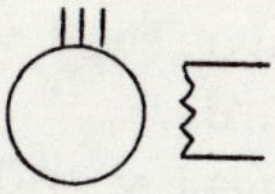
3 ø Synchronous motor

Wound-rotor motor or generator

Switches

Single-pole, single-throw

Single-pole, double-throw

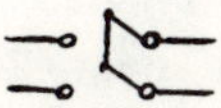
Double-pole, single-throw

Double-pole, double-throw

Pushbutton—make

Pushbutton—break

Single-pole

3-way

4-way

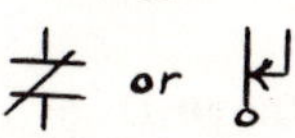

Contactor—closed

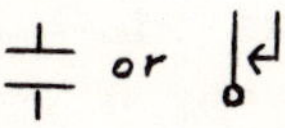

Contactor—open

FORMULAS

1. Area of rectangle = Length X Width

$A = L \times W$

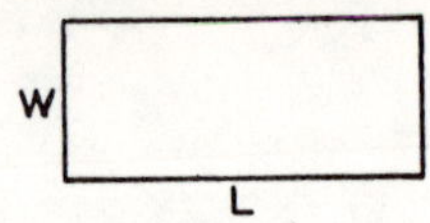

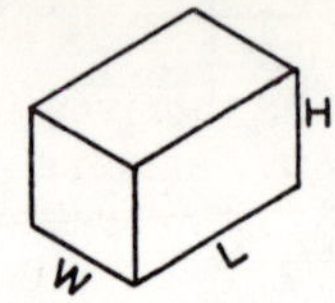

2. Volume of rectangular solid = Length X Width X Height

$V = L \times W \times H$

3. Area of circle = $.7854 \times (D)^2$

$A = .7854D^2$

or

$A = \pi R^2$

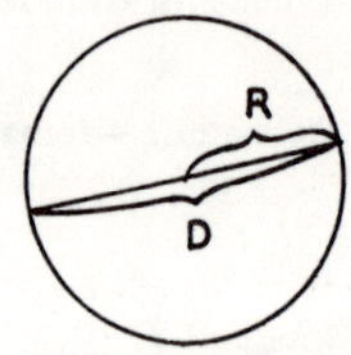

4. Circumference of circle = $\pi \times D$

$C = \pi D$

5. Volume of cylindrical vessel = Area of base X height

$V = .7854D^2h$

or

$V = \pi R^2h$

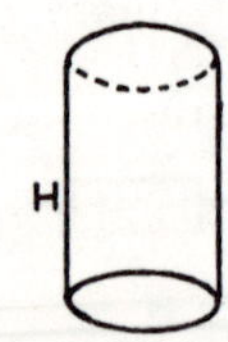

6. The following rule will be found useful for figuring problems with right-angle triangles.

$c^2 = a^2 + b^2$	$c^2 = 25$
$c^2 = (4)^2 + (3)^2$	$c = \sqrt{25}$
$c^2 = 16 + 9$	$c = 5$

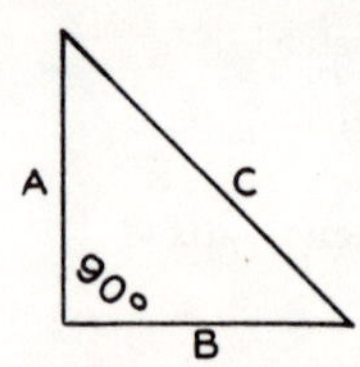

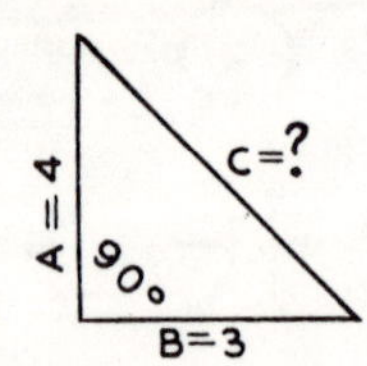

7. $°F. = \frac{9}{5}C. + 32$

$°C. = \frac{5}{9}(F. - 32)$

° Absolute = °C. + 273

° Absolute = °F. + 460

Taper formula for lathe work

$$\text{Taper per inch} = \frac{D - d}{L}$$

$$\text{Taper per foot} = \left(\frac{D - d}{L}\right) 12$$

Where D = diameter at large end
d = diameter at small end
L = length in inches

ABBREVIATIONS

A-area
A. C. -alternating current
amp. -ampere
B. H. P. -brake horsepower
B. T. U. -British Thermal Unit
C. -Centrigrade; circumference
cu. ft. -cubic foot
cu. in. -cubic inch
D or d-diameter
D. C. -direct current
F. -Fahrenheit
ft. ;' -foot
ft. per min. -feet per minute
G. R. & R. -General Rules & Regulations, U. S. C. G.
H. P. -high pressure; horsepower
hr. -hour
I. C. E. -Internal Combustion Engine
I. H. P. -indicated horsepower
in. ;" -inch
I. P. -intermediate pressure
Kw. -kilowatts
lbs. -pounds
lbs. per sq. ft. -pounds per square foot
lbs. per sq. in. -pounds per square inch
L. P. -low pressure
M. E. P. -mean effective pressure
min. -minute
p. f. -power factor
p. s. i. -pounds per square inch
R or r-radius
R. P. M. -revolutions per minute
sec. -second
S. H. P. -shaft horsepower
S. P. M. -strokes per minute
sq. ft. -square feet
sq. in. -square inches
T. D. C. -top dead center
T. S. -tensile strength
v. -volt
vol. or V-volume
W. or W. P. -working pressure
° -degrees

MEASURES

2 pts. = 1 qt. (quart)
4 qts. = 1 gal. (gallon)
1 gal. fresh water = 8. 3 lbs.
1 gal. = 231 cu. in.
1 cu. ft. fresh water = 62. 5 lbs.
1 cu. ft. salt water = 64 lbs.
1 B. T. U. = 778 ft. lbs. work
1 lb. coal = 13,000-15,000 B. T. U.
1 lb. oil = 18,000-20,000 B. T. U.
1 ton fresh water = 36 cu. ft.
1 ton salt water = 35 cu. ft.
1 ton fresh water = 268 gals.
1 gal. salt water = 8. 5 lbs.
7. 5 gals. oil = 1 cu. ft.
1 gal. fuel oil = 8 lbs.
1 bbl. oil = 42 gals.
Atmospheric pressure = 14. 7 lbs.
12 in. = 1 ft.
144 sq. in. = 1 sq. ft.
1728 cu. in. = 1 cu. ft.
1 H. P. = 33,000 ft. lbs.
π = 3. 1416
1 nautical mile or 1 knot = 6080 ft.
1 short ton = 2000 lbs.
1 long ton = 2240 lbs.
Pressure in lbs. per sq. ft. = 62. 5 X height of column in ft.
Pressure in lbs. per sq. in. = . 434 X height of column in ft.

THE BASICS OF AUTOMATION

This section includes only a brief treatise on the basic principles of automation. Its purpose is to give an insight as to the general plans and workings of some of the devices used in ship automation.

In the progress of modern shipbuilding it became necessary to increase their size and speed in order to be competitive with other countries (with lower wages) and earn money for steamship companies.

A vessel of the 1940's was a 10,000-ton vessel traveling at a speed of 10 knots with a crew of 45. A modern vessel of the late 1960's is 20,000 tons (or more) traveling at 25 knots with a crew of 35. Mathematically if we multiply the cargo capacity by the speed and divide by the number of men in the crew we get ton-knots per man.

$$\text{Old ship} \qquad \frac{10,000 \times 15}{45} = 3,333 \text{ ton knots per man}$$

$$\text{New ship} \qquad \frac{20,000 \times 25}{35} = 14,285 \text{ ton knots per man}$$

Thus we see that one new vessel can accomplish approximately four times as much as an older vessel.

In order to operate a larger vessel it becomes necessary to install a more powerful engine that can be operated by a smaller crew. The only way to accomplish this is to automate, which means to install equipment to do some of the work that would ordinarily be done by the engine room crew including engineers, firemen, watertenders, oilers, etc.

Shore plants were automated years ago. Then the question arises—why were ships so long in becoming automated? The answer is quite simple. Shore plants operate at a constant speed with small variations in load changes. A ship's propulsion plant is quite the opposite. The plant is operated at various speeds when the ship is being maneuvered also during which time the main propulsion unit is often reversed. It may also have to operate at slow speeds for long periods of time. It can readily be seen that under such circumstances the ship's plant cannot be compared with the shore plant. When the bridge control is moved, a complex chain of events takes place. Steam flow to the turbine must be adjusted. The fuel oil pressure must be adjusted accordingly which also means that the air pressure (forced draft blower), fuel oil temperature, etc., must also be adjusted. The speed of the feed pump must be changed and the vacuum must be maintained. All these things must be carefully planned out so that the plant will continue operating at its best efficiency and still not become involved with continual troubles which would necessitate change to manual control.

An automated system must have three essential characteristics. These can be compared with human capacities as shown in Fig. 1.

The nerve center of an automated engine room from which the engineer controls, interprets and takes corrective action is in a central location called the Central Operation Station (COS).

Ability to make decisions	Ability to carry out these decisions with physical action	Ability to evaluate and correct its own performance
Brain	Muscles	Nervous system
Computer	Mechanical devices	Feedback

Figure 1

Figure 2 represents a general schematic idea of a COS. The COS permits a one man operation of the power plant through bridge control of the propeller from full ahead to full astern, a centralized area for observing and operating the major components of the system and a simplified plant operation.

The bridge console contains very few devices, only the essential displays and a single operating lever to maneuver and communicate with the engine room.

The heart of the COS is the engine room console. It centralizes all normal watch duties. Information displayed on the usual three main sections of the console—propulsion—boiler and services—is presented to the engine room watch stander in three forms-

1. **Continuous or demand display**
2. **Monitor and alarm**
3. **Logged data on a typewriter print-out**

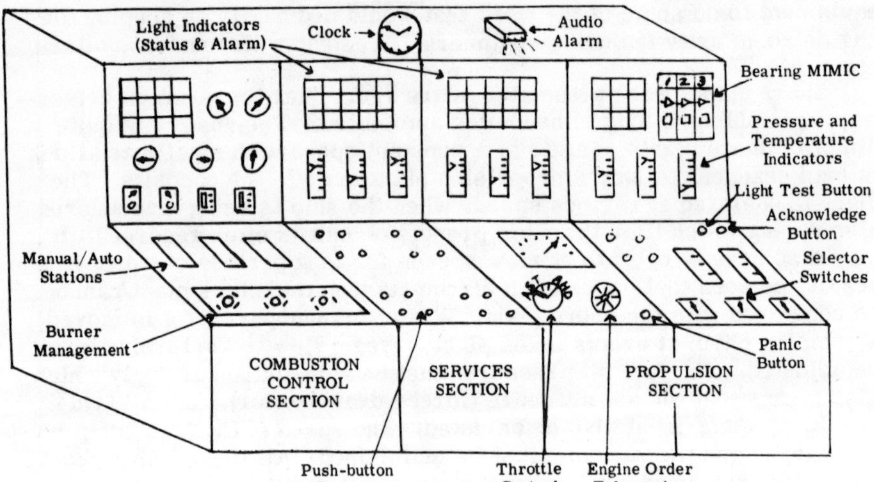

Figure 2

Figure 3 is a basic block diagram of the engine room and bridge console control.

Besides the engine-room console and the bridge console the COS includes the necessary sensors (such as thermometers, pressure gages, level indicators, etc.); transducers (devices for transmitting temper-

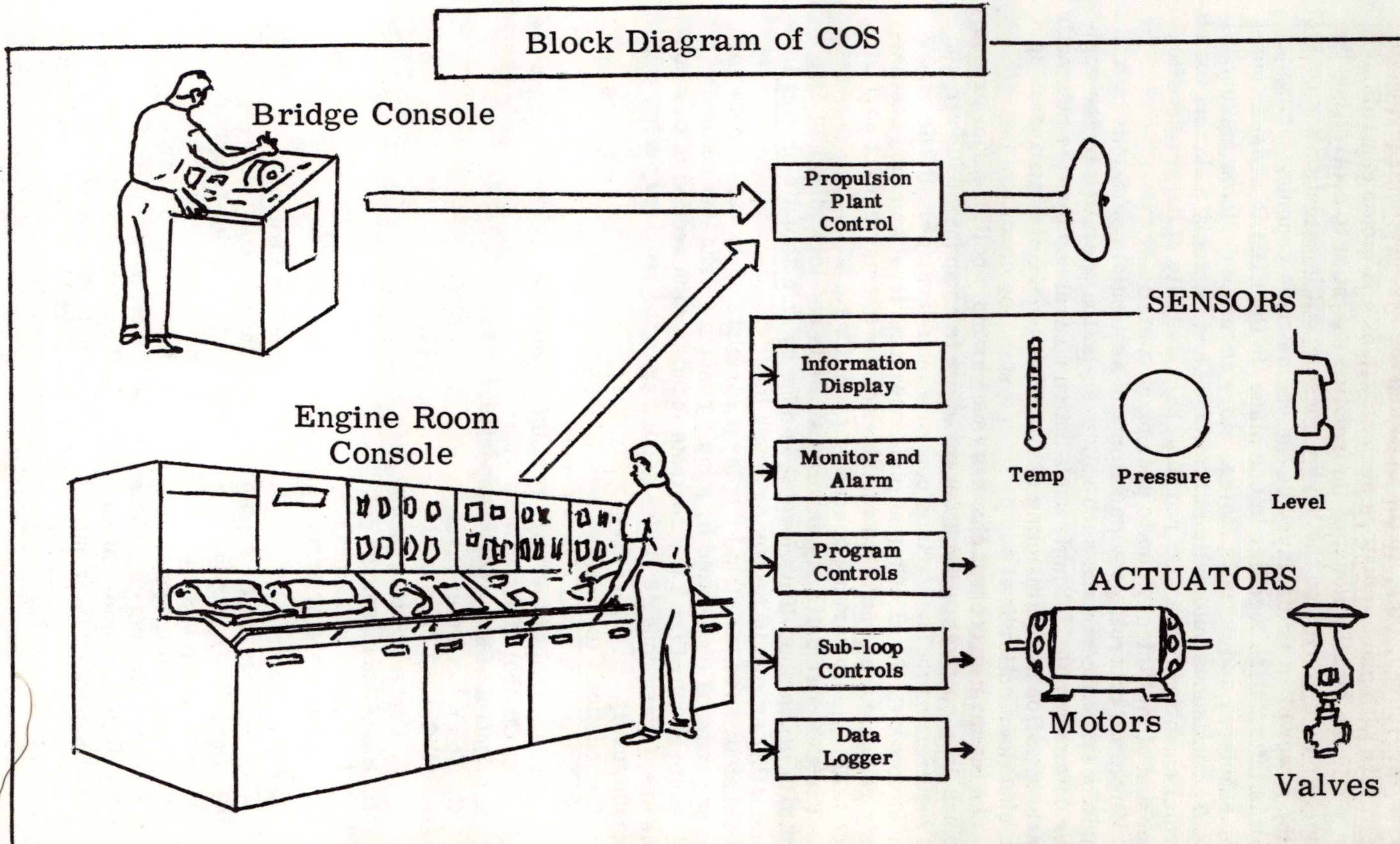

Block Diagram of COS
Bridge Console
Propulsion Plant Control
Engine Room Console
SENSORS
Information Display
Monitor and Alarm
Program Controls
Sub-loop Controls
Data Logger
Temp
Pressure
Level
ACTUATORS
Motors
Valves

Figure 3

atures, pressures, etc., to the COS) and the systems engineering to coordinate with plant actuators such as valves and motor controls.

While the small amount of information essential to controlling the propeller is displayed on the bridge console, all information for control of the power plant is displayed on the engine-room console. Bringing all this information together in one place (in the COS) increases operator efficiency and permits faster, more accurate decision making.

By continuous automatic surveillance and immediate alarms for off-normal conditions, the COS reduces the possibility of dangerous situations and of inefficiency due to poor plant operation.

Program controls provide automatic sequential operations by a single operating command such as a push-button operation by the engineer. For example, a single push-button control can provide automatic synchronization and paralleling of a turbine generator. Also start up and shut down boiler feed pumps from a stand-by condition.

As the master control changes (such as main boiler steam pressure) the sub-loop controls will cut out or adjust to compensate for air and oil requirements. The various controls (sub-loop) permit the engineer to set the most efficient plant operating level and then continuously regulate the controlled variable to provide stable plant performance under widely varying operating conditions.

The COS automatically records all bells and logs all other important data of the plant. This logging is done by an automatic printer thus eliminating errors in recopying rough logs.

In addition to the visual display a complete up-to-the-minute log can be produced on demand by simply pushing a print-out button.

Referring to Fig. 2 the controls found in each section of this console are: (every console varies according to the required design and manufacturer)

Combustion Control

A. Control of main steam pressure
   1. Control of fuel-oil pressure to burners
   2. Control of air to burners
B. Control of superheat temperature
C. Control of water level
   1. Control of feedwater regulating valve
   2. Control of feed pump speed
D. Burner management
   1. Control of burners
   2. Fail-safe protection

Services Control

A. Ship's service air supply
B. Control air supply
C. Fire pump system
D. Fresh water system
E. Salt water service system
F. Generator information
G. Evaporator information

Propulsion Control

A. Throttle control
B. Engine order telegraph
C. Shaft speed tachometer
D. Main lube oil pumps
E. Monitoring systems
F. Temperature controls
G. Main vacuum control
H. Condensate system control

CONSOLE DEVICES

Indicating Lights and Alarms. These lights are red and will light up if an alarm condition exists. They will flash on and off until the operator acknowledges by throwing the acknowledge button. When he does this the light will remain lit until the operator remedies the condition at which time the red light will go off and the green light will go on. When the alarm conditions exists an alarm will sound. This alarm stops when the acknowledge button is thrown.

Acknowledge Button. Depressing this button will silence the audio alarm and stop the indicating light from flashing. Do not depress the acknowledge button until you know what is alarming.

Status Indicating Lights. These are lights that indicate whether a piece of machinery is running or stopped or if a valve is open or closed. They do not flash. No alarm is sounded. Generally the running or open condition is a green light and the stopped or closed condition is an amber light.

Push-button Stations. These are start and stop positions for motors or open or closed positions for motor operated valves.

Selector Stations. These are switches which permit a variety of selections which the operator may require in order to keep the plant functioning. For example—if the main circulator breaks down the operator can switch to a fire pump or general service pump. If the #1 fuel-oil pump stops the operator can switch to the #2 pump, etc.

Continuous Display Indicators. These are pressure, temperature, liquid level, liquid flow or position indicators that show the varying conditions at the COS at all times.

Demand Display Indicators. By depressing a button or dialing an address the information requested is displayed at the COS.

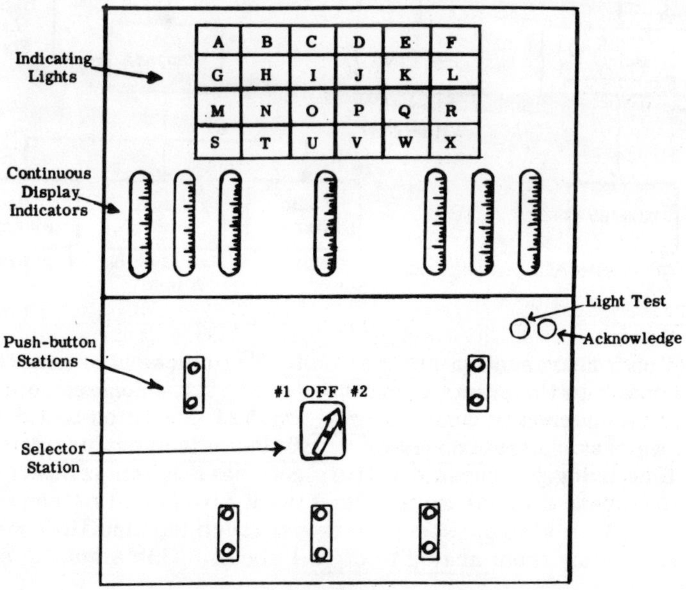

Figure 4

Figure 4 represents a typical service section on a COS. The boxes lettered A to X each represent a service unit. Suppose A and B represented the two salt-water pumps serving the sanitary system. If pump #1 (A) was running a green light on the lower section of the board (not shown) would be lit. If this pump stopped, the light in section A would start blinking and the audio alarm would sound. The operator would look to see which indicator was at fault. He would depress the acknowledge button to stop the audio and visual signals. He would then push the start button for pump #2 and the off button for pump #1.

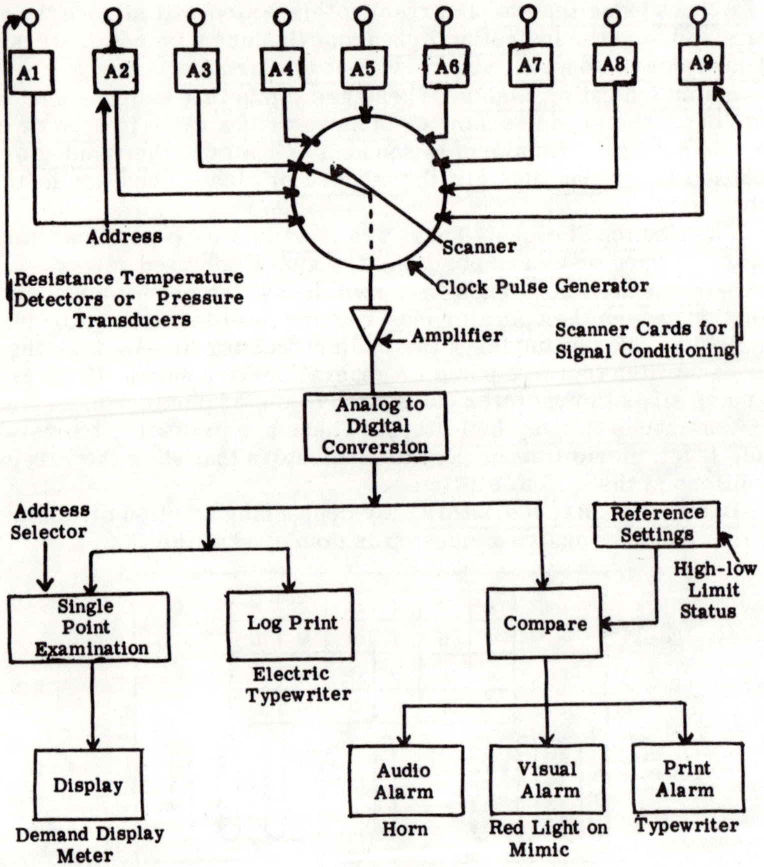

Figure 5

Figure 5 represents a scanner unit. The clock-pulse generator makes contact to the points being monitored. Each contact point is connected to an address or scanner card which is connected to a detector such as a pressure transducer or a resistance temperature detector (wheatstone bridge). These detectors note bearing temperatures, various pressures, etc. At each contact point a reading is given on the scanner card. It also passes a current through the amplifier and converts the reading from analog to digital signal. This signal is sent to a comparing unit. For example—suppose contact #1 is connected to the forward H.P. turbine bearing for a temperature checkout. The comparing element is set for the prescribed temperature—let us assume this to be 100°F. If the incoming reading is higher than this preset amount, an audio alarm and visual alarm are set off and a print alarm types out the reading and the time. If at any time the engineer wishes to check any single point in the system, he uses the demand display.

In the old engine room the fireman-watertender had to maintain the proper water level in the boiler. He did this manually (by hand) by closing or opening the boiler feed-check valve (controller) and thereby increasing or decreasing the flow of feedwater to the boiler.

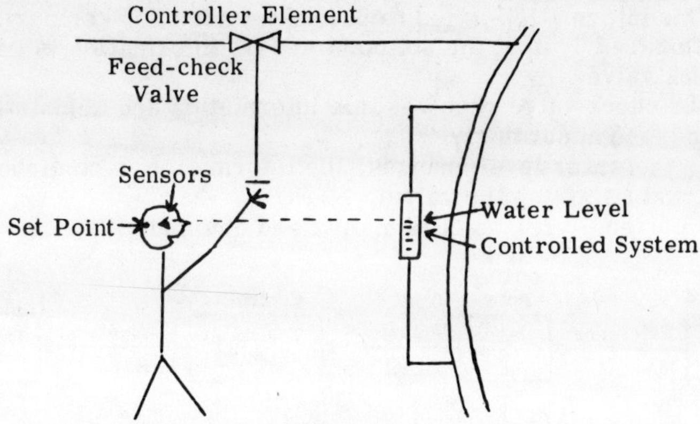

Figure 6

In Fig. 6 the fireman-watertender looks at the water level in the gage glass (controlled system). His eyes are sensors. His eyes send word to the brain (mixing point) which sends word to his hands (actuator) to open or close the check valve (controller) accordingly.

The greatest disadvantage of this manual type control is that the fireman-watertender cannot sense small variations in the water level and his manual control of the water level will vary greatly. These variations in plant control are inefficient and represent an actual loss of money.

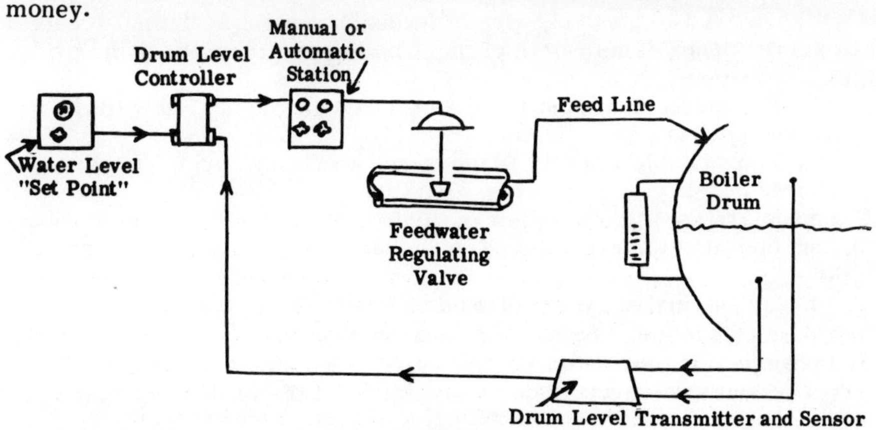

Figure 7

The automatic feed systems such as the Bailey or General units were a big improvement over the manual type of operation. These units can maintain a prescribed water level; however, if the feed pump should stop there is no alarm or immediate indication to warn the engineer of a serious situation. This situation is eliminated in the automated system.

Feedback. As the name implies—information is fed from the device being controlled back to the device which will make a correction in order to maintain the desired level of operation, proper temperature, pressure, etc.

Automatic Control with Feedback. Figure 7.

1. A level sensor interprets the water level in the boiler.
2. This information is fed back to a set point for comparison. Here it is compared against the set point and its information is relayed to the check valve.
3. The check valve receives this information and adjusts itself to open or close accordingly.
4. As the water level changes, the information is continually adjusting and trimming the valve.
5. A perfect water level is maintained automatically.

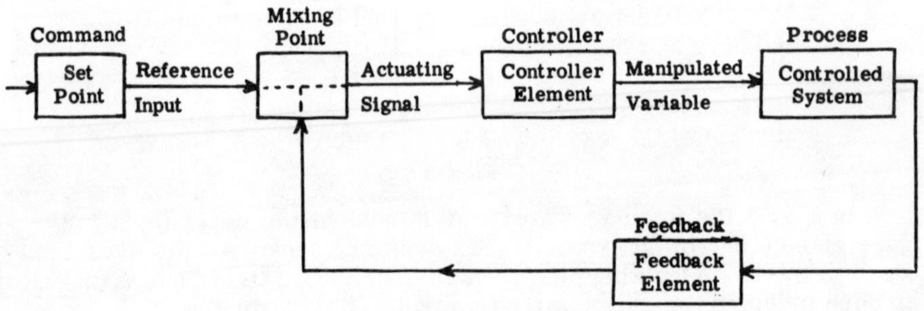

Figure 8

Figure 8 shows in block form how the feedback forms a closed-loop chain of action between the output of the Controlled System and the input to the control element.

There are two general types of feedback control systems—regulators and servo-mechanisms. Both of these units operate on the same principle:

1. The mechanical position is measured and compared with the desired position.
2. The variable quantity is measured and compared with the desired quantity.

If the two are not alike in either of the foregoing conditions the feedback system operates the control system and adjusts them to make them alike.

Every automated system should have a back-up system (usually manually controlled). Know your back-up systems; they are important. For example—if your automatic feedwater regulator became inoperative, your back-up system would be the operation of the feed-check valve by hand.

At this point let us go a little further into the workings of sensing devices. They are the heart of any automated system. They form the first link in a series or chain of control of equipment. It would be very difficult to bring information from source of origination to the COS if it were not for sensing devices.

In measuring a physical property we can use a thermometer to register temperatures, a tachometer to register speed or a Bourdon tube to register pressure. The operation of any of these instruments can be used to register directly on a recording device or an indicator. They also perform these operations much faster and more accurately than the human sensors—sight, feeling, etc.

Most sensing devices usually provide for:

1. Movement of a mechanical member.
2. Change in a force or pressure.
3. Change in an electrical resistance.
4. Change in light intensity.
5. Change in conductivity.

These sensors observe what is happening. They check or note plant conditions such as pressures, temperatures, etc.

One of the most commonly used mechanical devices for measuring pressure is the Bourdon tube. The free end of the tube is usually attached by gears and linkage to the pointer to show pressure. However, it can also be attached by means of linkage to:

1. A pressure switch to start or stop a pump.
2. A modulator to vary the speed of a pump.

SUMMARY TO THIS POINT

The input or command is changed to the reference input by the reference elements. A portion of the controlled variable is converted to the same type of signal as the reference input by the feedback element. The corrected signal from the controlled variable is the feedback signal. The reference input and feedback signal are compared in the mixing point to produce the actuating signal which tells the control element what to do. The control element is the device for changing the manipulated variable which governs the operation of the controlled system. The output of the controlled system is the controlled variable.

Thus, all closed-loop systems are basically alike because they contain the same basic functional elements. The system must contain some means of measuring the output of the system and a means of comparing the actual output with the desired output. The results of this comparison are used to determine what changes must be made in the operation of the system. Certain pieces of equipment must be used in the system to make any corrective actions that become necessary.

Inasmuch as various temperatures and pressures must be transmitted to the COS, some mechanical devices are necessary. A pressure transducer is a sensor which converts a pressure, temperature or liquid level into an electrical current. This current is then transmitted to an ammeter which is graduated in PSI and thereby displays information as a pressure indicator.

The trade name used by the General Electric Co. for their pressure transducers is GEMACS (General Electric Measurement and Control).

The principle of operation of a transducer is shown in Figs. 9A, B and C.

Figure 9A shows the COS. It is necessary for the engineer to know what steam pressure he is carrying in the main steam line. This pressure is transmitted from the pipeline to the COS by a pressure transducer.

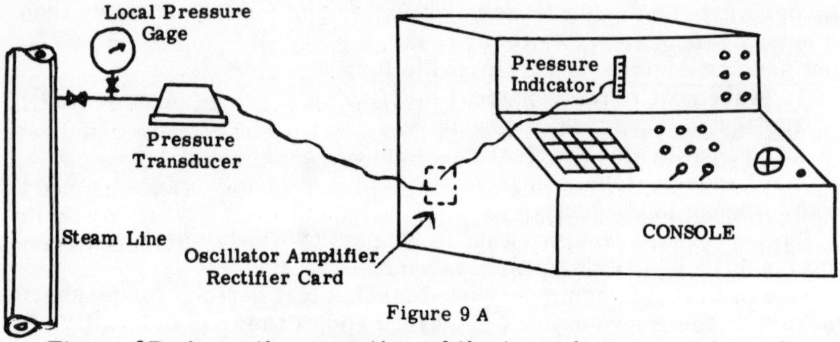

Figure 9 A

Figure 9B shows the operation of the transducer.

1. The mechanical force of the pressure bears down on a balancing beam.
2. The movement of the beam changes the air gap of the position error indicator.
3. The change in the air gap will change the amount of inductance of the electromagnet.
4. A change in inductance will vary the impedance of an oscillator circuit.
5. The oscillator output is enlarged in an amplifier.
6. The amplifier output is rectified to DC current by use of diodes.
7. The rectified current is in series with a coil in the null balance motor.
8. An increasing pressure will cause a greater current to flow through the coil, pulling the beam down and thereby balancing it.
9. The current is transmitted to an ammeter.
10. The ammeter is graduated in PSI and displays the signal as a pressure indicator.

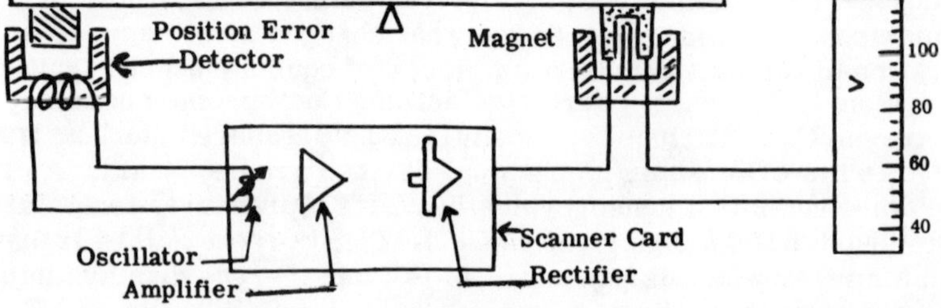

Figure 9 B

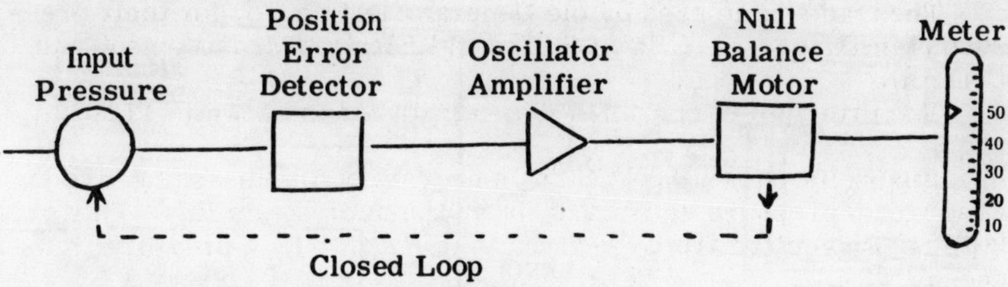

Figure 9 C

Figure 9C shows the system in block form.

A GEMS Level Switch converts a liquid level to an electrical signal. This unit shown in Fig. 10A consists of a hermetically sealed glass-enclosed magnet switch which is placed in a brass tube enclosure. A round plastic float with a permanent magnet enclosed, floats around the brass tube and rises and lowers with the liquid level. When the liquid level rises, the plastic float moves up and the magnet attracts the contacts and makes or breaks the circuit.

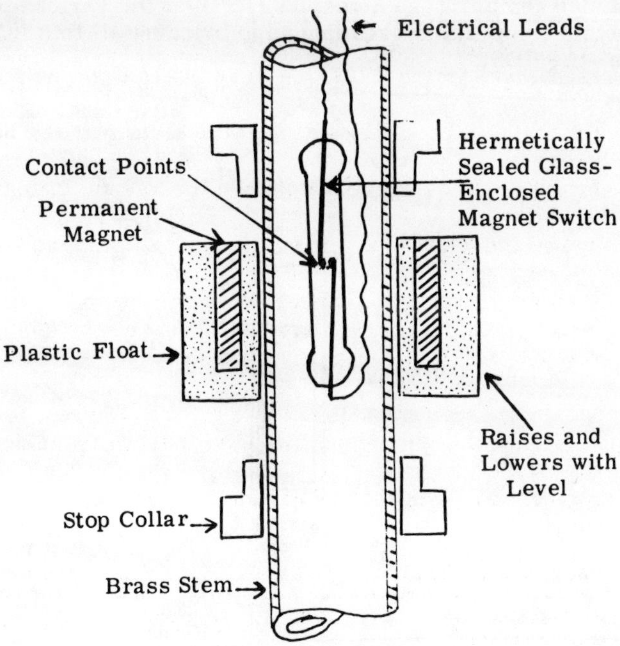

GEMS LEVEL SWITCH

Figure 10 A

Figure 10B shows a GEMS level switch located in the atmospheric drain tank and the Logic Annunciator Card attachment to the COS.

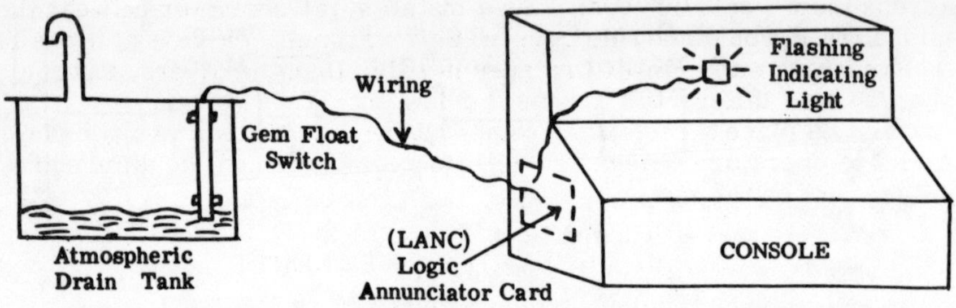

Figure 10B

REMOTE TEMPERATURE DETECTION

Whenever we wish to determine a temperature locally we may use a mercury thermometer or a dial thermometer. However, this would require walking around the entire fire and engine room in order to watch, log and correct any faults. In order to automate we need a means of gathering all of these readings in one place. This is accomplished by using resistance temperature detectors (RTD) at the various places one wishes to obtain temperature readings and bringing all this information to one place—the COS.

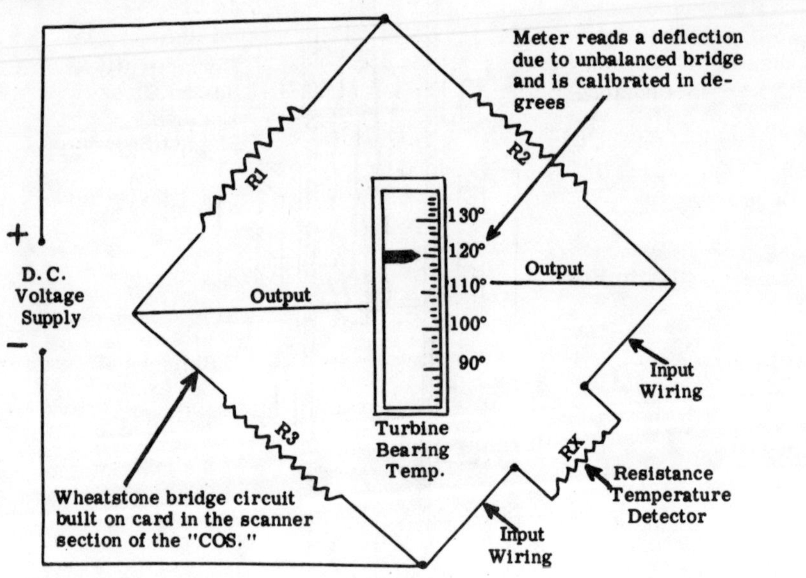

Figure 11A

Certain metals have a characteristic that as their temperature changes their resistance changes at a fairly uniform rate.

The Wheatstone bridge, Fig. 11A, is an electrical configuration consisting of four resistors. R1 and R3 are connected in series. R2 and RX are connected in series. R1 and R3 are in parallel with R2 and RX. If all four resistors were equal in value, then the voltage drop across them would be equal. If we install a galvanometer between their midpoints it would show no current flow or a null. However, if one of the resistors was greater or lower in value than the others, it would show up as a differential across the now unbalanced circuit.

Let us place a resistance temperature detector in the place of RX. As its temperature changes so will its resistance and the differential will reflect on the meter.

ASSEMBLY OF COMPONENTS

1. Resistance temperature detector (RX)
2. Scanner card (Wheatstone bridge)
3. D.C. electrical supply (from COS)
4. Galvanometer (temperature indicator)

Generally the material used for RTD's whose temperatures are below 300°F. is Nickel. Platinum is generally used for temperatures of 0° to 1000°F.

Adjustment. In order to properly balance the bridge and allow for the length of the leads in the cable between the RTD and the card, potentiometers or variable resistors are installed on the card.

Figure 11B shows some of the instrumentation and control symbols used in an automated plant. Planning the design for an automated plant is a tremendous job. Hundreds of these symbols are used to locate temperatures, pressures, etc., at various points in the plant. All of these devices must be correlated and brought back to the COS where the engineer can see what is happening at all times.

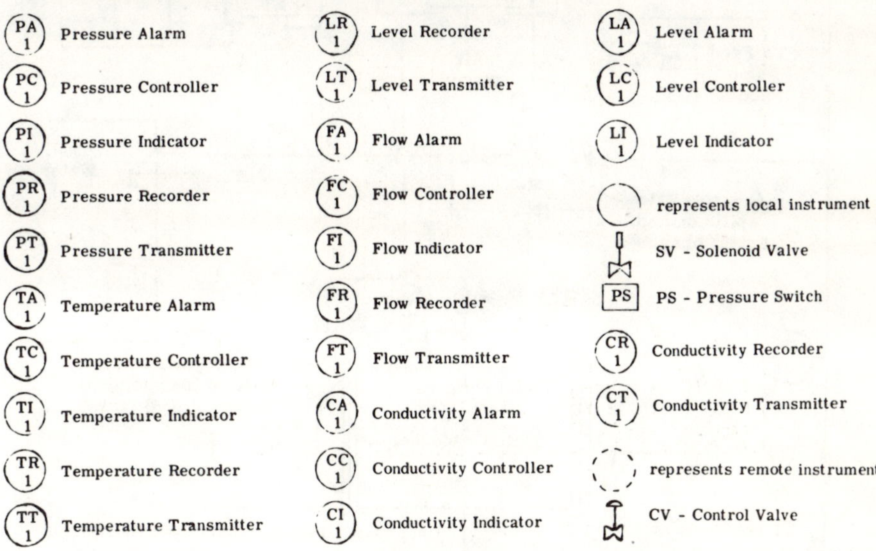

Figure 11B

Figure 12 shows a RTD connected to a main bearing to give a temperature reading at the COS.

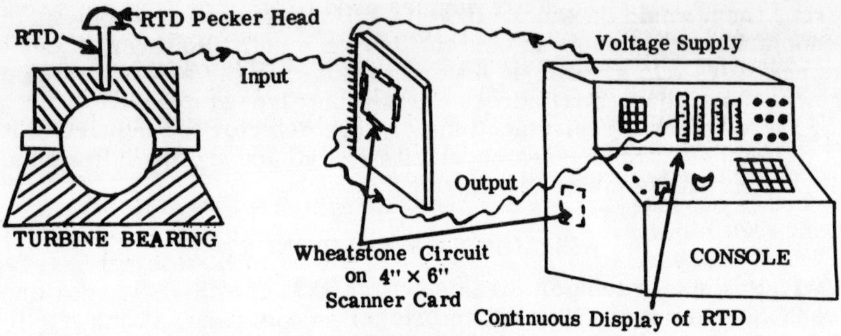

Figure 12

Figure 13 shows the control locations of level transmitters and controllers, temperature and pressure indicators, etc., as connected to a deaerating feed tank.

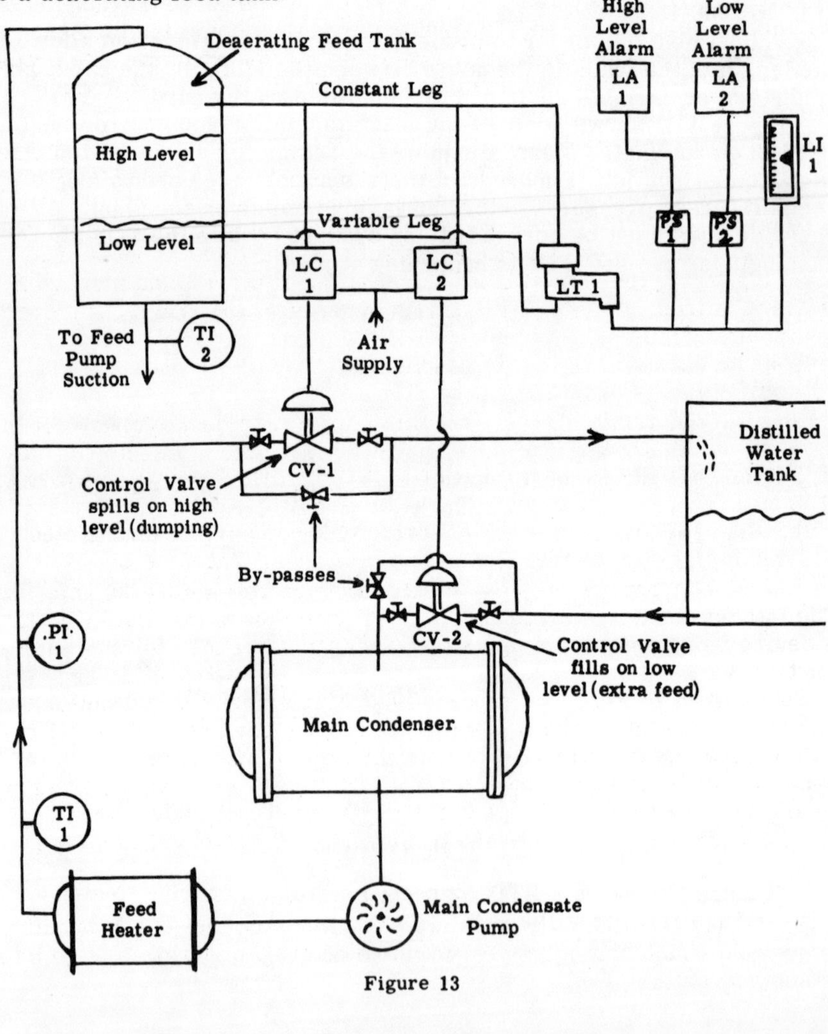

Figure 13

Computer Logic Gates are devices used to start or stop sequences of machinery operation under certain required conditions. Figure 14A represents an AND gate as shown on an automated blueprint. Figure 14B shows this gate as made into an electrical circuit simply for the purpose of explaining its operation. From this diagram it can be seen that both switches A and B must be closed (series circuit) in order to produce a closed circuit and give results at C.

Let us suppose that A represents the switch that operates the forced draft blower and B the switch that controls the water level in the boiler. C represents the fuel oil valve to the burners which opens when energized (when switches A AND B are closed). If either switch A or B is open, C will not energize. If the blower is operating, switch A will be closed. If the boiler water level is correct, switch B will be closed. We have now completed the circuit; oil valve C will open and allow oil to flow to the burners. We would not want oil to flow to the burners if either switch A or B were open.

In an automated system there are no actual physical movements of switches as just described. Diodes are used to accomplish the same results (see dictionary section). The diode steers the current to proper channels. When a computer is designed these diodes either produce electric flow or no electric flow.

There are two aids that are always used in conjunction with logic gates; they are Boolean Symbols and Truth Tables. The Boolean symbol for an AND gate is a dot (•). Thus A•B means A AND B. (Figure 14A)

Figure 14C is the Truth Table for an AND gate. If there is no input into either A or B then there is no output at C. If there is input at A but no input at B there is still no output at C. If there is no input at A but there is input at B there is still no output at C. However, if there is input at A and input at B then there will be output at C. Comparing these Truth Tables with the electrical symbol diagram should make this quite clear. A AND B must be 1 in order that C be 1.

Figure 15A represents an OR gate. The Boolean symbol for OR is (+). Figure 15B shows this type of gate as made into an electrical circuit. From this diagram we can see that the closing of either switch A OR B will complete the circuit (parallel circuit). Again, the Truth Table, Fig. 15C can be clearly understood if compared with the electrical symbol diagram. A OR B must be 1 in order that C be 1.

Figure 16A represents a NOT gate. Figure 16B shows the gate as made into an electrical circuit. The NOT gate means that the input to the device being operated must not be energized in order to have an output.

Suppose FS in Fig. 16B represents a Flame Scanner in a boiler. When the scanner sees the flame the normally closed contact (NCC) remains closed and the oil valve (OV) in the second circuit being in a closed circuit is kept open by energy. If the flame goes out the scanner de-energizes, the NCC will open, the switch in the oil valve circuit will open breaking the electrical circuit. The oil valve will close on spring pressure.

Figure 16C is the Truth Table for a NOT gate.

There are also NAND gates and NOR gates (see dictionary at end of this section) which can be studied in more detail if one wishes to go into this subject matter.

It is often necessary to use various combinations of these gates in order to obtain the required results for an automated system.

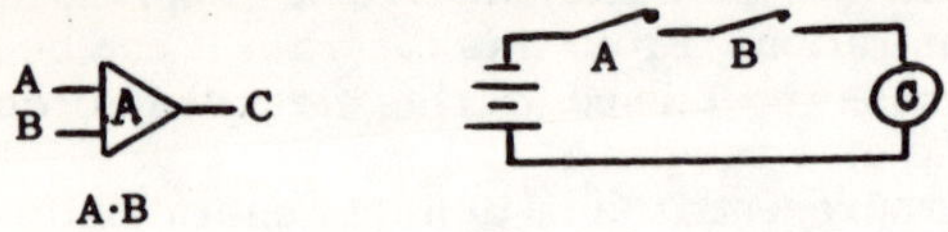

(A) (B)

INPUT		OUTPUT
A	B	C
0	0	0
0	1	0
1	0	0
1	1	1

Truth Table
(C)

Figure 14

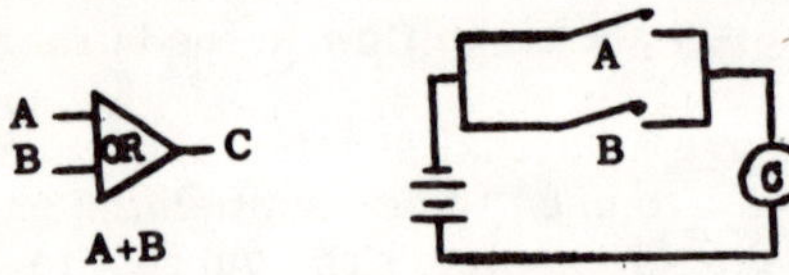

(A) (B)

INPUT		OUTPUT
A	B	C
0	0	0
0	1	1
1	0	1
1	1	1

Truth Table
(C)

Figure 15

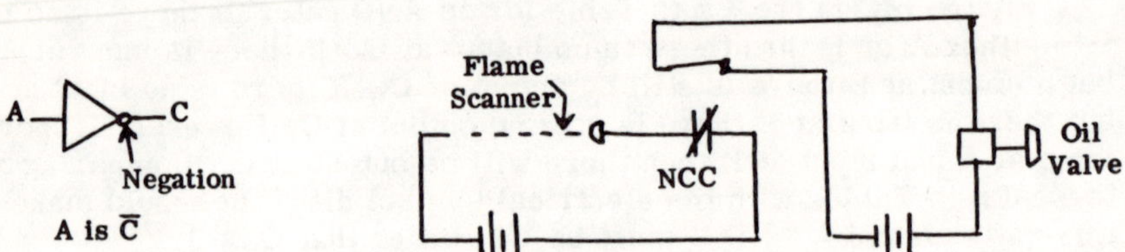

(A) (B)

INPUT	OUTPUT
A	C
1	0
0	1

Truth Table
(C)

Figure 16

MULTIPLE-CHOICE QUESTIONS & ANSWERS

QUIZ ON AUTOMATION

1. The nerve center or brains from which the engine room is controlled is called the:
   - A. Computer Room (CR)
   - B. Main Engine Station (MES)
   - C. Central Operating Station (COS)
   - D. Power Plant Station (PPS)

2. The entire housing which contains all the display information is called the:
   - A. log desk
   - B. console
   - C. burner housing
   - D. controller

3. A modern automated ship uses:
   - A. more men and has more speed than an older ship
   - B. less men and has less speed than an older ship
   - C. less men and has more speed than an older ship
   - D. more men and has less speed than an older ship

4. When something goes wrong in an automated plant:
   - A. an alarm sounds
   - B. an alarm sounds and an indicating light will flash
   - C. a light will flash
   - D. the fires will go out

5. A 4" × 6" board containing an arrangement of electronic logic circuits is called a:
   - A. card
   - B. deck
   - C. page
   - D. computer

6. In order to silence an alarm one must depress or throw the:
   - A. stop button
   - B. push button
   - C. alarm button
   - D. acknowledge button

7. A type of computer that uses continuous signals and gives continuous information is called a:
   - A. single computer
   - B. analog computer
   - C. digital computer
   - D. continuous computer

8. A closed loop chain of action between the output and the input of a controlled system is called:
   - A. double back
   - B. feedback
   - C. continuous loop
   - D. series control

9. In a controlled system another name for the command signal is:
   - A. set point
   - B. mixing point
   - C. control point
   - D. variable point

10. If a controlled system fails to operate a manual means may be used; this is called the:
   A. #2 system
   B. auxiliary system
   C. back-up system
   D. re-set system

11. Ship propulsion plants were longer in being automated (compared to shore power plants) because:
   A. shore plants are cheaper to install and operate
   B. ship plants are more complicated due to required variety of operating needs
   C. operating personnel were hard to train
   D. all of the above

12. The part of the human body that corresponds to the computer in an automated plant are the:
   A. hands
   B. ears
   C. eyes
   D. brains

13. A transducer is a device which:
   A. changes resistances in circuits
   B. transmits information to the bridge control
   C. converts a physical measurement to an electrical equivalent
   D. any of the above

14. Valves and motors that are controlled automatically are called:
   A. actuators
   B. sensors
   C. transducers
   D. none of the above

15. The automatic printer:
   A. records all telegraph bells
   B. logs important data (temperatures, pressures, etc.)
   C. eliminates errors in recopying rough log
   D. all of the above

16. An instant reading can be produced on demand by:
   A. using the typewriter
   B. dialing an address
   C. pushing the print out button
   D. B or C

17. The monitoring system that tests the plant for readings of temperatures, pressures, etc., is called a:
   A. demand display unit
   B. scanner unit
   C. comparing unit
   D. loop back

18. A sensing device usually provides for a measurement of:
   A. pressure
   B. temperature
   C. liquid level
   D. all of the above

19. One of the most common devices used to transmit a mechanical movement is a:
   A. Alpha tube
   B. Bourdon tube
   C. Venturi tube
   D. Vacuum tube

20. When a system has a means of comparing the actual output with the desired output and the results of their comparison are used to determine corrective changes in the system it is referred to as:

A. a continuous system C. a closed loop system
B. an open loop system D. any of the above

21. RTD stands for:

A. Return Time Delay
B. Resistance Temperature Drop
C. Resistance Temperature Detector
D. A or B

22. The metals used as a linear resistance in a Wheatstone bridge, for temperature indicating results are:

A. silver and brass C. nickel and platinum
B. bronze and brass D. vanadium and stainless steel

23. The reference input and feedback signals are compared in the:

A. controller B. mixing point C. set point D. process point

24. CV stands for:

A. Control Vent C. Console Vent
B. Control Voltage D. Control Valve

ANSWERS TO QUIZ

1. C	4. B	7. B	10. C	13. C	16. D	19. B	22. C
2. B	5. A	8. B	11. B	14. A	17. B	20. C	23. B
3. C	6. D	9. A	12. D	15. D	18. D	21. C	24. D

DICTIONARY

FUNDAMENTALS OF AUTOMATION

Acknowledge. The silencing of an Audio Alarm after the engineer has determined the malfunction. This is accomplished by depressing the acknowledge pushbutton for that section of the Console that is displaying the warning.

Actuating Signal. An electrical or Pneumatic signal whose strength is determined by the Set Point and Feedback. This signal carries information to the control device to regulate its position.

Actuator. An actuator is a device that upon receiving a control signal, can carry out the necessary operations of a control system. The signal allows the actuator to produce a linear or rotary mechanical movement. Examples of actuators are power cylinders, electric motors, diaphragms,etc.

Address Selector. Consists of rotary switches, which the engineer dials, and thereby selects an address. The address can be the location of a specific bearing temperature. He will then have a Digital Readout on a Demand Display Meter.

Amplifiers. By use of vacuum tubes or transistors, low level signals are enlarged or increased in size to a higher level or higher voltage. A ten-millivolt signal amplified 100 times will produce a one-volt signal.

Analog Computer. A type of computer that has the ability to use electrical circuits that are the equivalent of a mathematical formula. This causes a process to react according to the input. The electrical circuits are analogous to Physical Laws. The inputs are usually an electrical voltage of variable intensity. The output can control a process and also display information on a sliding scale. The analog allows for continuous calculation.

Analog. Digital Conversion. Analog to Digital conversion is that section of a computer capable of taking an Analog signal such as a voltage and converting that voltage into a given quantity of Digital Units. From this point on the signal is Digital and can be treated with Digital Logic.

Annunciator Card. An Electronic Computer Card whose function is to set off the Alarm light on the flashing bus and trigger the Audio circuit upon receiving a confirming signal.

Attemperator. A desuperheater located in the steam drum of a boiler. It is used for controlling superheat temperature. (See Superheat Temperature Control.)

Audio Alarms. Various types of noise signals used to alert the engineer to a condition in the engine room, i.e.:

1. Central Operations Console Alarm
2. Combustion Control Console Alarm
3. Engine Order Telegraph Alarm
4. Power Failure Alarm
5. Dial Telephone Alarm
6. Sound Powered Phone Alarm
7. Smoke Detecting Alarm
8. General Alarm
9. Cargo Vent Failure Alarm
10. Carbon Monoxide Alarm
11. CO_2 Release Alarm

These consist of buzzers, gongs, bells, sirens, horns and klaxons, each having a distinct sound.

Automation. Any or all of the combined processes used aboard ship that have the ability to control work, notify the operator of its status and warn of malfunctions. This allows less people to control more machinery in a more accurate and efficient manner.

Automatic Control. A system in which the value of a process is compared against a desired value and corrective action taken to correct the deviation without the use of human help.

Autorotation. A signal system used in throttle control. Following a 120 second count with the propeller shaft stopped, a signal goes out to crack the throttle and rotate the shaft. As soon as the shaft moves, the throttle shuts off. When the shaft stops again, the count starts again. This system prevents warping of the shaft during the periods of maneuvering mode and is especially useful when in Bridge Control.

Availability. Prior to the process of lighting-off a boiler burner, certain conditions must be proven before permitting light-off. These are:
1. prove burner is in place
2. prove minimum atomizing steam
3. prove minimum of fuel oil pressure
4. prove minimum of fuel oil temperature

If the above conditions are met the burner indicator will read Burner Available.

Averaging Relay. A pneumatic relay receiving two different inputs and whose output is the average of both units. (See Totalizer.)

Back Up. A provision in any automated system that will permit the operator to use a valve or control, or another system in case of a failure of the system in use. An engineer should try to learn as many Back Up systems as possible in the event of a malfunction.

Band Width. Represents the sensitivity of the controller. (See Proportional Band.)

Bellows. Used extensively in pneumatic control devices, the basic principle being that an increasing signal will force open the accordion-like leaves of a bellows which will move a beam. The movement of the beam toward or away from an air restriction nozzle will cause supply air to bleed to a greater or lesser degree to the atmosphere resulting in a proportional signal pressure.

Binary Numbers. A mathematical system where each column of numbers doubles in value as compared to our decimal system where each column increases by a factor of 10. This system permits the use of ONES and ZEROS to state a number and forms the basis of Digital Logic.

Bits. In a digital computer the voltage signals in the form of square waves travel through circuitry and represent digits or quantities for mathematical purposes. (Bits is an abbreviation for BInary digiTS.)

Bridge Control. A throttle control system whereby a throttle lever connected to a potentiometer can be used for controlling the speed and direction of the main propulsion turbine through the sophistication of an Analog Computer. The procedure for using such a control is for the bridge to ring BRIDGE on the engine order telegraph. The wheelhouse then adjusts the throttle to match the Valve Position Indicator. The engine flips the Throttle Mode Selector to Bridge. With the bridge throttle now taking over and being in the matching position, there will be a Bumpless Transfer and no change in engine speed during transition. The engine throttle will track any bridge throttle movements so that the engineer is aware of their movements. In case of a problem the engine room can recapture control at once by flipping the Mode Selector to Engine. No matching is required since the engine throttle is always tracking.

Brushless Exciters. A modern system that uses silicon controlled rectifiers to convert AC into the proper amount of DC for excitation of large AC generators. When producing DC by means of rectification instead of with an exciter, brushes are eliminated.

Bumpless Transfer. When an operator wishes to transfer from Automatic Mode to Manual Mode he matches the present automatic signal going to the controller and then switches to Manual. The controller will not see any difference. He may now elect to lower or raise the signal on a manual or open loop basis. (See Bridge Control for bumpless transfer of the throttle.)

Burner Management. The entire combination of electrical and pneumatic controls and circuitry involved in controlling, firing, protecting, indicating and warning and shutting down of an automated burner system. (See Purge, Flame Scanner, Low Firing Position, Timers, Permissives, Availability.)

Cards. A printed wiring board approximately 4 by 5 inches in size, containing electronic components such as resistors, transistors, capacitors and potentiometers. The end of the card has 50 contact points for external connections to other Cards and Terminal Boards. The Cards are modular in concept and may contain from two to sixteen similar circuits per Card. The Cards have a copper laminate on one or both sides and are made of glass epoxy to resist moisture and warping. The pins are tin plated for proper electrical contact and corrosion resistance. After component assembly, a conformal coating such as silicon is applied to protect the Card against dirt, moisture, fumes and other environmental contamination. Hundreds of these Cards are installed in a row and column manner to form a PAGE and are the Logic Elements or BRAINS of a computer.

Cascade Control. A control system in which one controller alters the set value of one or more other controllers. An example is a Boiler Master Controller which receives a signal input proportional to Boiler Steam Pressure. It then sends out signals to the Air Damper Controller and the Fuel Oil Regulator Controller.

Central Operating System. Usually referred to as the COS. The concept of designing and operating a marine power plant from a central station containing all the necessary devices to allow the engineer complete control of the plant.

Chopper. A device for periodically interrupting a DC signal thereby making it into a square wave signal so that it may be amplified in an AC amplifier. The chopper can be accomplished by a vibrating reed or a transistorized multivibrator circuit. After amplification it is usually rectified back into DC. This process of chopping eliminates DRIFT, an undesirable characteristic of a DC amplifier.

Closed Loop. In an automatic process the closed loop is formed by the Sensor which feeds back information of the final process. This Feedback or Closed Loop, returns into a Mixing Point or comparison point where it compares itself against a reference or Set Point signal. Any deviation between these two signals will alter or adjust the control mechanism.

Combustion Control. The entire process of controlling the steam pressure, superheat temperature, air flow, fuel flow, feedwater flow and other devices related to the production of steam for the power plant.

Command Signal. The Command Signal is the Set Point or Reference Input signal for the desired level the operator wishes an automatic system to maintain.

Component Failure. Generally in the assemblage of electronic components there is a high rate of failure. This is known as the high Infancy Mortality Rate. The manufacturers have a "burn in" period or "soak" test of 500 hours. During this test the mortality rate falls off rapidly and levels off to an extremely low failure rate. Surviving components will now survive for the full life expectancy. As the wearout period occurs, as in old age, the rate of failure increases rapidly. Drawing this curve graphically it would appear in the shape of a bathtub and is often referred to as the "bathtub curve."

Computer. In the pure sense of the word a computer is a machine that can calculate. In the practical sense we look upon the computer as being able to monitor a power plant, totalize, compensate, record, alarm, adjust, and feed information visibly, audibly and graphically. Computers are divided into two classifications. They are either ANALOG or DIGITAL.

Conductivity. The ability of a material to allow electrons to flow. In automation we use conductivity cells in certain controls and processes. For example:

1. oil detection of the contaminated system
2. salinity indication of feedwater system
3. evaporator malfunction and trip
4. total dissolved solids of boiler water

Console. The entire housing of the computers, controls, and visual signals from which the engineer controls the power plant.

Continuous Controller. A controller that is capable of modulating itself through its entire range in order to take corrective action.

Continuous Display. Any instrument that shows temperature, pressure, level, position, speed, etc., at all times.

Controller. The device in an automatic system that controls the final process.

Control Valve. Any valve in an automatic system that responds to a signal to control the flow of a medium that is required in a process. In instrumentation charts they are designated as CV.

Cybernetics. Taken from the Greek word meaning "steersman." Ampere later coined the word "cybernetique." Today's meaning is the study of Communications and Controls.

Data Logger. That section of a computer in which is mounted a typewriter whose keys are controlled electronically. The data logger is capable of producing a printed record of Off Limit Conditions. It can also produce a printed log of plant Sensors at predetermined intervals or on demand.

Dead Band. A zone within which a change of value of an Input Signal does not cause a change in the Controller. In throttle control this represents the amount that you can move the throttle ahead or astern and not cause the steam valve to move. If the dead band is too small, the steam valve may hunt. If the dead band is too large, speed regulation will be poor.

Decade Box. An electronic instrument consisting of precision resistors arranged in columns of 9 each. One row has 1,000 ohm resistors, the next row 100, the next row 10 and the final row 1/10 ohm resistor. By connecting them to switches the operator may select any amount of resistance and connect it into a circuit for the purpose of setting Transducers or Resistance Temperature Detectors. (See Zero Adjustment and Span Adjustment.)

Delta "P". This represents the pressure differential across a regulating valve. For example: if the feedwater pressure on the inlet side of a feedwater regulating valve is 900 psi and the pressure on the outlet side is 830 psi the Delta "P" is 70 psi. (See Differential Pressure Transmitter.)

Demand Display. A device on the console that does not give out information until such time as the engineer dials an address or pushes a button demanding the information. This allows for one device to gather information from many points into one unit.

Deviation. The difference between the Measured Value as sensed by the Feedback Loop and the Set Point.

Diaphragm Actuator. A pneumatic operator. It consists of a diaphragm operating against a spring. The diaphragm exerts a force on a push rod opening or closing a valve in proportion to the strength of the signal it receives.

Differential Pressure Transmitter. A transmitter capable of measuring a pressure drop across a valve. In controlling the speed of a feed pump the differential pressure transmitter senses a greater drop across the valve as it is closing. The valve is closing because the water level in the boiler is rising. The transmitter now signals the governor of the feed pump to slow down.

Digital Computer. A type of computer that works with numbers or digits by counting "Bits" of electrical pulses. It can perform mathematical problems fed into it. It makes use of the Binary System of numbers and two states of logic namely the Zero State and the One State. The complex circuitry contains Flip Flows, Matrixes, Multiplexes, Memory Elements, Decoders, Counters, Gates, Pulse Multipliers, Clock Generators, and other electronic circuit configurations to complete the job.

Digital Display. A device that reads out temperatures, pressures, levels, etc., in digital form as opposed to reading it out on Analog devices such as a pressure gage, thermometer or Eye Hye.

Diode. An electronic device that allows electrons to flow in one direction but is highly resistant to current flow in the opposite direction. (Similar to a check valve in a piping system.)

Discontinuous Controller. This is a two step controller and represents the simplest type of control. An example would be a contact type thermostat for a start-stop operation of a house air conditioner or home heating system.

Edgewise Meters. With the advent of numerous indicators in a central area, large dial type gages took up too much room. Most of the instruments on the modern automated vessel are edgewise meters with movements usually ranging from 0 to 50 milliamps. They contain a Zero Adjustment on the front and a Span Adjustment on the back.

Electronics. The use of low voltage and low power signals in the field of communications and computers for the purpose of control, logic, computation and transmission of information.

Encapsulated Relay. An electrical relay usually consisting of many relay contactors enclosed in a plastic enclosure. Its terminals are similar to a vacuum tube and plug in. The relays can control numerous electrical circuits and are much more reliable since they are protected from moisture, oxygen and pitting.

E/P Converter. A transducer that converts an electrical signal into a pneumatic signal for the purpose of pneumatically controlling a process.

Fail Safe. The thought and logic in the design of an automated system so that in the event of a failure it will fail in such a manner as to not cause any damage. An example would be the master fuel oil valve which is a two position valve held open by a solenoid valve which controls air to the diaphragm. If the ship's power is lost the solenoid not having any power de-energizes and shuts the air off the master valve. If the air system fails the lack of air will shut off the master valve. Therefore:

1. Solenoid is energized to open—fail safe electrically
2. Valve requires air supply to open—fail safe pneumatically.

Feedback. A Sensor value of a final process that is fed back to the input of the Mixing Point or Comparing Element for the purpose of automatically controlling a process.

Flame Out. When the photo electric cell or flame scanner fails to see the light intensity of a burning fire, it will shut down the burner oil valve and air register and alarm the engineer that the burner has a Flame Out.

Flame Scanner. A photo electric cell in an automated oil burner that is energized by light energy from a fire. The scanner converts the light energy into electrical energy and performs the function of keeping the fuel oil valve open as long as it "sees" the fire. Once it has been converted into electrical energy it can, through relays, keep the engineer notified of its status by showing a light indicator that it is BURNING or set off an alarm that it has failed—FLAME OUT.

Flashing Bus. Flashing is the condition of an indicator light as soon as a malfunction has occurred. It will blink on and off rapidly to attract your attention. After you have Acknowledged that you are aware of the condition, it will then burn steady. After the malfunction has cleared, the light will go out.

Flip Flops. An electronic circuit configuration using two transistors to form a counter for making calculations of numbers by receiving Bits.

Flow Device. A Sensor or instrument capable of measuring the flow of water, steam, air or gas in a system or process.

Frequency. Frequency is the number of cycles or oscillations per second. We generally think in terms of AC current as 60 cycles per second. In automation however we deal with clock pulses of 20 kc or 20,000 cycles per second. The new terminology is the use of the word HERTZ which is used to replace the words "cycles per second." When we say 20 Kilo Hertz we mean 20,000 cycles per second. We also find a varying frequency occurring in an oscillating tank of a Transducer. (See Pressure Transducer and Position Error Detector.)

Fuel-Air Ratio Controller. A pneumatic device for trimming the amount of oil that will burn with a given proportion of air. It is adjustable or "biased" by the engineer. Once set it will maintain this relationship of oil consumed with a proportional quantity of air.

Function Generator. An operational amplifier used in an Analog Computer of a throttle control system. It is capable of producing a cubic output from the input reference signal. Since turbine power requirements is proportional to the cube of the ship's speed, and since it is desirable to have a good linear operating characteristic of ship's speed to throttle position, it is necessary to have a Function Generator in the control system.

Gates: AND—OR—NOT. These are circuits used in a Digital Computer.

1. AND gate. Requires two or more circuits to be closed in order to trigger an Output. This is similar to two switches in series. Instead of thinking in terms of the word closed use the term One State. We can say that if Input "A" is One AND Input "B" is a One, then Output "C" will be a one. From this we derive a Truth Table.

Input "A"	Input "B"	Output "C"
0	0	0
1	0	0
0	1	0
1	1	1

2. OR gate. Requires any one of many circuits to be closed in order to trigger an Output. This is similar to two or more switches in parallel that will operate a circuit. From this we derive the following Truth Table.

Input "A"	Input "B"	Output "C"
0	0	0
1	0	1
0	1	1
1	1	1

3. NOT gate. This gate will produce no output if there is a logic input and will produce a One output if there is no logic input. This can be thought of in terms of normally open or normally closed relays. This is also called an Inverter. (See NAND Logic.)

GEMACS. A trade name used by the General Electric company which basically stands for General Electric Measurement And Control. The GEMAC is a Pressure Transducer for converting pressure into an electrical signal. When the GEMAC is used as a Differential Pressure Transducer it can then be used to measure a level. (See Pressure Transducer, Position Error Detector, Null Balance Motor.)

GEMS. A trade name of the GEMS company which produces level switches. These switches consist of an encapsulated switch in a brass rod which is stroked by a magnet in the proximity of the switch. The magnet is attached to a float surrounding the brass rod. Information as to high or low level status of vented tanks is most useful in automation.

Group Control. A system whereby all the motor controls of the engine room are grouped into one large master cubicle. This allows for repair, start, stop, indication information, and resetting of protective devices from a centrally located area. The unit is usually located close to the COS.

Hardware. All of the major components and parts that physically make up a computer. (Mechanical, magnetic, electrical and electronic devices.)

High Limit Status. A device found in the Monitoring System of a Digital Computer consisting of an X-Y Matrix. As the monitoring system contacts an address it permits that address to enter into a column on the Matrix. That particular horizontal column contains an arrangement of Diodes. The original sensing element's Analog voltage has already been converted into an equivalent BIT count. The vertical rows of the Matrix are arranged in Binary count. Starting from right to left the count columns are 1, 2, 4, 8, 16, 32, 64, 128, 256. Let us trace a typical sensor such as the RTD of the main thrust bearing whose address is C-3. The RTD has sent out an Analog voltage the equivalent of its particular temperature which is converted into digital BITS. Let us say that

300 BITS is equal to 180 degrees. We have set our High Limit Status Matrix to alarm us at 179 degrees which digitally would be 298 BITS. There would be installed in that address's horizontal row, Diodes in the following columns 256, 32, 8, 2 (equals 298). As the digits flow through the Matrix and since we have more than the measured 298, the alarm will flash which means that the bearing temperature is over the preset limit. The Low Limit Status works the same way for digits below a preset minimum.

Hunting. When an automatic system is highly sensitive or has a small Proportional Band, hunting is likely to occur. Instead of remaining at a steady state or steady level, it will cycle or fluctuate above and below the set point.

Hydraulic Operator. A device when signalled uses a hydraulic medium to operate against a plunger to control a valve or physically move another object with a power cylinder.

Ignitors. An electrical device that will cause a high energy spark to jump across a gap in order to light off a burner.

Indicators. Any device on the console offering information is an Indicator.

Input. Energy or information either electrical or pneumatic that enters into a control system or Computer Card for further correction or evaluation is an Input.

Interfacing. In the designing of an automated power plant it is necessary to plan out various signals between different manufacturers so that these electrical or pneumatic signals are introduced to operate from each other's devices. This is a complicated process and planning stage that requires many meetings and the cooperation between vendors and the designers and builders at the shipyard.

Interpage Wiring. The wiring that leaves the back of the Page of a console and enters other Pages of the same console.

Interlock. An arrangement whereby an operation of one part or mechanism automatically brings about or prevents the operation of another.

Instrumentation. All of the various forms of indicators that give information of pressure, temperature, flow, speed, position, conductivity, etc. For the purpose of delineation the following applies:

First letter	Second letter
P - pressure	A - alarm
T - temperature	C - controller
F - flow	I - indicator
L - level	R - recorder
C - conductivity	T - transmitter

From the above table one can see that LA would be a Level Alarm. PC is a Pressure Controller. FR is a Flow Recorder. Three other common classifications are: CV—Control Valve; SV—Solenoid Valve; PS—Pressure Switch.

Inverter. A type of Logic Gate that will convert a logic ONE to a ZERO or a logic ZERO to a ONE. Also called a NOT Gate. A NOT gate logic applied to an AND circuit will produce a NOT AND or better known as NAND logic.

Jumper Wire. A wire used to bypass circuitry that has failed to operate. This must be done with extreme caution.

Lead and Standby. A selector arrangement on the console for controlling pumps or motors. When you select pump #1 to run it becomes the Lead pump. Pump #2 becomes the Standby pump. All valves should be lined up on all units. A pressure switch now holds pump #2 from running. If the pressure falls below a preset minimum the standby pump starts. The standby pump will also start upon electrical overload or failure of the Lead motor. It is good practice to periodically change over making pump #2 the Lead and #1 the Standby.

Level Switch. A switch designed to notify the operator of a high or low level condition. (See GEMS.)

Level Transmitter. A feedback device that measures level by comparing the differential between a constant leg and a variable leg. Level transmitters are used in feedwater control systems and also for shutting off the master oil valve if the boiler water level reaches a point of low-low. (See Three Element Regulator, Permissive.)

Manual Control. The operator switches off the Feedback Loop and now controls the process himself by raising or lowering the control signal.

Master Controller. A master unit sending out more than one signal to other controllers in a Cascade system or complex automatic control system such as the boiler master which receives the equivalent of steam pressure and sends out signals to the damper control and fuel control.

Memory Element. The ability of a computer to store Digital information which can be referred to for evaluation and computation.

Minimum Firing Rate. A design into the combustion control system so that when the master controller goes to a minimum signal it will not shut off the fires but allows a minimum regulator to supply a small air signal maintaining a fire still visible to the Flame Scanner.

Mimic. A mimic is a device on the console that is imitating graphically or diagrammatically a particular process or arrangement. It may have within it Indicator Lights, showing Status information of valves and pumps that are in operation. It may also have pushbuttons or Selectors for controlling these units. Another type in use is the bearing mimic which imitates the physical relationship of the main turbine, reduction gears and propeller shaft bearings.

Mode Selector. A selector switch enabling the operator to select a Mode of operation. Throttle Control has a Mode Selector for bridge-engine direct control. A plant Mode Selector allows for Port Mode, Maneuvering Mode and Full Power Mode control. When selection is made, various circuitry is brought into play to place the plant or throttle into that desired condition.

Modulating Valve. A valve in a control system that is capable of adjusting through a range of partially open to full open. It is controlled by a Continuous Controller and is usually diaphragm operated and has a positioner for accurate setting.

Monitoring System. That portion of the computer that continuously tests all of the pertinent sensors of the plant and will alarm and notify the operator of any Off Limit conditions through the High and Low Limit Matrixes.

Motor Operated Valve. Often referred to as an M.O. valve. An electric motor operates the valve from a remote station or the valve can automatically open or close in a programmed process. The opening and closing of valves by motor lends itself readily to automation. Large valves thus operated save much physical exertion. These valves may also be operated manually by pushing down a lever which disconnects the motor drive (this could be referred to as an emergency back-up system).

NAND Logic. NAND is an abbreviation for NOT AND. This logic is an inverted AND gate. This means that if A is ONE and B is ONE then C is ZERO.

Nuisance Alarm. When an alarm intermittently keeps going off and the operator has determined that the alarm is at fault but the system is working properly, he calls this a nuisance alarm.

Null Balance Motor. A magnetic coil on a Pressure Transducer which receives electrical energy of sufficient strength to counter the force exerted on the other end of the beam.

ON-OFF Action. In a control system where a signal is used to operate a Two Position Valve or solenoid valve, the control is either ON or OFF. The valve is either OPEN or CLOSED.

Open Loop. An Open Loop is a manual process whereby information is NOT fed back to the control system. A man is required to sight visually or read display information and then make an adjustment.

Oscillator Circuit. In electronics an oscillator circuit or "tank" is used to produce a given frequency. In a Pressure Transducer we find that the Position Error Detector will vary the inductance of an oscillator tank, thereby changing its frequency in proportion to the force applied to the beam.

Optimizing. The design and logic built into a computer that makes it capable of adjusting plant conditions to a more efficient control than is humanly possible.

Output. Energy or information either electrical or pneumatic that leaves a control device or a circuitry Card is called an Output.

Page. A section in back of the computer containing rows and columns of Cards of electronic circuitry. The Page is hinged and swings clear like a door exposing further electronic equipment within the console cubicle.

Pneumatic. Referring to the use of AIR for signal information and control in a process.

Pneumatic Relay. A pneumatic device receiving one or more input signals and whose output is designed to be capable of adding, subtracting or averaging of signals. (See Totalizer.)

Permissive. Permission or proof within a control system allowing a process to start or continue. The loss of a permissive will shut down the system. In burner management the master fuel oil valve can only remain open if: 1) the forced draft blower is running: 2) there is proper water level in the boiler. Either of these permissives will prevent original light-off or shut the master oil valve during operation.

PNP Transistor. A transistor whose barrier layers are sandwiched in an arrangement of negative material surrounded by positive material on each side.

Positioner. A device on a control valve that mechanically relates the position of the valve in order to inform and control the air to the diaphragm that it is in a given position. This allows for more accurate valve control.

Position Error Detector. A device in a Pressure Transducer which causes a change of inductance in an oscillating circuit when a beam forces an iron core deeper into the flux path of a coil. (See Pressure Transducer.)

Position Feedback. In a control system where the final process is a physical movement, the device in moving to its position also moves a potentiometer which feeds back an electrical equivalent of its position. This information is then compared with the Reference Setting or Command Signal. (See Throttle Control.)

Potentiometer. Basically a small rheostat or an adjustable resistor, allowing for a change in reference setting, position feedback or signal strength adjustment.

Power Positioner. A pneumatically operated actuator for controlling the movement of forced draft inlet vanes or windbox dampers.

Pressure Feedback. Generally referred to as a pressure transmitter or transducer controlling a feedback loop. In throttle control, pressure feedback may be used in positioning the throttle.

Pressure Switch. A device that will convert a pressure into an ON-OFF electrical signal.

Pressure Transducer. A device capable of converting pressure into an electrical equivalent. Pressure exerted on a beam moves a Position Error Detector which causes a change in inductance of an oscillator circuit. This varying oscillation is then amplified and rectified and balances the beam via a Null Balance Motor. The resulting signal may be used on an Analog Edgewise Meter for continuous display of a pressure indicator. The signal may also be converted into a digital equivalent for monitoring and alarming.

Printed Wiring Board. A small card approximately 4" by 5" containing resistors, capacitors, transistors, potentiometers and external connections. The circuitry is printed and the components fitted into holes and then group soldered. (See Cards.)

Process. The automated controls and equipment in use to accomplish a specific job. For example:
1. throttle control
2. burner management
3. steam pressure control
4. feedwater control
5. superheat temperature control

Process Lag. Delay in response of the final process to a change in value of the control valve.

Programming. Building logic steps into a computer so that when a permissive is met, or the operator starts a mode or system, a series of related functions will take place due to the programming of these items.

Proportional Band. Represents that range of values of deviation as compared with the full operating range of a signal of a controller. It is expressed in percentage of controller scale; therefore, the wider the proportional band the less sensitivity the controller will have.

Propulsion Section. That section of the console containing the control devices, indicators and alarms for the main turbine. The parts usually found in this section are:
1. throttle control lever
2. throttle mode selector
3. plant mode selector
4. pressure, temperature and level displays
5. trip button
6. shaft speed indicator
7. engine order telegraph
8. throttle override switch

Proximity Switch. A contactless type switch working on the principle of a magnet on the moving leg and a coil on the fixed leg. When the two units are in proximity a ferro resonant effect provides a snap action of response for ON or OFF. They are now in use as push-buttons or as limit switches the advantage being the ability to switch without wear and tear on contacts.

Purge. The changing of air in the boiler forcing out old gases and explosive fumes. Generally four changes of air are needed to clear a boiler (usually 30 to 40 seconds). In the automated system the purge must be made and then proven prior to permission for lighting-off.

Push Button Station. Often called a PB station. Simply an electrical pushbutton used to start and stop a motor or open and close a valve. Many modern PB stations have the indicating light built into them indicating the status condition of the unit you are controlling.

Readout. When the engineer desires a set of pressures or temperatures on a Demand Display device or on a Data Logger, he either dials an address or pushes a button calling for a readout on the information that he wishes.

Recorders. Graphically penned devices which can record for a 24-hour period leaving a permanent record or history of operating conditions for a particular process. Usually used for pressures, temperatures, flows, levels, speeds, etc.

Rectifier. An electrical device or circuit capable of allowing an AC current to be converted into a DC output.

Reference Signal. In throttle control this signal is established by the potentiometer attached to the throttle lever or handwheel. The signal has a polarity and voltage level in proportion to the handwheel's position. This signal now represents the Command Signal or reference signal which now enters into the Function Generator. (See Throttle Control, Summing Amplifier, Function Generator.)

Regenerative Cycle. The steam cycle of a turbine power plant whereby steam is extracted or bled off the turbine after it has done work. This steam is then utilized to heat feedwater and other plant requirements thereby saving live steam and offering greater plant efficiency.

Reheat Cycle. With greater horsepower requirements on modern vessels designers are using a reheat cycle to increase plant efficiency. The steam is extracted from the main turbine after it has done some work. It is then reheated, its temperature increased and then brought back to the turbine to do more work. The regenerative cycle may still be used in a reheat plant.

Relay. An electrical device consisting of a coil that when energized will open or close one or more contactors. These contacts or switches may be in circuitry for controlling process or logic. A relay may be of another nature. (See Pneumatic Relay.)

Remote Indicator. In instrumentation, any device that indicates pressure, temperature, level or flow in an area away from the point of attachment.

Resistance Temperature Detector. Often referred to as RTDs. Located in large quantities throughout the turbines and reduction gears, lube systems and steam flow paths. These units have a linear relationship between temperature and resistance and when installed in the circuit with a Wheatstone bridge, will affect a voltage output in proportion to the temperature that it senses. Once the signal is into the console it can be treated as Analog and also be converted to Digits for monitoring and alarming purposes.

Retracts. Soot blowers in a modern plant that must clean the superheater and screen tubes. The elements would become warped if they lay in the path of the flue gas. They are designed to enter into and then retract from the boiler by motor or pneumatic controls.

Rotator. A rotating visual light to attract the attention of the engineer. Used in conjunction with an alarm.

Scanner. The sensing Hardware and signal conditioning is the scanner system of the monitoring system of the plant.

Sealing Relay. Used in conjunction with a Push Button Station. Since the pushbutton is only a momentary switch, when it is depressed it energizes a relay coil. One of the coil's contacts now closes paralleling the pushbutton station. When the PB is released to open circuit, the relay now keeps the circuit closed.

Selector Station. A station on the console from which the operator can choose an automatic or manual process control or a mode of operation, or select the pump or motor he desires for Lead or Standby.

Sensors. All of the devices installed in the path of flow to measure or Sense pressure, temperature, level, etc.

Sequential. When two or more devices will perform a function in an orderly manner, and are so arranged that "A" occurs first, followed by "B" and then "C" are said to be sequential operated devices.

Service Section. The section of the console that contains the controls, displays and alarms for the auxiliary machinery such as main and auxiliary circulators, air compressors, saltwater service pumps, fire pumps, etc.

Servo Motor. A motor capable of producing linear or rotary motion which acts as a slave for another device.

Set Point. In an automated system this represents the Command or Reference Signal and is the point at which an operator wants to maintain a desired pressure, temperature or level.

Shaft Stopped Alarm. When the main shaft has stopped, a counter in an electric circuit counts up to 160 pulses each of one second duration. It will then alarm the engineer that the shaft has stopped. The alarm clears on shaft rotation and the counter will start counting again as soon as the shaft stops again.

Signals. Any message sent, either electrical or pneumatic, for the purpose of measurement, control or logic.

Silicon Controlled Rectifiers. Often referred to as SCRs. They contain an electronic configuration of transistors, diodes and related circuitry that can control the "firing angle" of phases of AC current. These are now converted into DC pulses. Advancing the firing angle allows larger DC pulses through and retarding the angle only permits small pulses through. In so doing we can readily control the speed and movement of a small DC motor or the current of an exciter. In modern ships it is used in Throttle Control and Brushless Exciters.

Simulate. In automation we refer to the ability of being able to falsely duplicate one or more conditions in order to see if our logic circuits are responding. Therefore we simulate certain conditions to see how they are affecting the Output.

Software. The collection of all programs for a computer.

Solenoid Operated Valve. A two-position valve that will open or close by a solenoid coil. In smaller valves, the solenoid coil energizes and attracts magnetically a core attached to the valve stem, raising it to the open position. In larger valves, the solenoid controls air pressure which can open or close the valve by operating against a diaphragm.

Solid State. Several years ago vacuum tubes were used in electronic circuits. In modern circuitry Diodes and Transistors have replaced the tubes and the electrons no longer travel through a vacuum but through solid material such as germanium, indium and silicon, therefore all of this is called Solid State.

Span Adjustment. An adjustment on a potentiometer so that a meter or a transducer will reach its proper maximum value from a given maximum input. This setting allows the device to properly "track" the input.

Speed Error Correction. In throttle control a speed error correction will allow the system to obtain faster response and more accurate positioning. If the throttle lever is moved from full ahead to slow the speed error circuit will cause the valve to "over close." As the propeller slows down speed error is gradually reduced. The valve will open slowly and stop to a position when the propeller reaches the desired speed.

Speed Feedback. Speed or shaft RPM information that is fed back in an automatic process. In throttle control a tachometer sends speed feedback information into the Summing Amplifier and while the vessel is in the maneuvering mode it helps to hold the speed to the throttle setting. If it is desired to have a main turbine to travel at a given or governed speed, speed feedback could be used, and the reference would be the engineer's setting for full speed.

Spill and Fill Valves. Two valves arranged by deaerating level control so that when the deaerating feed tank is high, condensate will spill through a valve into the distilled water tank. When the deaerating feed tank is low, it will open a valve taking on extra feed from the distilled water tank to the main condenser.

Static Switching. A means within a computer whereby electronic circuitry will allow switching of current paths without the use of conventional switch-type contacts. Since this involves no moving parts such as is found in regular switches, it is called static switching and is both reliable and less apt to fail.

Status. The present condition of a specific piece of machinery or control. For example: indicators that show that a pump is running or stopped; a valve that is open or closed; a gear that is engaged or disengaged. These are status indicating lights and are generally lit up in green.

Steam Pressure Control. A master controller in a Cascade System in which steam pressure controls air and fuel being supplied to the boiler.

Subtracting Relay. A pneumatic relay receiving two different inputs and whose output represents the mathematical difference of both signals.

Summing Amplifier. Used in Throttle Control. It is an amplifier which compares the various signals applied to its input terminal and then produces an output which is directly proportional to the algebraic sum of the input signals. The inputs are: Reference Signal from the throttle handwheel's potentiometer; Speed Feedback from the turbine shaft and Position Feedback from the valve's actual position. The Summing Amplifier's output is fed into the SCR package which will control the DC motor of the oil relay valve.

Superheat Temperature Control. An automated system in which the superheat temperature is sensed and fed back to control valves which allow a portion of the superheated steam to enter a desuperheater or attemperator.

Supply Disturbances. These are disturbances which may occur in the power plant and will have an affect in the process of steam production. Examples:

1. feedwater pressure
2. fuel oil flow
3. feedwater temperature
4. fuel oil temperature

Synchro Motor. A motor containing three stator windings and two rotor leads which when connected to another synchro will follow and move exactly as the master.

Tacho Generator. A small generator whose output is proportional to speed. Used in Throttle Control systems and feed pump and generator information regarding speed and control.

Terminal Boards. Insulated boards containing screws to which external circuits of the engine room are introduced and connected. These are then connected internally to the console logic circuitry.

Thermocouple. A heat sensor consisting of two dissimilar metal wires connected together at one end and terminating at an electrical coil at the other end. As it senses an increase or decrease in heat, a current will flow between the dissimilar wires. This can be amplified and read off as a temperature. The thermocouple is an excellent sensor for high temperature.

Thermodynamics. The study of characteristics and behavior of liquids and gases. In the marine power plant we are concerned with these properties relating to water and steam.

<u>Three Element Regulator.</u> A feed control system in which steam flow compares itself against feedwater flow and the resulting output trims up the signal of the Drum Level Transmitter for ultimate control of the feedwater regulating valve.

<u>Throttle Control.</u> The complex automated management of the throttle through the use of an Analog Computer. Some of the requirements are:

1. controlling throttle from engine room or bridge
2. displaying of shaft speed and throttle position
3. steam and water level overrides
4. speed error correction
5. position or pressure feedback
6. speed feedback
7. shaft stop alarm or autorotation
8. wrong direction alarm
9. turning gear interlocks

<u>Throttle Override.</u> A section of throttle control which takes into account the steam pressure and boiler water level. Steam pressure falling below a preset minimum will start to close the throttle and at a lower preset minimum will shut it off entirely. High and low water levels will prevent further opening of the throttle valve. The engineer may eliminate these overrides by flipping a selector marked Throttle Override Lockout.

<u>Time Delay.</u> In motor control a time delay device will allow a motor to accelerate. In level control it can be used to prevent an alarm from ringing because of a variation in level due to the rolling of the ship. This can be accomplished by adding capacitors to a circuit which require time to change before allowing the circuit to alarm. Ten (10) second time delays are common on boiler levels, deaerator tank levels, atmospheric tank levels, main condenser hotwell levels, etc.

<u>Timers.</u> Timers are electrically activated spring wound mechanisms capable of measuring a small period of time such as 30 seconds. When the timer has run for that length of time its contacts then activate a circuit to start or stop an operation. Examples used in burner management are:

1. purge timers
2. light-off timers
3. cancel low-fire timers

<u>Totalizer.</u> A pneumatic relay capable of receiving one or more signals and delivering an output proportional to the inputs. The output can be the result of adding, subtracting or averaging.

<u>Transducers.</u> A device capable of converting a pressure, temperature or level into an electrical equivalent. Converts energy from one form to another. (See Pressure Transducer, E/P Converter.)

<u>Transmitters.</u> Pneumatic devices that sense pressure, temperature and level and convert them into a pneumatic equivalent for display, evaluation and control.

Trip Button. Sometimes called the Panic Button. Usually enclosed in a plastic housing to prevent accidental tripping. Can be designed to trip the main throttle, generator or a feed pump from the COS.

Truth Tables. Tables derived in Digital Computer logic that state all the possible conditions that can exist for a given circuit or Gate.

Turn Down Ratio. Represents the ability of a burner to handle and burn fuel oil from its maximum delivery to its minimum delivery. If a burner is capable of burning 5 gallons per minute maximum and can still maintain a fire and burn 1/2 gallon per minute, then it has a turn down ratio of 10 to 1.

Unidirectional Switch. If a switch has three or more selections it is sometimes necessary to prevent the operator from going back on a process, therefore the switch only allows movement in one direction.

Valve Range. Comparing the minimum controllable flow to the maximum controllable flow of a valve. (See Turn Down Ratio.)

Volt Ohmmeter. Usually called a VOM. An instrument widely used for measuring resistance, voltage and milliamperes and checking circuitry for continuity.

Wheatstone Bridge. An electrical bridge configuration of 4 resistors in an arrangement that a Null exists across the bridge if all 4 resistors are equal. The variation of one resistor will upset the circuit and proportionally alter the null. (See Resistance Temperature Detectors.)

Word. A set of Bits comprising the smallest addressable unit of information in a programmable memory.

Word Length. The number of Bits in a word.

Wrong Direction Alarm. An arrangement between the engine order telegraph and the throttle handwheel whereby if the telegraph is in the AHEAD position and the throttle handwheel is in the ASTERN position an alarm is sounded.

Zero and One State. The only two states that can exist in a Digital Computer. The One state contains energy. It is energized or ON. The Zero state does not contain energy. It is de-energized or OFF. Ones and Zeros form the logic of Binary Numbers and by this means the computer can perform fantastic calculations.

Zero Adjustment. Instrumentation circuitry in order to be properly adjusted must have a starting reference. The manufacturer specifies the amount of resistance to be injected with a Decade Box. A VOM is placed on the specified testing output point. The Zero Adjust potentiometer is turned until the specified voltage is attained. The device has now been Zero Adjusted. (See Span Setting for further adjustment.)